Lecture Notes in Earth Sciences 92

W0232283

Springer-Verlag Berlin Heidelberg GmbH

Per Christian Hansen   Bo Holm Jacobsen
Klaus Mosegaard (Eds.)

Methods and
Applications
of Inversion

With 111 Figures and 13 Tables

 Springer

Editors

Per Christian Hansen
Dept. of Mathematical Modelling
Building 321 Technical Univ. of Denmark
2800 Lyngby, Denmark

Klaus Mosegaard
Department of Geophysics
University of Copenhagen
Juliane Maries Vej 30
2100 Copenhagen O., Denmark

Bo Holm Jacobsen
University of Aarhus
Department of Earth Sciences
Finlandsgade 8
8200 Aarhus N., Denmark

"For all Lecture Notes in Earth Sciences published till now please see final pages of the book"

Cataloging-in-Publication data applied for

Die Deutsche Bibliothek - CIP-Einheitsaufnahme

Methods and applications of inversion: interdisciplinary elements ofmethodology, modeling, and computational algorithms / ed.: Per Christian Hansen

(Lecture notes in earth sciences; Vol. 92)
ISBN 978-3-540-65916-7   ISBN 978-3-540-48866-8 (eBook)
DOI 10.1007/978-3-540-48866-8

ISSN 0930-0317
ISBN 978-3-540-65916-7

© Springer-Verlag Berlin Heidelberg 2000
Originally published by Springer-Verlag Berlin Heidelberg New York in 2000

The use of general descriptive names, registered names, trademarks, etc. in this publication does not imply, even in the absence of a specific statement, that such names are exempt from the relevant protective laws and regulations and therefore free for general use.

Typesetting: Camera ready by author
Printed on acid-free paper   SPIN: 10528995   32/3142-543210

Preface

These proceedings consist of 20 refereed papers, covering most of the talks presented at the Second Interdisciplinary Inversion Conference, held August 1998 in Copenhagen. The conference brought together researchers working in many different areas of inversion, and the topics of the talks ranged from theoretical aspects and algorithmic issues to industrial applications.

The present proceedings reflect this variety of topics at the conference, and thus demonstrates the inherent interdisciplinary nature of inversion research.

In spite of the variety of topics covered, the different papers in the proceedings share a common interest in *inversion methods*. Some papers give insight into the early stages of algorithm development, other papers discuss various properties, advantages, and disadvantages of state-of-the-art algorithms, while yet other papers demonstrate the power of inversion methods in large-scale applications.

The editors would like to thank both the authors and the referees for their contributions and their hard work during the preparation of the proceedings.

January 2000, Lyngby, Aarhus, and Copenhagen

Per Christian Hansen
Dept. of Mathematical Modelling
Technical University of Denmark

Bo Holm Jacobsen
Dept. of Earth Sciences
University of Aarhus

Klaus Mosegaard
Department of Geophysics
Niels Bohr Institute
University of Copenhagen

Organization

The Second Interdisciplinary Inversion Conference IIC98 was held August 12–14, 1998 in Copenhagen. It was organized by the Danish Interdisciplinary Inversion Group, sponsored by the Danish Natural Science Research Council.

Program Commitee

Conference Chair:	Klaus Mosegaard (University of Copenhagen, DK)
Members:	Bo Holm Jacobsen (Århus University, DK)
	Per Christian Hansen (Technical University of Denmark)
Secretaries:	Charlotte Barnkop (University of Copenhagen, DK)
	Henriette Hansen(University of Copenhagen, DK)

Referees

Ann-Charlotte Berglund
Per Christian Hansen
Bo Holm Jacobsen
Per Knudsen
Kevin Kreider
Klaus Mosegaard

Jacob M. Pedersen
Frank P. Pijpers
Carl E. Rasmussen
Peter Skjellerup
Curtis R. Vogel

Table of Contents

List of Speakers

Bjarne Andresen
Ørsted Laboratory
University of Copenhagen
Copenhagen, Denmark

Torben Bach (current address)
National Survey and Cadastre
Copenhagen, Denmark

Christophe Barnes
Institut de Physique du Globe de Paris
Paris, France

Svend Berntsen
Dept. of Mathematical Sciences
Aalborg University
Aalborg, Denmark

Miguel Bosch
Facultad de Ingeniería
Universidad Central de Venezuela
Los Chaguaramos, Caracas, Venezuela

Marwan Charara
Institut de Physique du Globe de Paris
Paris, France

Hugues Djikpéssé
Institut de Physique du Globe de Paris
Paris, France

Flemming Effersø
Dansk Geofysik A/S
Aarhus, Denmark

Lars Eldén
Department of Mathematics
University of Linköping
Linköping, Sweden

Leiv-J. Gelius
Department of Geophysics
University of Oslo
Oslo, Norway

Lykke Gemmer
Department of Earth Sciences
Aarhus University
Aarhus, Denmark

Freeman Gilbert
Inst. of Geophys. and Planetary Phys.
9500 Gilman Drive
La Jolla, California, USA

Mårten Gullikson
Department of Computing Science
Umeå University
Umeå, Sweden

Per Christian Hansen
Dept. of Mathematical Modelling
Technical University of Denmark
Lyngby, Denmark

Martin B. Hansen
Dept. of Mathematical Sciences
Aalborg University
Aalborg, Denmark

Bo Holm Jacobsen
Department of Earth Sciences
Aarhus University
Aarhus, Denmark

Jesper Munk Jensen
Department of Earth Sciences
Aarhus University
Aarhus, Denmark

X

Amir Khan
Department of Geophysics
University of Copenhagen
Copenhagen, Denmark

Uwe Koppelt
Institute of Geophysics and Geology
Leipzig University
Leipzig, Germany

Olwijn Leeuwenburgh
National Survey and Cadastre
Copenhagen, Denmark

Guilhem Moreaux
Department of Geophysics
University of Copenhagen
Copenhagen, Denmark

Mette Munkholm
Z&S GeoScience
Copenhagen, Denmark

Ingelise Møller
GEUS
Copenhagen, Denmark

Frank P. Pijpers
Institute for Physics and Astronomy
Aarhus University
Aarhus, Denmark

Otmar Scherzer
Industrial Mathematics Institute
Johannes Kepler Universität
Linz, Austria

Gabriel Strykowski
National Survey and Cadastre
Copenhagen, Denmark

Stig Syndergaard (present address)
Institute of Atmospheric Physics
University of Arizona
Tucson, Arizona, USA

Lisbeth Engell Sørensen
University of Bergen, Parallab
Bergen, Norway

Curt R. Vogel
Department of Mathematical Sciences
Montana State University
Bozeman, Montana, USA

Solvej M. P. Waagepetersen
Department of Earth Sciences
Aarhus University
Aarhus, Denmark

Per-Åke Wedin
Department of Computing Science
Umeå University
Umeå, Sweden

Static Light Scattering
Minimum Contrast Estimation for Point Processes

Kim E. Andersen[1] and Martin B. Hansen[1,2]

[1] Department of Mathematical Sciences, Aalborg University,
Fredrik Bajers Vej 7E, DK-9220 Aalborg Ø, Denmark
[2] MaPhySto*
Centre for Mathematical Physics and Stochastics

Abstract. Approximate formulas for the mean scattering intensity are only known for a limited number of point process models. One approach to handle this problem is to invert the integral operator relating the scattering intensity to the pair correlation function for the point process. However, this is only feasible when the pair correlation function is known explicitly. This is usually not the case for Gibbs (or Markov) point processes wherefore simulation techniques become important. This paper investigates to what extent Monte Carlo optimization of objective functions can be utilized in light scattering.

1 Introduction

A major problem in the field of material research is resolving the structure of dissimilar types of diluted substances. One of the experimental methods used for structural investigations of amorphous structures is electromagnetic radiation by matter. Electromagnetic radiation can interact with matter in a number of ways, e.g. by absorption or by scattering. In absorption, the thermal motion of the molecules in the solution is increased due to an elevated energy concentration in the dilution. In scattering, the electrons may start to oscillate acting as a minute antenna and reemit some energy as fluorescence or phosphorescence in directions other than the incident beam. Placing detectors at different scattering angles makes it possible to measure the scattered intensity giving the observer informations on the atomic scale structure. The experimental setup is sketched roughly in Fig.1.

In the present paper we study point processes for modelling a structure of randomly distributed point scatterers. Although, some interesting problems arise both from a theoretical and applied statistical point of view not many statisticians seem to have considered stochastic approaches to light scattering, see Sonntag et al. [16], Hanisch [9] and Stoyan et al. [18]. Following their lines, a way of making inference for parametric point process models is to utilize the following approximate relation between the mean scattering intensity $I(h)$ and

---

* Funded by a grant from The Danish National Research Foundation

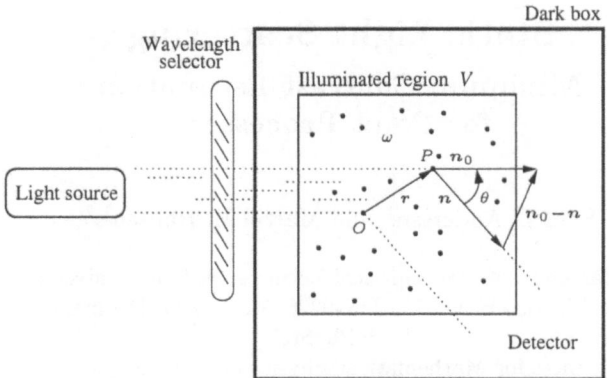

Fig. 1. Schematic representation of the experimental setup. The incident wave impinges on two elements of a point pattern Φ, at O and P. Vector r denotes the displacement of P from O. The unit vectors n_0 and n indicates the direction of the incident wave and the scattered wave at scattering angle θ. The length of $n_0 - n$ is seen to be $2\sin(\theta/2)$.

the pair correlation function function $g(r)$ (which is an important second order-characteristics for point processes, see Stoyan and Stoyan [17], Sec. 14.4)

$$I(h) = 1 + \lambda \int_0^\infty 4\pi r^2 \frac{\sin(hr)}{hr}(g(r) - 1)\,dr \ . \tag{1}$$

By inverse Fourier transformation of this integral equation it is possible to obtain an expression for $g(r)$ as a function of the measured intensity $I(h)$, i.e.

$$g(r) = 1 + \frac{1}{2\pi^2 \lambda r} \int_0^\infty h(I(h) - 1)\sin(hr)\,dh \ .$$

If the pair correlation function for the model parameterized by ψ is given explicitly one can fit it to the experimental pair correlation function $\hat{g}(r)$ by minimizing the objective function

$$\int_0^\infty \{\hat{g}(r) - g_\psi(r)\}^2\,dr \ , \tag{2}$$

with respect to ψ. This approach has the disadvantage that one has to solve an ill-posed inverse problem, and does not give any ideas on how to proceed when the pair correlation function is not given explicitly, as is the case for Gibbs (or Markov) point processes. We will consider how one by Monte Carlo methods in some cases can get around these drawbacks.

In Sec. 2 general properties of point processes and static light scattering are presented and approximate scattering intensity formulas are reviewed for special types of point processes. Parameter estimation by Monte Carlo optimization of objective functions are presented in Sec. 3 and an approximation of the expected scattering intensity is derived. The performance of the proposed estimation procedure is studied in Sec. 4.

2 Preliminaries

2.1 Point Processes

The illuminated region V is assumed to be a known bounded subset of $\mathbb{R}^2$. The illuminated structure is assumed to be a point pattern ω contained in V. Consequently $\omega = \{\omega_1, \ldots, \omega_n\}$ where the number of points n is not fixed and each ω_i is a point in V. The point pattern ω is assumed to be a realization of a point process Φ in V and to have density $f(\omega)$ with respect to the distribution μ of the homogeneous Poisson process with intensity 1 on V (Kingman, [12]).

Readers should not be terrified by the measure-theoretic language used here. The probability measure π corresponding to the density $f(\omega)$ should be understood as a probability measure on the subsets of the space Ω of all finite point configurations. The probability of an event $B \subset \Omega$ is then given by

$$\pi(B) = \int_B f(\omega) d\mu(\omega) \quad , \tag{3}$$

where μ is the probability measure of the homogeneous Poisson process with intensity rate 1, which for a subset $B \subset \Omega$ can be defined by

$$\mu(B) = \exp(-\mathrm{Vol}(V)) \left[\mathbb{1}_B(\emptyset) + \sum_{n=1}^{\infty} \frac{1}{n!} \int_V \cdots \int_V \prod_{i=1}^{n} \mathbb{1}_B(\omega_i) d\omega_1 \cdots d\omega_n \right] \quad ,$$

where $\mathrm{Vol}(V)$ denotes the volume of V and $\mathbb{1}_B(\cdot)$ is the indicator function. Hence,

$$\pi(B) = \exp(-\mathrm{Vol}(V)) \left[\mathbb{1}_B(\emptyset) + \sum_{n=1}^{\infty} \frac{1}{n!} \int_V \cdots \int_V f(\omega) \prod_{i=1}^{n} \mathbb{1}_B(\omega_i) d\omega_1 \cdots d\omega_n \right] \quad ,$$

for $B \subset \Omega$. For a discussion on this issue, see e.g. Daley and Vere-Jones [2].

Specific models with density w.r.t. the distribution of the homogeneous Poisson point process with intensity 1 are scattered around in the literature. However, it is instructive briefly to list two examples. Firstly, we consider the homogeneous Poisson point process Φ which has density

$$f_\lambda(\omega) \propto \lambda^{n(\omega)} \quad , \qquad \lambda > 0 \quad , \tag{4}$$

where $n(\omega)$ is the number of points in ω and λ is the intensity of Φ. An example based on the homogeneous Poisson point process is the model of Strauss [19], which has density

$$f_{\beta,\gamma}(\omega) \propto \beta^{n(\omega)} \gamma^{s(\omega,R)} \quad , \qquad \beta > 0 \quad , \tag{5}$$

where $s(\omega, R)$ is the number of inter-point distances no greater than the hard-core distance $R > 0$. Additionally, the interaction parameter $\gamma \in [0, 1]$ controls the degree of repulsion. Particularly, if $\gamma = 1$ then the Strauss point process becomes a homogeneous Poisson point process on V with intensity β, whereas taking $\gamma = 0$ induces the hard-core model, in which the constituent points are prohibited to be closer than the hard-core distance R. The Strauss model is comprehensively discussed in many papers, see e.g. Strauss [19], Kelly and Ripley [11], and Stoyan et al. [18]. Strauss [19] suggested $\gamma > 1$ as a model for clustering, however, Kelly and Ripley [11] showed that (5) is not integrable then. Simulated point patterns for the Strauss point process are shown in Fig. 2. The examples are simulated by rejection sampling (Ripley, [14]). If the number of points $n = n(\omega)$ is assumed to be known, we talk about a conditional Strauss point process with n points and the density becomes $f_\gamma(\omega) \propto \gamma^{s(\omega, R)}$. This density is integrable for all $\gamma > 0$.

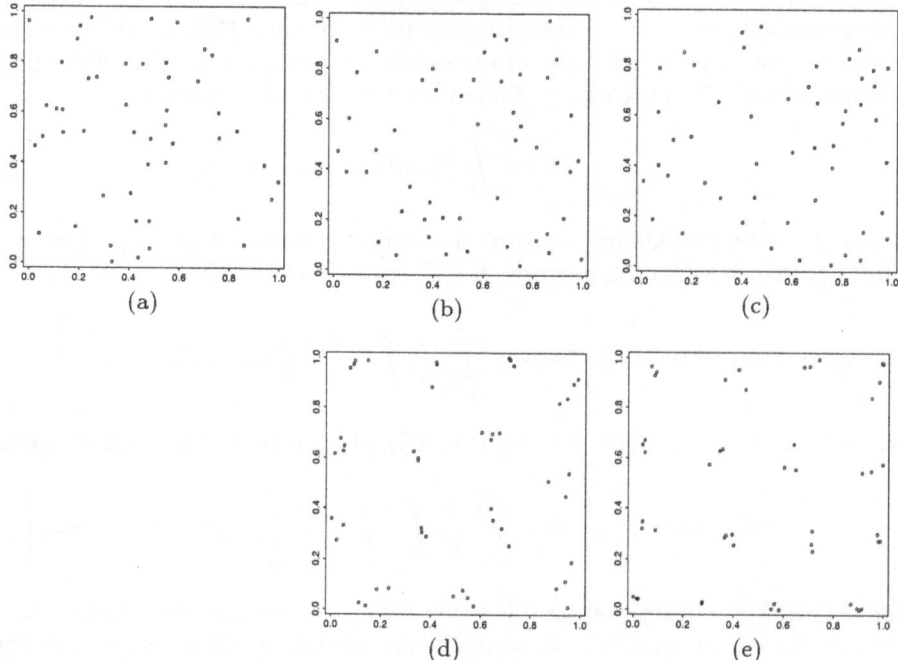

Fig. 2. Simulated point patterns for a conditional Strauss point process with 50 points in the unit square. (a) $\gamma = 1$; (b) $\gamma = 0.05, R = 0.05$; (c) $\gamma = 0.01, R = 0.05$; (d) $\gamma = 0.01, R = 0.25$; and (e) $\gamma = 0.05, R = 0.25$.

2.2 Scattering Analysis of Point Processes

In the following we model the intensity function for the scattering of electromagnetic radiation by random point patterns. Hereby we neglect the shape and size of the studied particles, see Hanisch [9]. Following Beeman et al. [1], the total scattering amplitude from a point process Φ in $V \subseteq \mathbb{R}^2$ can be modelled as

$$A_\Phi(\boldsymbol{h}) = \sum_{\boldsymbol{x} \in \Phi} \exp(-i\langle \boldsymbol{h}, \boldsymbol{x} \rangle) \ , \tag{6}$$

where i denotes the imaginary unit, $\boldsymbol{h} = (h_1, h_2) \in \mathbb{R}^2$ denotes the scattering vector, and $\langle \cdot, \cdot \rangle$ denotes the usual inner product. The scattering vector is defined by directions of the incident beam $\boldsymbol{n}_0$ and the detector $\boldsymbol{n}$ as $\boldsymbol{h} = 2\pi(\boldsymbol{n}_0 - \boldsymbol{n})/\lambda_I$, where λ_I is the wave length of the incident light, see Fig. 1. The magnitude of $\boldsymbol{h}$ is found as $h = \|\boldsymbol{h}\| = 4\pi \sin(\theta/2)/\lambda_I$.

The observed scattering intensity I_Φ at scattering vector $\boldsymbol{h}$ is given as the modulus of the amplitude of A_Φ, i.e.

$$I_\Phi(\boldsymbol{h}) = A_\Phi(\boldsymbol{h})A_\Phi^*(\boldsymbol{h}) = \sum_{\boldsymbol{x} \in \Phi} \sum_{\boldsymbol{y} \in \Phi} \cos(\langle \boldsymbol{h}, \boldsymbol{x} - \boldsymbol{y} \rangle) \ . \tag{7}$$

From this, the mean scattering intensity $I(\boldsymbol{h})$ is defined as $I(\boldsymbol{h}) = \mathbb{E}_\pi I_\Phi(\boldsymbol{h})$, whenever is exists which is the case for the Poisson point process and the conditional Strauss point process. As we will see in the following section, the assessment of the mean scattering intensity plays an essential role in our proposed estimation procedure.

It is often reasonable to assume that Φ is isotropic, that is, the distribution of Φ is invariant under rotation (implicitly assuming that V is so large that edge effects can be ignored). This leaves I_Φ as a function of $h = \|\boldsymbol{h}\|$ alone. Hanisch [9] gives for a large illuminated region V the following approximate formulas for the mean scattering intensity for two types of isotropic point processes.

1. Let Φ be a homogeneous Poisson point process with a small intensity λ, then $\mathbb{E}_\pi I_\Phi(h) \simeq \lambda \text{Vol}(V)$.
2. Let Φ be a hard-core point process with hard-core distance $R > 0$ and small intensity β, then

$$\mathbb{E}_\pi I_\Phi(h) \simeq \beta \text{Vol}(V) \left[1 - \frac{4\pi\beta}{h^2} \left\{ \frac{\sin(Rh)}{h} - R\cos(Rh) \right\} \right] \ . \tag{8}$$

The Strauss model introduced in (5) has no known approximating formula for the mean intensity. Accordingly another approach is required.

3   Parameter Estimation by Monte Carlo Optimization

In the physics context V is the irradiated 2-dimensional region including the point pattern φ of interest. Now, assume φ is a realization of a point process Φ with distribution π_ψ which depends solely on an unknown parameter ψ in a parameter space Ψ. If it is believed that φ is a realization of e.g. a homogeneous Poisson point process, then the unknown parameter is the intensity λ implying $\Psi = [0; \infty)$. Moreover, let $I_\varphi(\boldsymbol{h}_i)$ denote the observed scattering intensity at the ith detector for the realization φ and let $\mathbb{E}_{\pi_\psi} I_\Phi(\boldsymbol{h}_i)$ denote the corresponding mean scattering intensity. The problem is now to fit the parameter ψ via minimization of the following objective function

$$Q(\psi \mid \varphi) = \sum_{i=1}^m \left\{ I_\varphi(\boldsymbol{h}_i) - \mathbb{E}_{\pi_\psi} I_\Phi(\boldsymbol{h}_i) \right\}^2 \ . \tag{9}$$

This is minimized by the value $\widehat{\psi}$ that satisfies

$$\widehat{\psi} = \arg\min_{\psi \in \Psi} Q(\psi \mid \varphi) \ ,$$

which is usually termed the minimum contrast estimate $\widehat{\psi}$ of ψ. In this study we are using least squares estimates as we are dealing with simulated data. These are not influenced by measurement noise and outliers. In real applications one might have to use a more robust norm. Furthermore, it may be beneficial to use a norm which down weigh unreliable detectors. The precise form of Q is unknown and indeed intractable. Accordingly, we compute the objective function by Monte Carlo techniques. We now briefly introduce parameter estimation by Monte Carlo optimization of object functions, see Diggle and Gratton [3] and Geyer and Thompson [7]. Confer Geyer [5] for a recent review on methods for Monte Carlo optimization.

In order to find the minimum contrast estimate $\widehat{\psi}$ of ψ we must be able to assess the value of $\mathbb{E}_{\pi_\psi} I_\Phi(h_i)$ for any given value of $\psi \in \Psi$. When the distributional properties of Φ are known, the mean scattering intensity at h_i is for a given $\psi \in \Psi$ approximated by

$$\mathbb{E}_{\pi_\psi} I_\Phi(h) = \int_\Omega \sum_{x \in \varphi} \sum_{y \in \varphi} \cos(\langle h, x - y \rangle) f_\psi(\varphi) \, d\mu(\varphi) \tag{10}$$

$$\simeq \frac{1}{N} \sum_{i=1}^{N} \sum_{x \in \varphi_i} \sum_{y \in \varphi_i} \cos(\langle h, x - y \rangle) \ , \tag{11}$$

where $\varphi_1, \ldots, \varphi_N$ is a sample from π_ψ. With this sample from Φ we can evaluate $Q(\psi \mid \varphi)$ for one $\psi \in \Psi$, however, the evaluation of (9) would require an entirely new sample $\varphi_1, \ldots, \varphi_N$ for each value of ψ. Diggle and Gratton [3] therefore suggested to evaluate Q on a grid and then smooth the resulting values by e.g. a kernel smoother in order to overcome the rough approximation of the objective function in (11). The minimum contrast estimate is then the value of ψ that minimizes the smooth approximation. This many-sample method is obviously very computationally demanding.

We shall now study an estimation procedure that exploits only one sample to evaluate $Q(\psi \mid \varphi)$ for any $\psi \in \Psi$. Gibbs point processes are normally only specified up to proportionality as was seen in (4) and (5). Moreover, the simulation techniques we are going to utilize in order to evaluate $Q(\psi \mid \varphi)$ will only require knowledge up to proportionality. Hence, we will work in the following framework. Let $\overline{\mathcal{F}} = \{\overline{f}_\psi : \psi \in \Psi\}$ be a family of unnormalized densities with respect to the measure μ. Then $\overline{f}_\psi$ is a nonnegative function on Ω with finite and non-zero normalizing constant

$$c(\psi) = \int_\Omega \overline{f}_\psi(\varphi) d\mu(\varphi) \ ,$$

and $f_\psi(\varphi) = \overline{f}_\psi(\varphi)/c(\psi)$ is a probability density w.r.t. μ. Now, let $\overline{f}_\psi$ be the unnormalized density of Φ with normalizing constant $c(\psi)$, and define a reference value $\eta \in \Psi$ so $\overline{f}_\eta(\varphi) = 0$ implies $\overline{f}_\psi(\varphi) = 0$ for all $\psi \in \Psi$ and for μ-almost all $\varphi \in \Omega$. Rewriting (10), we find

$$
\mathbb{E}_{\pi_\psi} I_\Phi(h) = \int_\Omega \sum_{x \in \varphi} \sum_{y \in \varphi} \cos(\langle h, x - y \rangle) \frac{\overline{f}_\psi(\varphi)}{c(\psi)} \frac{\overline{f}_\eta(\varphi)/c(\eta)}{\overline{f}_\eta(\varphi)/c(\eta)} \, d\mu(\varphi)
$$

$$
= \frac{c(\eta)}{c(\psi)} \int_\Omega \sum_{x \in \varphi} \sum_{y \in \varphi} \cos(\langle h, x - y \rangle) \frac{\overline{f}_\psi(\varphi)}{\overline{f}_\eta(\varphi)} f_\eta(\varphi) \, d\mu(\varphi)
$$

$$
= \frac{\mathbb{E}_{\pi_\eta} \sum_{x \in \Phi} \sum_{y \in \Phi} \cos(\langle h, x - y \rangle) \frac{\overline{f}_\psi(\Phi)}{\overline{f}_\eta(\Phi)}}{\mathbb{E}_{\pi_\eta} \frac{\overline{f}_\psi(\Phi)}{\overline{f}_\eta(\Phi)}}. \tag{12}
$$

An approximation of $\mathbb{E}_{\pi_\psi} I_\Phi(h)$ is obtained by drawing a sample $\varphi_1, \ldots, \varphi_N$ from π_η and replacing the expectation by the empirical approximations

$$
\mathbb{E}_{\pi_\psi} I_\Phi(h) \simeq \frac{\frac{1}{N} \sum_{i=1}^N \sum_{x \in \varphi_i} \sum_{y \in \varphi_i} \cos(\langle h, x - y \rangle) \frac{\overline{f}_\psi(\varphi_i)}{\overline{f}_\eta(\varphi_i)}}{\frac{1}{N} \sum_{i=1}^N \frac{\overline{f}_\psi(\varphi_i)}{\overline{f}_\eta(\varphi_i)}}. \tag{13}
$$

We note that for every ψ the approximation depends on the sample but different weights $\overline{f}_\psi(\varphi_i)/\overline{f}_\eta(\varphi_i)$ are used. These are called importance weights and this way of approximating is often termed importance sampling. This single-sample method is very efficient, as it uses the same sample for all evaluations of $Q(\psi \mid \varphi)$ in contrast to the much more computationally burdensome many-sample method. Hence the objective function $Q(\psi \mid \varphi)$ can be evaluated for any $\psi \in \Psi$ and the minimizer is therefore easily found. Note, a smoothing is not needed now. One should notice that there exists a variety of alternatives and hybrids of the abovementioned methods, see the references in Geyer [5].

Sampling from a distribution π_η having density f_η w.r.t. μ without any knowledge about the normalizing constant can be done by e.g. rejection sampling (Ripley, [15]) and the Metropolis-Hastings algorithm (Metropolis et al., [13]; Hastings [10]). A description of these methods and the requirements for the approximations in (13) to be valid is out of the scope of the present paper. For an introduction and some of its many ramifications in applications see Gilks et al. [8].

It should be stressed here, that the minimum contrast estimation method is also applicable for situations where more than one parameter is unknown.

4 Simulation Study

In this section we investigate the two Monte Carlo minimum contrast estimation procedures for two specific types of point processes. In the study we simulated observed scattering intensities at 40 detectors with angles equidistant placed on the positive half circle, $\theta_i = 180i/41$ for $i = 1, \ldots, 40$. For the incident light beam we used a wavelength $\lambda_I = 630$ nm. The observation window V was chosen as $1\text{m} \times 1\text{m}$.

4.1 Homogeneous Poisson Point Process

Let φ be a realization of a homogeneous Poisson point process with an unknown intensity λ, then we can exploit the approximate formula for the expected intensity to estimate λ. In the Poisson case the object function $Q(\lambda \,|\, \varphi)$ becomes

$$Q(\lambda \,|\, \varphi) \simeq \sum_{i=1}^{40} \{I_\varphi(h_i) - \lambda \text{Vol}(V)\}^2 \ ,$$

which is seen to be minimized for

$$\widehat{\lambda} = \frac{1}{40 \cdot \text{Vol}(V)} \sum_{i=1}^{40} I_\varphi(h_i) \ . \tag{14}$$

Consequently we can compare the performance of the Monte Carlo minimum contrast estimator with the direct estimator.

To assess the performance of the Monte Carlo minimum contrast estimator we simulated three independent homogeneous Poisson point processes with intensity $3.0 \cdot 10^{-3} \text{cm}^{-2}$, see Fig. 3 for the patterns and the corresponding scattering intensities. From the computed scattering intensities we are able to find an immediate estimate of the intensity by (14). The achieved intensity estimates are given in Table 1.

Table 1. Estimated intensities for three realizations of a homogeneous Poisson point pattern with intensity $\lambda = 3.0 \cdot 10^{-3} \text{cm}^{-2}$ (units are in 10^{-3}cm^{-2}).

Sample	Direct estimation	Many sample estimation	Single-sample estimation Reference intensity		
			2.60	3.00	3.40
Poisson 1	2.72	4.80	2.70	2.83	3.05
Poisson 2	5.60	4.60	5.38	2.78	3.15
Poisson 3	4.07	4.80	3.80	2.83	2.98

The intensity was also estimated by the many-sample and single-sample estimation procedures described in Sec. 3. In the former method we evaluated the objective function on a grid of intensity values in the interval $[1.4 \cdot 10^{-3} \text{cm}^{-2}; 7.0 \cdot 10^{-3} \text{cm}^{-2}]$ with an equal spacing of $0.2 \cdot 10^{-3} \text{cm}^{-2}$ using a sample size of $5,000$. Hereby a total of $145,000$ samples were simulated to obtain the objective function, which was then kernel smoothed with a Gaussian kernel with a suitably

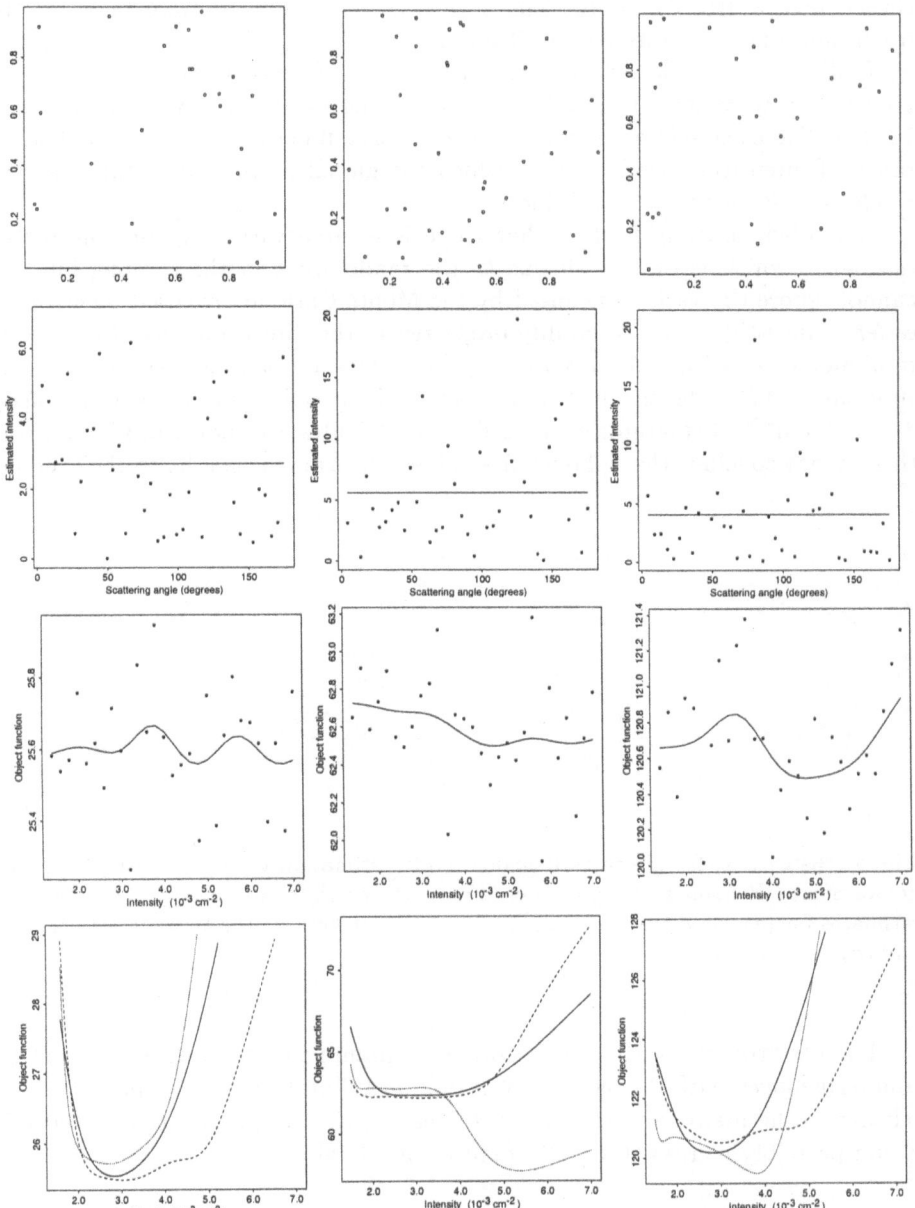

Fig. 3. Parameter estimation by Monte Carlo minimization of $Q(\lambda \,|\, \varphi)$ for three Poisson point processes with intensity $\lambda = 3.0 \cdot 10^{-3}\,\mathrm{cm}^{-2}$. From top to bottom are shown: The three simulated Poisson point patterns with intensity $\lambda = 3.0 \cdot 10^{-3}\,\mathrm{cm}^{-2}$; their corresponding observed scattering intensity (horizontal line indicates mean scattering intensity); the smoothed many-sample objective function; and the single-sample objective function with reference intensities: $2.6 \cdot 10^{-3}\,\mathrm{cm}^{-2}$ (dotted line), $3.0 \cdot 10^{-3}\,\mathrm{cm}^{-2}$ (solid line), and $3.4 \cdot 10^{-3}\,\mathrm{cm}^{-2}$ (dashed line)

chosen bandwidth of $1.5 \cdot 10^{-3} \mathrm{cm}^{-2}$. The attained minima were taken as the minimum contrast estimates, see Table 1.

Utilizing the single-sample method we used reference intensities $\lambda_1 = 2.6 \cdot 10^{-3} \mathrm{cm}^{-2}$, $\lambda_2 = 3.0 \cdot 10^{-3} \mathrm{cm}^{-2}$, and $\lambda_3 = 3.4 \cdot 10^{-3} \mathrm{cm}^{-2}$ with sample size $10{,}000$. For each reference value, the objective function was evaluated for a range of intensities, see Fig. 3, in which the global minima were taken as the single-sample estimates, see Table 1.

It is obvious from Table 1 that there is a large variability present in the estimates, which is first of all due to the randomness in the sample, but one cannot ignore the variation caused by the Monte Carlo approximation error. In order to investigate the variability properties of the direct and the Monte Carlo minimum contrast estimator based on the single-sample objective function, we simulated $1{,}000$ independent homogeneous Poisson point patterns with intensity $3.0 \cdot 10^{-3} \mathrm{cm}^{-2}$. Histograms of the achieved estimates are shown in Fig. 4. From this we may conclude that all reference values perform reasonable, on the average.

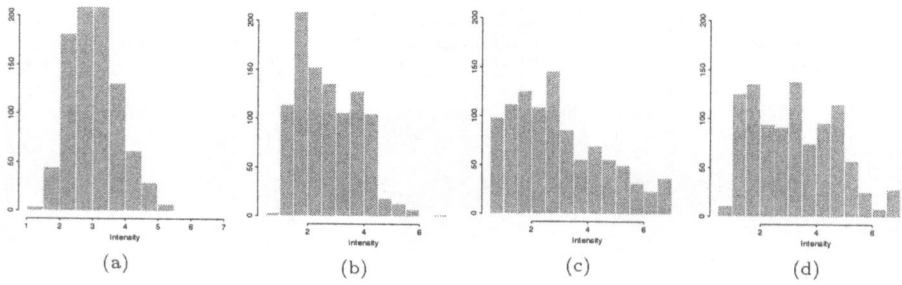

Fig. 4. Histograms for direct and single-sample estimation of the intensity for $1{,}000$ homogeneous Poisson point processes. From left to right are shown: The obtained estimates for (a) the direct method; (b) $\lambda_1 = 2.6 \cdot 10^{-3} \mathrm{cm}^{-2}$; (c) $\lambda_2 = 3.0 \cdot 10^{-3} \mathrm{cm}^{-2}$; and (d) $\lambda_3 = 3.4 \cdot 10^{-3} \mathrm{cm}^{-2}$.

The performance of the single-sample minimum contrast estimator must be counterbalanced with the fact that it is only based on very few samples ($10{,}000$) relative to the many-sample method. Increasing the sample size to e.g. $100{,}000$ would probably increase the performance considerably.

4.2 Conditional Strauss Point Process

The same experiment was conducted for three independent realizations of a conditional Strauss point process with 30 points, hard-core distance $R = 0.25$ m, and interaction parameter $\gamma = 0.05$. Figure 5 depicts the three independent point patterns with their corresponding scattering intensities. The point processes were simulated by means of rejection sampling (Ripley, [14]).

We pursued the many-sample and single-sample methods described in Sec. 3 to obtain estimates of γ. The objective function based on the many-sample method was computed on an equidistant grid in the interval $[0.01; 0.10]$ with spacing 0.005 resulting in totally 95.000 samples. The achieved intensities were then smoothed with a Gaussian kernel with bandwidth 0.05, see Fig. 5. The minimum contrast estimates are shown in Table 2 for the results. Note, that one of the minimum contrast estimates are obtained at the end points of the interval.

Table 2. Estimated interaction parameters for three conditional Strauss point patterns with interaction parameter $\gamma = 0.05$

Sample	Many sample estimation	Single-sample estimation Reference interaction		
		0.030	0.050	0.070
Strauss 1	0.010*	0.028	0.077	0.062
Strauss 2	0.040	0.035	0.061	0.061
Strauss 3	0.040	0.039	0.068	0.046

* minimum attained at the boundary

The single-sample estimation performance was assessed by evaluation of $Q(\gamma \,|\, \varphi)$ for three reference values $\eta_1 = 0.03$, $\eta_2 = 0.05$, and $\eta_3 = 0.07$. The obtained objective functions are shown in Fig. 5. We investigated the achieved estimates for $1,000$ simulated Strauss point patterns, see Fig. 6 for the results. From this we see that using 0.05 as reference value we obtain the best estimate on the average.

5   Conclusion

In the present paper we have illustrated how Monte Carlo methods can be employed to avoid an unstable inversion procedure and to make inference when the pairwise correlation function is unknown.

The Monte Carlo minimum contrast estimator based on the single-sample technique seem to be a promising method compared with the computationally demanding many-sample method. However, the obtained single-sample estimates seem to be more or less dependent on the choice of reference value, nevertheless, choosing the reference parameter wisely we obtain encouraging results. Comparing to Geyer [6], who also performs a single-sample Monte Carlo study based on the true parameter as reference value, we obtain promising results. In applications this is impossible and one has to find ways to "search" for a good reference point. One possibility is first to minimize the object function based on the many-sample method with a rough grid and choose the parameter which minimizes the object function as a good reference point. But other more sophisticated methods have been suggested, e.g. trust regions (Geyer and Thompson, [7]) and inverse logistic regression (Geyer, [6]).

From an inferential point of view it will be very interesting to investigate the variability of the estimators. When inference is based on a Monte Carlo

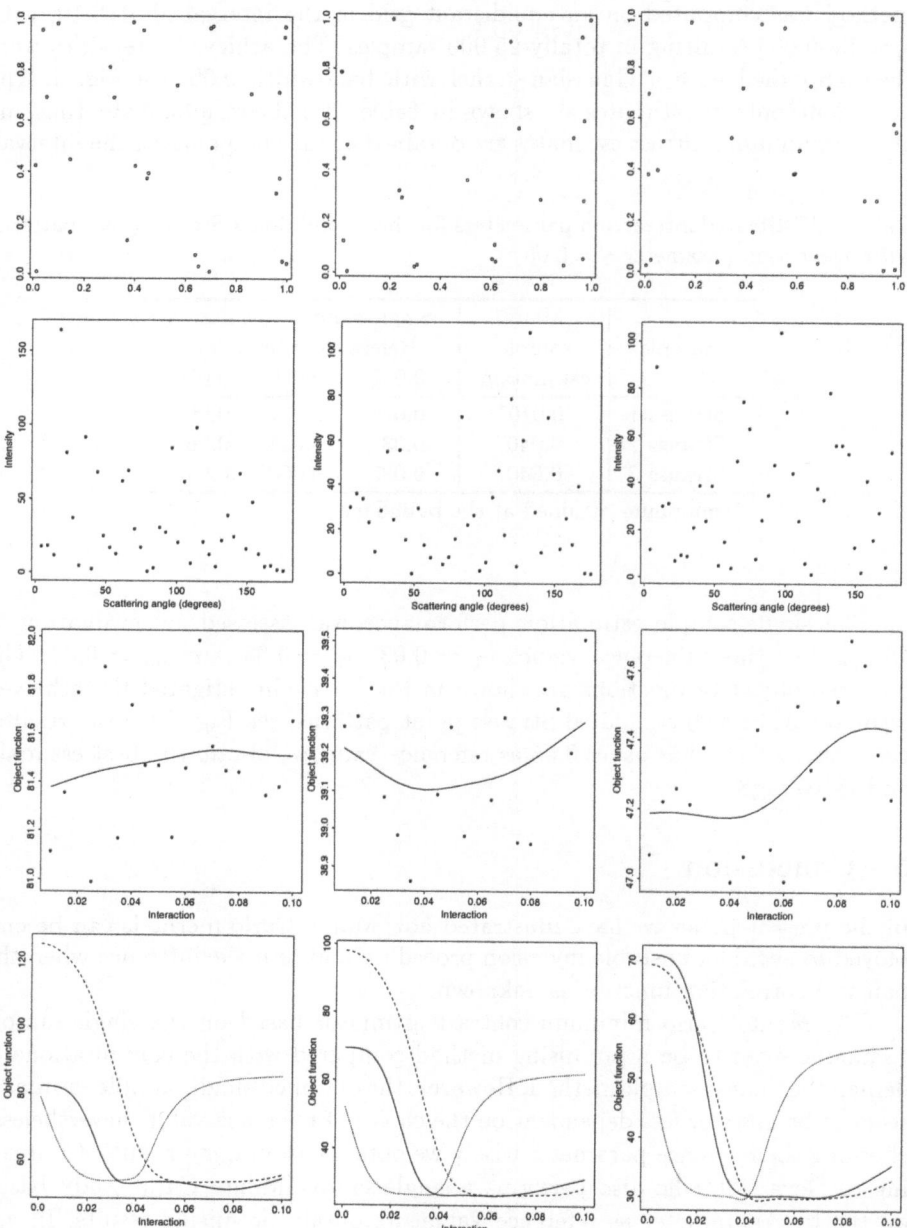

Fig. 5. Parameter estimation by Monte Carlo minimization of $Q(\lambda \mid \varphi)$ for three independent conditional Strauss point processes with known hard-core distance. From top to bottom are shown: The three simulated Strauss point patterns with interaction parameter $\gamma = 0.05$; their corresponding total scattering intensity; the many-sample object function; and the single-sample object function with reference intensities: 0.03 (dotted line), 0.05 (solid line), and 0.07 (dashed line)

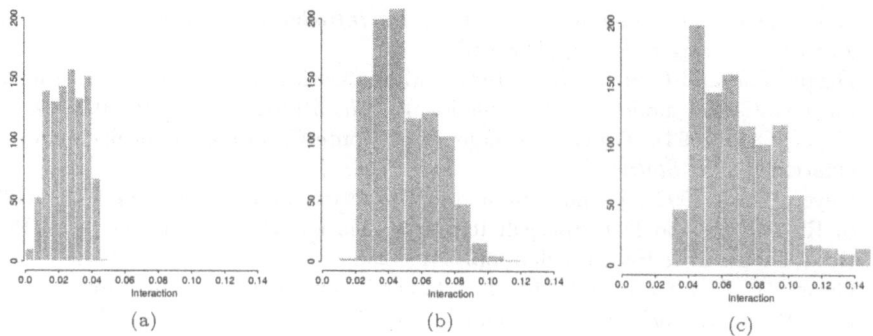

Fig. 6. Histograms for single-sample estimation of γ for $1,000$ conditional Strauss point processes with known hard-core distance. From left to right are shown: The obtained estimates for (a) $\eta_1 = 0.03$; (b) $\eta_2 = 0.05$; and (c) $\eta_3 = 0.07$

procedure there are always two uncertainties in play. One is the Monte Carlo approximation error and one is the variability caused by the sample. For the case of Markov Chain Monte Carlo Maximum Likelihood for exponential families a quite developed asymptotic analysis exists (Geyer, [4]). This has to some extent been carried over to optimal decisions in sequential control problems (Geyer, [5]). It would be very interesting to see if this works in the present setting.

From a physical point of view we are working with an extremely diluted solution of point scatterers ($30 \text{ m}^{-1} = 3 \cdot 10^{-17}\text{nm}^{-1}$). This is, of course, unrealistically low but working with more realistic dilutions will create an computational burden, which will be a challenge to solve.

Above we have focused on the many-sample and single-sample approach for parameter fitting. But there exist a collection of other methods. For a comprehensive reference list see Geyer and Thompson [7].

Given the fact that the techniques seems promising on simulated examples it will be very exiting to see how it can be used to make inference for real solutions containing a random ensemble of point scatterers.

Finally, it is our feeling that with a slight change of object function and the type of Gibbs fields we are considering, these methods should be applicable in many areas of science.

Acknowledgement

The work was supported by the European Union's network Statistical and Computational Methods for the Analysis of Spatial Data. ERB-FMRX-CT96-0095.

References

1. Beeman, W.W., Kaesberg, P., Anderegg, J.W. and Webb, M.B. (1957), Size of particles and lattice defects, *in* S. Flügge, ed., 'Handbuch der Physik', Vol. XXXII, Springer-Verlag, Berlin, pp. 321 – 442.

2. Daley, D.J. and Vere-Jones, D. (1988), *An Introduction to the Theory of Point Processes*, Springer Verlag, New York.

3. Diggle, P.J. and Gratton, R.J. (1984), 'Monte Carlo methods of inference for implicit statistical models (with discussions)', *J.R. Statist. Soc.* B, **46**, 193 – 227.

4. Geyer, C.J. (1994), 'On the convergence of Monte Carlo maximum likelihood calculations', *J.R. Statist. Soc.* B, **56**, 261 – 274.

5. Geyer, C.J. (1996), Estimation and optimization of functions, *in* W.R. Gilks, S. Richardson and D.J. Spiegelhalter, eds, 'Markov chain Monte Carlo in Practice', Chapman & Hall, London, pp. 241 – 258.

6. Geyer, C.J. (1999), Likelihood inference for spatial point processes, *in* O.E. Barndorff-Nielsen, W.S. Kendall and M.N.M. van Lieshout, eds, 'Stochastic Geometry: Likelihood and Computation', Chapman and Hall, London, pp. 141 – 172.

7. Geyer, C.J. and Thompson, E. (1992) , 'Constrained Monte Carlo maximum likelihood for dependent data (with discussions)', *J.R. Statist. Soc.* B, **54**, 657 – 699.

8. Gilks, W.R., Richardson, S. and Spiegelhalter, D.J. (1996), *Markov chain Monte Carlo in Practice*, Chapman & Hall, London.

9. Hanisch, K.-H. (1984), 'Scattering analysis of point processes and random measures', *Math. Nachr.* **117**, 235 – 245.

10. Hastings, W. (1970), 'Monte Carlo sampling methods using Markov chains and their applications', *Biometrika* **57**, 97–109.

11. Kelly, F.P. and Ripley, B.D. (1976), 'A note on Strauss's model for clustering', *Biometrika* **63**, 357 – 360.

12. Kingman, J.F.C. (1993), *Poisson Processes*, 3 edn, Oxford University Press.

13. Metropolis, N., Rosenbluth, A.W., Rosenbluth, M.N., Teller, A.H and Teller, E. (1953), 'Equation of state calculations by fast computing machines', *J. of Chem. Phys.* **21**, 1087 – 1092.

14. Ripley, B.D. (1979), 'Simulating spatial patterns: dependent samples from a multivariate density. Algorithm AS137', *Appl. Statist.* **28**, 109 – 112.

15. Ripley, B.D. (1987), *Stochastic Simulation*, John Wiley & Sons, New York.

16. Sonntag, U., Stoyan, D. and Herman, H. (1981) , 'Random set models in the interpretation of small-angle scattering data', *Phys. Stat. Sol. (a)* **68**, 281 – 288.

17. Stoyan, D. and Stoyan, H. (1994), *Fractals, Random Shapes and Point Fields*, John Wiley & Sons, Chichester.

18. Stoyan, D., Kendall, W.S. and Mecke, J. (1995), *Stochastic Geometry and its Applications*, 2 edn, John Wiley & Sons, New York.

19. Strauss, D.J. (1975), 'A model for clustering', *Biometrika* **63**, 467 – 475.

20. Younes, L. (1988), 'Estimation and annealing for Gibbsian fields', *Annales de l'Institut Henri Poincaré. Section B, Probabilites et statistique*, **24**, 269 – 294.

Thermodynamic Geometry Determines Optimal Temperature Profile in Distillation Column

Bjarne Andresen[1] and Peter Salamon[2]

[1] Ørsted Laboratory, University of Copenhagen
Universitetsparken 5, DK-2100 Copenhagen Ø, Denmark

[2] Department of Mathematical Sciences, San Diego State University
San Diego, CA 92182-7720, U. S. A.

Abstract. General principles of inversion in connection with thermodynamic length optimizations on discrete systems are used to specify the optimal temperature of each tray in a distillation column and consequently the amount of heat to be added or withdrawn in order to maintain that temperature. The thermal efficiency of the distillation column may be improved considerably by permitting such heat exchange on every tray rather than only in the reboiler and the condenser.

1. Introduction

Inversion in a broad sense means determining the interior workings, maybe even the optimal interior workings, of some system based on externally observable quantities. Those observables may either be insufficient to fully determine the interior profile, or several interior profiles may result in the same exterior observables, or both. Stochastic processes, and in particular those based on thermodynamic geometry as the distance measure, have proven very useful for this purpose. In the present study the objective is to use geometric optimization to obtain the interior temperature and concentration profiles in a distillation column for given exterior concentrations (feed and products).

Geometric aspects have always played a prominent role in optimizations. One of the early results, cast in the language of finite-time thermodynamics but actually quite universal [1], is a general bound on the entropy produced in bringing a thermodynamic system from a given initial to a given final state through a sequence of steps. The system is required to follow a prescribed equilibrium path specified by N intermediate equilibrations with the environment. For an arbitrary sequence of reservoirs this bound provides the lower limit on the entropy generated by this process. The derivation also shows how these N intermediate reservoirs may guide the system in the least dissipative fashion from the initial to the final state. Such a process has become known as a *horse-carrot process* and the associated bound as the *horse-carrot theorem* since the system (the 'horse') is coaxed along a sequence of states by successive contacts with generalized baths (the 'carrots'). The intensities of these baths (temperature, pressure, etc.) define successive states in a sequence of equilibria. Replace 'entropy' by the objective function of choice and the thermodynamic equilibrations by the relevant dynamics and the statement still applies for any abstract system.

Fractional distillation [2] is a very old process used to separate a mixture of compounds into its components of specified purity by making use of differences in boiling point. The liquid and vapor phases of the mixture are brought into equilibrium with one another at successive points along a column that is heated at the bottom and cooled at the top. In trayed columns these equilibrations occur on the trays, i.e. at a finite number of discrete points, while packed columns provide continuous contact between liquid and vapor. In this gradient of temperature, the more volatile components will preferentially migrate to the top of the column and the less volatile components to the bottom. In the remainder of this paper we consider only binary separation for simplicity and trayed columns for their correspondence to the step processes mentioned above. It is then the goal of this paper to extend the general results for step processes to the steady state separation inside a distillation column in order to minimize the entropy production.

The internal distribution of temperature and mole fractions of light and heavy components in such a conventional distillation column where heat is added and withdrawn only in the reboiler and the condenser, respectively, is fixed exclusively by the laws of energy and mass conservation. In most columns that leads to an S-shaped curve of temperature versus tray number with most of the variation occurring near the end points of the column, connected with a flat stretch around the feed point. A qualitatively similar picture emerges for the mole fractions. This temperature profile implies that the major part of the entropy production in the distillation process occurs near the ends of the column and is thus not uniformly distributed. Even more importantly, if the number of trays is increased, essentially only the middle flat section is extended while the segments of rapid variation are unchanged. This means that dissipation does not approach zero as the number of trays goes to infinity.

The principle of constant thermodynamic speed for continuous systems, or equivalently equal thermodynamic distance for discrete systems, aims at minimizing total dissipation in the column by distributing it evenly among the trays whatever their number. The result of the optimization, for processes not too far from equilibrium, is that the minimum of the entropy production is achieved when the thermodynamic distance between each pair of trays is kept constant, i.e. they are equidistant in that measure. This distance is controlled by adjusting the tray temperatures appropriately. For any other distribution of the total length along the column dissipation is larger. Obviously such freedom of adjustment does not exist in a conventional adiabatic column. Rather, it is necessary to allow individual heat exchange with each tray to maintain it at the desired temperature.

This mode of operating a distillation column may reduce the entropy production (or loss of exergy) by anywhere from 25% to a factor of 4 or more depending on the particular mixture treated and the size of the column. In general the larger the column, the larger the saving.

In section 2 below, we define the thermodynamic geometry. Section 3 derives the optimal path for a staged process. Section 4 provides some comments on traditional distillation as a prelude to finding the optimal interior temperature profile of the column. This profile is obtained by geometric means in Section 5. Finally section 6 contains an example of the savings which are possible with this new temperature profile.

2. Thermodynamic Length

Thermodynamic geometry lives in the space of all the extensive thermodynamic variables of a system: energy, entropy, volume, amount of material 1, amount of material 2, etc. In this space Weinhold [3, 4] defined a metric that is the second derivative of one extensive quantity (usually entropy or energy) with respect to the other extensive quantities. For example, in the energy picture, the system is described by the internal energy $U(S, V, N_1, N_2, ...)$ expressed in terms of the other extensive quantities. The metric in this picture is

$$M_U = \left\{ \frac{\partial^2 U}{\partial X_i \partial X_j} \right\} \tag{1}$$

with the remainder of the extensive variables represented by X_j. Weinhold's purpose was the calculation of all the usual partial derivatives in traditional static thermodynamics at a particular point. However, any metric invites integration over a range of its variables, in this case leading to the definition of a thermodynamic length L [5] by the usual formula

$$L_U = \int dL_U = \int \sqrt{dX\, M_U\, dX}. \tag{2}$$

Boldface indicates vector quantities. The alternative mixed form of the differential length

$$dL_U = \sqrt{dY\, dX}, \tag{3}$$

where $Y = \partial U/\partial X$ is the vector of intensive quantities conjugate to X, may be useful if some of those elements happen to be constant, e.g. temperature or pressure.

Salamon and Berry [6] found a connection between this thermodynamic length along a continuous process path and the (reversible) availability lost in the process. Specifically, if the system moves via states of local thermodynamic equilibrium from an initial equilibrium state i to a final equilibrium state f in time τ, then the dissipated availability $-\Delta A$ is bounded from below by the square of the distance (i.e. length of the shortest path) from i to f times ε/τ, where ε is a mean relaxation time of the system. If the system proceeds entirely through a sequence of equilibrium states the bound can be strengthened to

$$-\Delta A = \frac{L_{U^2}\,e}{t}, \tag{4}$$

where L_U is the length of the *traversed* path from i to f. This will be the case for example if the process is endoreversible, i.e. the system is reversible in its interior while all irreversibilities are associated with its coupling to the environment [7]. Equality in Eq. (4) is achieved at constant thermodynamic speed $v = dL/dt$, assuming that the process proceeds slowly, i.e. is close to equilibrium with the environment at all times. At higher speeds corrections may be applied [8].

Viewed in the entropy picture $S(U, V, N_1, N_2, ...)$ an analogous expression exists for the total entropy production during the process:

$$\Delta S^u \geq \frac{L_S{}^2 \varepsilon}{\tau}, \tag{5}$$

where the length L_S is then calculated relative to the entropy metric

$$M_S = -\left\{ \frac{\partial^2 S}{\partial X_i \partial X_j} \right\}. \tag{6}$$

When expressed in identical coordinates these two metrics are related by [9] $M_U = -T\,M_S$, where T is the temperature of the system.

3. Optimization of a Step Process

First consider a single step [1] where a system described by the extensive variables X_i is coming to equilibrium with an environment at the intensities Y_i^0, the simplest possible horse-carrot process. The system intensities conjugate to X_i will, in the entropy picture, be denoted by $Y_i = \partial S/\partial X_i$. Specifically $X = (U, V, N, ...)$ and $Y = (1/T, -p/T, \mu/T, ...)$. Then a second order power series expansion of the system entropy S about equilibrium yields

$$S = S^0 + \sum_i Y^0{}_i(X_i - X^0{}_i) - \frac{1}{2}\sum_{ij}(X_i - X^0{}_i)M^0{}_{ij}(X_j - X^0{}_j), \qquad (7)$$

where the superscript zero denotes values at equilibrium with the environment. Then the change in entropy of the system becomes

$$\Delta S = S^0 - S = -\sum_i Y^0{}_i \Delta X_i + \frac{1}{2}\sum_{ij}\Delta X_i M^0{}_{ij}\Delta X_j. \qquad (8)$$

The corresponding change of entropy of the environment (reservoir) is

$$\Delta S^{en} = -\sum_i Y^{en}{}_i X^{en}{}_i, \qquad (9)$$

since its intensive quantities $Y^{en}{}_i$ are constant. At equilibrium $Y^0{}_i = Y^{en}{}_i$, and conservation of energy and matter relates $\Delta X_i = -\Delta X^{en}{}_i$, yielding a total change of entropy of the universe as a result of this small equilibration process

$$\Delta S^u = \Delta S + \Delta S^{en} = \frac{1}{2}\sum_{ij}\Delta X_i M^0{}_{ij}\Delta X_j \qquad (10)$$

or in matrix notation

$$\Delta S^u = \frac{1}{2}\,\Delta X\,M\,\Delta X. \qquad (11)$$

The superscript zero on M indicating an equilibrium value will be presupposed in the following.

Next consider N consecutive small steps. The nth step in this sequence produces the dissipation

$$\Delta S^{u\,n} = \frac{1}{2}(Ds^n)^2, \qquad (12)$$

where

$$Ds^n = \sqrt{\Delta X^n M_{s^n} \Delta X^n} \qquad (13)$$

to first order in the stepsize is the thermodynamic length of the step. Summing over all N steps in the process, we find using the Cauchy-Schwarz inequality

$$\Delta S^u = \frac{1}{2} \sum_{n=1}^{N} (D_S^n)^2 \geq \frac{1}{2N} \left(\sum_{n=1}^{N} D_S^n \right)^2 = \frac{L_S^2}{2N}, \tag{14}$$

where $L_S = \sum D_S^n$ is the thermodynamic length of the full N-step process. Similar to the continuous results Eqs. (4) and (5), equality is achieved only when all D_S^n are equal.

This bound and the optimal equality of all steps (in thermodynamic geometry), valid for many small steps, are the crucial results used in the following analysis of the distillation process.

4. A Classical Distillation Column

A conventional binary distillation column is constructed as sketched in Fig. 1 with feed entering at the steady state rate F around the middle of the column and being separated into distillate leaving the top at rate D and waste W (also called bottoms) leaving at the bottom. The mole fractions x_F, x_D, and x_W refer to the light component in those three streams. The separation is affected by a heat flow q_W entering the bottom at temperature T_W and a corresponding flow q_D leaving the top at the lower temperature T_D, thus creating vapor flows V and liquid flows L inside the column. In all large-scale distillation columns interior equilibration between vapor and liquid is achieved on trays where vapor bubbles through a thin layer of liquid.

The reversible separation (unmixing) of a feed stream F into its pure components ($x_D = 1$, $x_W = 0$) requires the power [2]

$$W_{rev} = -RT_D [x_F \ln x_F + (1-x_F)\ln(1-x_F)] F, \tag{15}$$

where R is the gas constant. Mass and energy conservation inside a conventional distillation column, on the other hand, requires a minimum heat flow

$$q_{min} = FR \frac{T_D T_W}{T_W - T_D} \tag{16}$$

as derived in any chemical engineering textbook [2]. Using the Carnot efficiency, this is equivalent to a theoretical power

$$W_{min} = \frac{T_W - T_D}{T_W} q_{min} = FRT_D. \tag{17}$$

Thus the thermal efficiency of a distillation column can never exceed

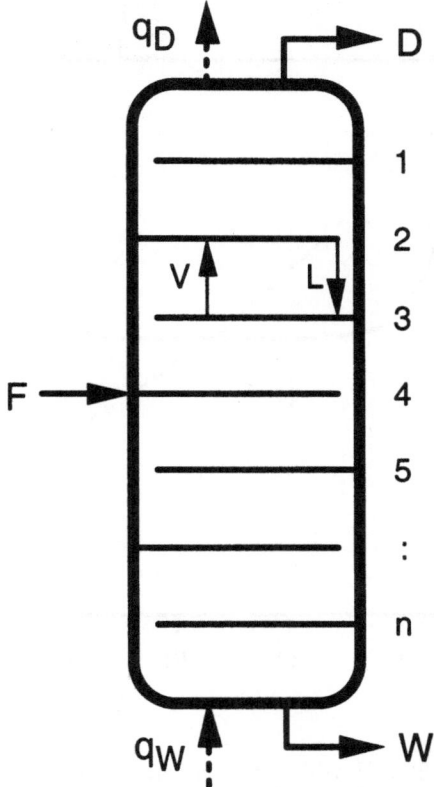

Figure 1. Sketch of a conventional distillation column with feed, distillate, and waste (bottoms) rates F, D, W, heating and cooling rates q_W and q_D, and tray numbers n.

$$\varepsilon = W_{rev}/W_{min} = -[x_F\ln x_F + (1-x_F)\ln(1-x_F)] \tag{18}$$

which has a maximum of ln2, about 70%, for an equal mixture feed, $x_F=0.5$. Deviation from this balanced feed as well as any non-idealities in the system will further reduce the efficiency. Considering that distillation consumes a sizeable fraction of the world energy demand, this is an unfortunate situation. Of course, a completely equivalent analysis can be carried out in terms of entropy produced during the separation.

The internal distribution of temperature and mole fractions of light and heavy components in a conventional distillation column where heat is added and withdrawn only in the reboiler and the condenser, respectively, is fixed exclusively by the conservation laws for energy and mass. In most long columns this leads to an S-shaped curve of composition versus tray number with most of the variation occurring near the end

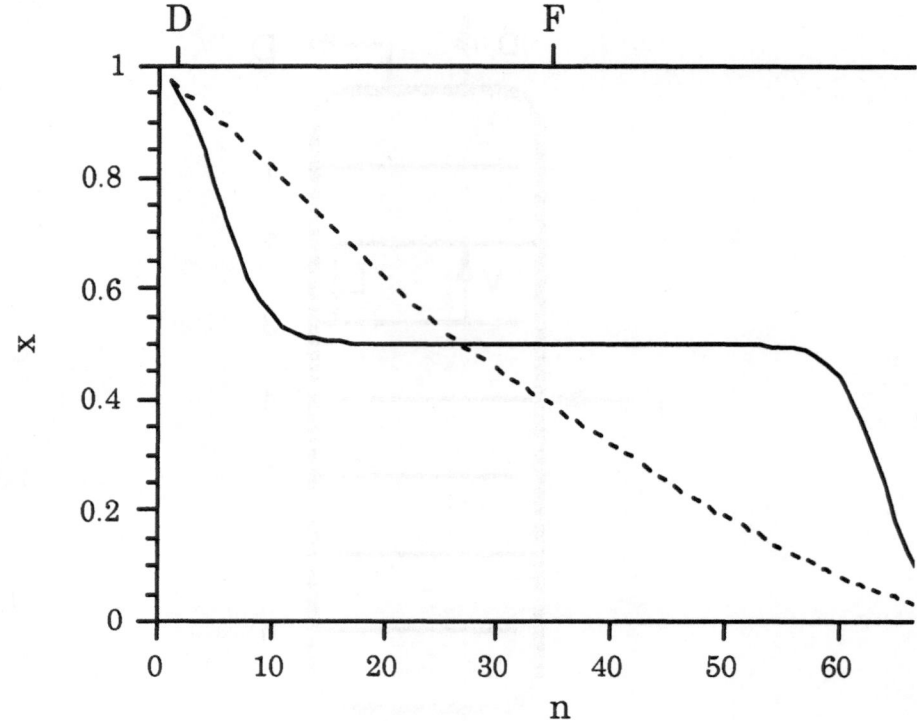

Figure 2. The liquid composition profile x as a function of tray number, counted from the condenser, for conventional (solid) and equal-thermodynamic-distance (dashed) separation of an ideal benzene-toluene system.

points of the column, connected with a flat stretch around the feed point (see Fig. 2). A qualitatively similar picture emerges for the temperature. This composition profile implies that the major part of the entropy production in the distillation process occurs near the ends of the column and is thus not uniformly distributed. Even more importantly, if the number of trays is increased, essentially only the middle flat section is extended while the segments of rapid variation are unchanged. This means that dissipation does not approach zero as the number of trays goes to infinity.

5. Optimal Temperature Profile

The principle of equal thermodynamic distance shows us how to minimize the total dissipation in a trayed column by distributing the dissipation evenly among the trays whatever their number [10, 11]. Above, a general quasistatic step process was optimized, i.e. a process composed of N discrete steps where the system equilibrates fully after each step. The standard description of a distillation column is exactly such a process where it is assumed that gas and liquid come to equilibrium at a particular temperature on each tray. Entropy is produced when the up- and down-moving flows encounter liquid on the next tray at slightly different temperature and composition. The concept of thermodynamic length defined in refs. [1, 6] not only provides a lower bound on dissipation, it also predicts which path will achieve that bound, namely operation with equal thermodynamic distance between the trays.

Two important consequences of this general result are immediate: dissipation (here entropy production) must be equally distributed along the column; and the total dissipation approaches zero as N, the number of trays, goes to infinity (Eq. (14)), i.e. the separation becomes reversible. Neither are satisfied in ordinary distillation columns.

In binary distillation 8 extensive quantities are involved on each tray. In the energy picture, these quantities are the entropy, volume, mole number of light component, and mole number of heavy component for each phase, i.e. for both vapor and liquid: $X = (S_V, V_V, N_{1V}, N_{2V}, S_L, V_L, N_{1L}, N_{2L})$. This leads to an 8×8 metric matrix

$$
\mathbf{M_U} = \begin{array}{cc} \quad V \quad\quad L \\ \begin{pmatrix} \mathbf{M_U}^V & \mathbf{0} \\ \mathbf{0} & \mathbf{M_U}^L \end{pmatrix} \begin{array}{c} V \\ L \end{array} \end{array} \tag{19}
$$

Fortunately a number of relations allows one to reduce the dimensionality of the problem dramatically. First of all, as already indicated by the zeroes, the two physical states are usually considered energetically noninteracting, at once making $\mathbf{M_U}$ block-diagonal. After a partial Legendre transform to the mixed intensive-extensive variables $X = (T, p, N_1, N_2)$ each of the submatrices $\mathbf{M_U}^V$ and $\mathbf{M_U}^L$ may be further block-diagonalized to

$$
\mathbf{M_U}\,(T, p, N_1, N_2) = \begin{pmatrix} -G_{TT} & G_{Tp} & 0 & 0 \\ G_{pT} & -G_{pp} & 0 & 0 \\ 0 & 0 & G_{N_1N_1} & G_{N_1N_2} \\ 0 & 0 & G_{N_2N_1} & G_{N_2N_2} \end{pmatrix}, \tag{20}
$$

where the subscripts indicate partial derivatives. Note that the dependent function is now Gibbs' free energy G corresponding to the variables (T, p, N_1, N_2).

Constant pressure in the column eliminates the second row and column. Next we recombine the molar quantities N_1 and N_2 into the total amount of material, $N = N_1 + N_2$ and the molar fraction of component 1, $x = N_1/(N_1 + N_2)$ (in the liquid phase, the letter y is traditionally used for the vapor phase). The purpose is once again to effectively eliminate one variable, N since it only describes the total scaling of the process. The metric matrix for the liquid is now down to

$$
M_U{}^L(T, x, N) = \begin{pmatrix} -G_{TT} & 0 & 0 \\ 0 & -\dfrac{G_{N_1N_2} N^2}{x(1-x)} & 0 \\ 0 & 0 & 0 \end{pmatrix} \tag{21}
$$

with an equivalent expression for the vapor. The square of the length element is then

$$
(dL_U)^2 = -G_{TT}{}^V dT^2 - \frac{G_{N_1N_2}{}^V (N^V)^2}{y(1-y)} dy^2 - G_{TT}{}^L dT^2
$$
$$
- \frac{G_{N_1N_2}{}^L (N^L)^2}{x(1-x)} dx^2
$$

$$
= \left[-G_{TT}{}^V - \frac{G_{N_1N_2}{}^V (N^V)^2}{y(1-y)} \left(\frac{dy}{dT}\right)^2 - G_{TT}{}^L \right.
$$
$$
\left. - \frac{G_{N_1N_2}{}^L (N^L)^2}{x(1-x)} \left(\frac{dx}{dT}\right)^2 \right] dT^2 \tag{22}
$$

since the temperatures in the two phases are equal. The vapor and liquid flows N^V and N^L are identified with the actual material flows V and L in the column at that particular tray (see Fig. 1).

The G_{TT} terms are related to the ordinary heat capacities of the vapor and liquid mixtures, respectively,

$$
G_{TT}{}^L = -\frac{L}{T} \left[x \, C_1{}^L + (1-x) \, C_2{}^L \right] , \tag{23}
$$

where C_i is the molar heat capacity at constant pressure of component i, with a similar equation for the vapor. The terms $G_{N_1N_2}$ are cross derivatives of the chemical potentials,

$$qG_{N_1N_2}{}^L = \frac{\partial \mu_{2^L}}{\partial L_1} = \frac{\partial \mu_{1^L}}{\partial L_2}, \tag{24}$$

i.e. derivatives of the heavy chemical potential μ_2 with respect to the light flow $L_1 = xL$ and vice versa. The vapor-liquid equilibrium conditions $\mu_1{}^V = \mu_1{}^L$ and $\mu_2{}^V = \mu_2{}^L$ have also been invoked. For ideal mixtures, $G_{N_1N_2} = -RT/N$, where R is the gas constant. Again analogous expressions exist for the vapor phase.

Above we have made use of the constancy of pressure to reduce the size of the problem in a fairly straightforward fashion. Mass conservation is applied in a more unusual way. In the traditional description of a distillation tray [2] the material flows of each of the two components entering and leaving a particular tray n are balanced:

$$V^{n+1} y_1{}^{n+1} + L^{n-1} x_1{}^{n-1} = V^n y_1{}^n + L^n x_1{}^n$$

$$V^{n+1} y_2{}^{n+1} + L^{n-1} x_2{}^{n-1} = V^n y_2{}^n + L^n x_2{}^n. \tag{25}$$

As above, V and L are the vapor and liquid flow rates leaving the tray indicated in the superscript, and x_i and y_i are the molar fractions of component i in the liquid and vapor phases, respectively. These quantities are pictured in Fig. 3.

Calculating the entropy produced in this mixing process of four streams is a bit cumbersome. Instead we consider the entropy produced in connection with the generation of a small bubble of vapor on tray n+1 at temperature T^{n+1} and its absorption on tray n. The vaporization creates no entropy since it is an equilibrium process; the vapor of composition $(y_1{}^{n+1}, y_2{}^{n+1})$ is in equilibrium with the liquid of composition $(x_1{}^{n+1}, x_2{}^{n+1})$ from which it is generated at that temperature. On arriving on tray n we split the process of assimilation of the bubble on that

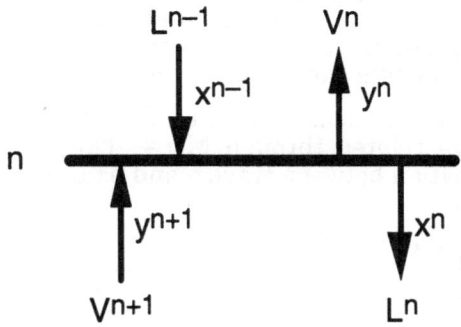

Figure 3. Definition of quantities around tray n.

tray into two parts: first exchange of only heat and work with the liquid-vapor system on tray n, followed by exchange of materials now at the proper temperature. In the first step the tray fluid acts as the heat and work reservoir for the equilibration, the type of simple horse-carrot process described in Sect. 3. The entropy production in this step is given to first order in the step size by Eq. (10) which, when expressed in the mixed intensive-extensive formulation of Eq. (3), has only two terms in the sum,

$$\Delta S^{u\,n} = \frac{1}{2}\left[\Delta\left(\frac{1}{T^n}\right)\Delta U^n \Delta\left(\frac{p}{T^n}\right)\Delta V^n\right]. \tag{26}$$

This is easier to calculate using the energy picture and the equivalence $M_U = -TM_S$ [9] so that

$$\Delta S^{u\,n} = \frac{1}{2T^n}\left[\Delta T^n \Delta S^n \Delta p \Delta V^n\right] \tag{27}$$

since the constancy of the pressure knocks out the second term. In this thermo-mechanical equilibration some of the vapor in the bubble (of composition y^{n+1}) will condense so that the bubble volume now contains some liquid of composition x^n plus some vapor of composition y^n due to the new temperature T^n. Introducing the constant pressure saturation heat capacity C_r of the two phase mixture, i.e. the effective heat capacity for gas in equilibrium with liquid [12], so that $\Delta S = C_r\Delta T/T$ makes

$$\Delta S^{u\,n} = \frac{1}{2}\frac{C_{r^n}}{(T^n)^2}(\Delta T^n)^2. \tag{28}$$

Once the temperature and the pressure in our bubble have equilibrated to the surroundings, we proceed to the second step during which each fluid phase can mix with the reservoir fluid. Since in a binary mixture T and p fix the composition, this second step occurs reversibly. The total entropy generation for this tray is thus given by Eq. (28) above.

On a strictly formal basis we could also recall that the energy and entropy metrics are related through $M_U = -TM_S$ [9] and thus quickly arrive at the dissipation between trays n and n+1, from eq. (22),

$$\Delta S^{u\,n} = \frac{1}{2}\left(\frac{\Delta L_{U^n}}{T^n}\right)^2 = \frac{1}{2}\frac{C_{r^n}}{(T^n)^2}(\Delta T^n)^2, \tag{29}$$

where C_r may be expressed as

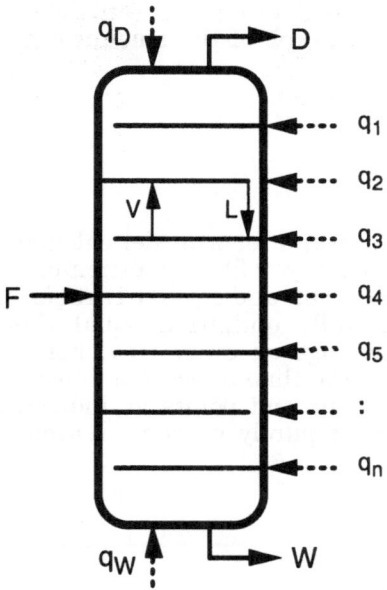

Figure 4. Sketch of an equal-thermodynamic-distance distillation column with feed, distillate, and waste (bottoms) rates F, D, W as in the traditional column Fig. 1, but with heating or cooling on all n trays.

$$C_r = V \left[y \, C_1^V + (1-y) \, C_2^V \right] - G_{N_1 N_2}^V \frac{TV^2}{y(1-y)} \left(\frac{dy}{dT} \right)^2$$

$$+ L \left[x \, C_1^L + (1-x) \, C_2^L \right] - G_{N_1 N_2}^L \frac{TL^2}{x(1-x)} \left(\frac{dx}{dT} \right)^2, \qquad (30)$$

and where all quantities are related to the particular tray n, and □ indicates the difference between trays n and n+1.

The computational procedure is to integrate Eq. (29) from the distillate temperature T_D to the reboiler temperature T_W (both of course given by the required product purities) to obtain the total thermodynamic column length L_S. The distance from one tray to the next must then be fixed at $D_S = L_S/N$ for optimal performance by adjusting the tray temperatures appropriately according to Eq. (13).

Obviously such freedom of adjustment does not exist in a conventional adiabatic column where heat is added and removed only in the reboiler and condenser, respectively (Fig. 1). Rather, it is necessary to allow individual heat exchange with each tray to maintain it at the desired temperature (see Fig. 4). This heat addition/removal is of course part of the energy balance used above. The result of the whole calculation is either a graph similar to Fig. 2 specifying the temperature of each tray in the column or a graph of the amount of heat added/removed at each tray.

6. Example

Figure 5 is such a graph for a model calculation on an ideal 50/50 benzene-toluene mixture (x_F=0.5). Separation of 1 mole of mixture into 99% pure products (x_D=0.99 and x_W=0.01) requires an exergy expenditure of 191 J in the fully optimized equal thermodynamic distance column with heat exchange on every tray. For comparison the exergy requirement is 842 J, more than a factor of four larger, in a conventional column heated only at the end points as indicated with the two filled circles. If for economic or purely practical reasons one does not wish to

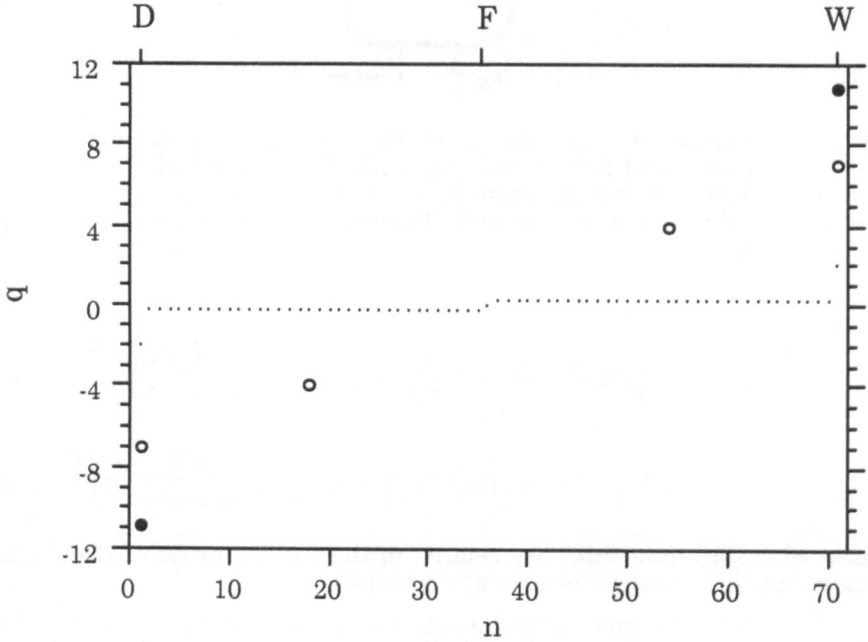

Figure 5. Amount of heat in J added per mole of benzene-toluene feed on the individual trays of a 71-tray column operated in the traditional fashion with heat added only in the reboiler and withdrawn only in the condenser (•); operated according to equal thermodynamic separation specifications with heat added or withdrawn on each tray (·); and operated with just 2 additional heat exchange points, optimally located (o).

mount heat exchangers on every tray in the column, the first step of improvement will be to add just two heat exchangers at appropriate locations, making a total of four positions of heat exchange. Such a column only needs 423 J to do the job, a saving by a factor of two compared to the conventional column.

It should be mentioned that the total amount of *heat* used to perform a certain equal-thermodynamic-distance separation is only marginally different from that required by a conventional column, but a large part of it is used over a much smaller temperature difference than T_W to T_D, leading to a correspondingly smaller *exergy* requirement. The gradual addition and withdrawal of heat lends itself to coupling with other sources and sinks of heat at the plant which might otherwise have been lost or degraded. This distillation system is the first application of the equal thermodynamic distance principle to a chemical system. However, the entire analysis is not restricted to a distillation column but applies equally well to any staged process, be it gas separation by diffusion, staged refrigeration, or batch chemical reactions.

Acknowledgements

The work reported in this outline has been done in collaboration with James D. Nulton and Gino Siragusa.

References

[1] J. Nulton, P. Salamon, B. Andresen, and Q. Anmin: "Quasistatic processes as step equilibrations", J. Chem. Phys. **83** (1985) 334.

[2] See for example C. J. King: "Separation processes", McGraw-Hill, New York (1971).

[3] F. Weinhold: "Metric geometry of equilibrium thermodynamics", J. Chem. Phys. **63** (1975) 2479.

[4] F. Weinhold: "Geometrical aspects of equilibrium thermodynamics", in D. Henderson and H. Eyring (eds.), *Theoretical chemistry, advances and perspectives*, Vol. 3 (Academic Press, New York) (1978).

[5] P. Salamon, B. Andresen, P. Gait, and R. S. Berry: "The significance of Weinhold's length", J. Chem. Phys. **73** (1980) 1001, 5407E.

[6] P. Salamon and R. S. Berry: "Thermodynamic length and dissipated availability", Phys. Rev. Lett. **51** (1983) 1127.

[7]   M. H. Rubin: "Optimal configuration of a class of irreversible heat engines. I", Phys. Rev. A **19** (1979) 1272.

[8]   B. Andresen, and J. M. Gordon: "On constant thermodynamic speed for minimizing entropy production in thermodynamic processes and simulated annealing", Phys. Rev. E. **50** (1994) 4346.

[9]   P. Salamon, J. Nulton, and E. Ihrig: "On the relation between energy and entropy versions of thermodynamic length", J. Chem. Phys. **80** (1984) 436.

[10] P. Salamon and J. Nulton: "The geometry of separation processes: A horse-carrot theorem for steady flow systems", Europhys. Lett. **42** (1998) 571.

[11] P. Salamon, B. Andresen, J. Nulton, and G. Siragusa: "The horse-carrot theorem and geometric optimization of distillation", (in preparation) (1999).

[12] J. S. Rowlinson: "Liquids and liquid mixtures", Plenum, New York (1969).

Inversion of Seismic AVO Data

Torben Bach[1], Thomas B. Espersen[1], Jacob M. Pedersen[1], Richard Hinkley[2],
William R. Pillet[3] and Klaus B. Rasmussen[1]

1 Ødegaard A/S, Kroghsgade 1, 2100 Copenhagen Ø, Denmark
2 Texaco Britain Limited, 1 Westferry Circus, Canary Wharf, London E14 4HA,
United Kingdom
3 Fina Exploration Minh Hai B.V., 157 Nguyen Van Troi Street, Phu Nhuan
District, Ho Chi Minh City, Vietnam

1. Introduction

The combination of AVO (Amplitude variation Versus Offset) analysis and seismic inversion gives new possibilities for fluid detection and lithology prediction. Inversion of seismic AVO data for physical properties enables direct lithological interpretation.

We use a 3D multi-trace globally optimised seismic inversion method. This method incorporates spatial variation of wavelet frequency and energy, the possibility for up-link to dip and continuity data for horizontal continuity, and use of stacking or migration velocities. These features make it possible to correctly invert very large data cubes. Unlike results obtained from conventionally weighted stack type AVO sections [Smith and Gidlow (1987)], the implementation of AVO data in actual seismic inversion results in inherent benefits from the inversion (signal-noise optimisation, wavelet removal and absolute values of physical properties) and thus makes the obtained petrophysical sections more reliable. The basic method and some case studies are presented.

2. Implementation

AVO data used for the inversion can be either intercept-gradient, angle stack or offset stack data. The implementation is based on the Shuey approximation [Shuey (1985)],

$$(1) \qquad R_P(\theta) = I + G \sin^2(\theta),$$

with implementation of a Gardner constant [Gardner, 1974] and a low-frequency V_p/V_s model estimated from well log data.

For intercept-gradient data the first step is calculation of acoustic and shear reflectivity sections, cf. Equations (2 and 3). Hereafter, an acoustic wavelet and a shear wavelet are estimated for use in the inversions. The inversion results are acoustic impedance and shear impedance, which can be used to calculate additional lithological identification volumes, such as Poisson's ratio, V_P-V_S ratio, Lamé constants etc. [Mavko, 1998].

(2) $$R_P = I$$

(3) $$R_S = \frac{I/(1+a) - G}{2(2V_S/V_P)^2} + \frac{a}{2}\frac{I}{1+a}$$

For the angle stack and offset stack data [Resnick, 1993] the concept of effective impedance in Equation (4) makes use of the convolutional model possible [Connolly, 1999]. For each angle stack an effective impedance log is calculated inserting the median angle for each angle stack in the Shuey approximation. Individual wavelets are estimated for each angle stack, ensuring that effects of frequency variation with offset and NMO effects present in the data are reduced in the inversion results, making comparison of the effective impedance at different angles possible. After the inversion has been performed additional sections based on the inversion results can be calculated (acoustic and shear impedance, Poisson's ratio etc.). The inversion of offset stacks follows the same concept, only varying the effective angle with depth compensating for the offset stack not having a constant median angle.

(4) $$Z_\theta = Z_p \exp([\log(Z_P) - 2\log(Z_S) + C]\sin^2(\theta))$$

The inversion method used is a commercial method developed to invert full stack data for acoustic impedance. As we can apply the same principle to the AVO data we have the special features from the commercial package:

- Global optimisation.
- Spatial variation of wavelet frequency and energy.
- Up-link to dip and continuity volumes for horizontal continuity constrain.
- Use of stacking or migration velocities.
- Well log constraints can be varied, making the inversion applicable both for exploration purposes in areas with little well control and for detailed reservoir characterisation in areas with high well control.

These features make the inversion method well suited for AVO inversion of very large data volumes.

The interpretation of the generated inversion results can be used to estimate lithologies. This is done either by interpretation of the inversion results directly or by comparing the relationship of the inverted properties with lithology in the well logs. This is where AVO inversion is superior to traditional inversion of full-stack data where only acoustic impedance is estimated.

Figure 1 shows a cross-plot of acoustic impedance versus density with empirical curves for some common lithologies [Castagna, 1993]. Contours of acoustic impedance are also shown. The plot illustrates that the acoustic impedance of sand and shale are similar, and that the sand-shale variation cannot be separated by acoustic impedance alone.

V_P versus Density

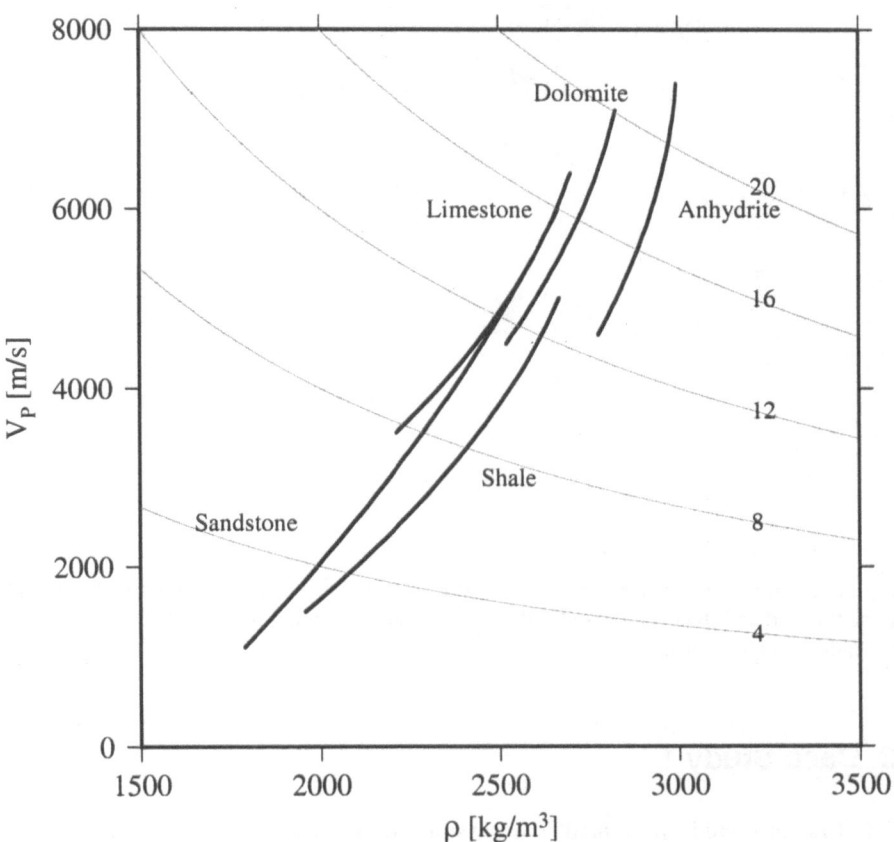

Figure 1. Empirical relations between acoustic velocity and density for major lithologies [Castagna, 1993]. Contours of constant acoustic impedance ($\times 10^6$) are overlaid. Sand and shale fall within the same acoustic impedance region making separation uncertain.

Figure 2 shows a cross-plot of acoustic impedance versus Poisson's ratio. Empirical relations for sand and shale are plotted, and general trends of varying pore fluid and compaction are indicated. The plot illustrates that by using a combination of acoustic impedance and Poisson's ratio it is possible to separate sand from shale. Furthermore, the plot indicates that water filled sand can be separated from gas filled sand.

Acoustic impedance vs. Poisson's ratio

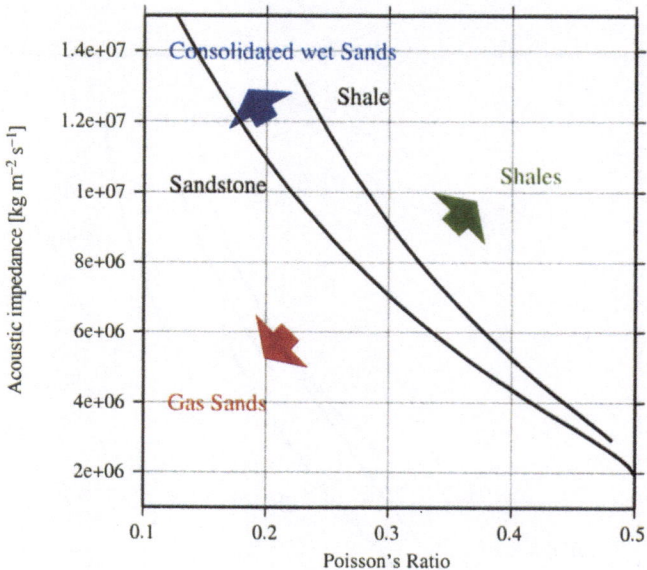

Figure 2. Empirical relation between acoustic impedance and Poisson's ratio. Empirical lines for sand and shale are plotted and general trends of varying pore fluid and compaction are indicated with arrows.

3. Case Study 1

The first case study presented is a possible reservoir situation from the Victory Field, West of Shetland. Seismic data were acquired and true amplitude processed by Texaco Britain Limited. AVO seismic attributes, intercept and gradient, were calculated from the processed data and subsequently inverted for acoustic impedance and shear impedance. Well log data were available for one well situated close to the possible reservoir. An additional well penetrating the reservoir was omitted from the inversion and used by Texaco Britain Limited for quality control of the results.

Figure 3 shows the intercept data (R_p), the calculated shear wave reflectivity (R_s), and the acoustic and shear impedance inversion results without addition of low-frequency information. The shear seismic shown has been calculated from Equation (3) utilising the intercept and gradient data, in combination with a low-frequency V_p/V_s model and a Gardner constant both estimated from log information.

The gas in the possible reservoir is indicated by a large drop in relative acoustic impedance (see Figure 3; left-hand side). However, from the intercept data alone it is not possible to determine whether the acoustic anomaly is caused by the presence of gas or a by a variation in the porosity.

35

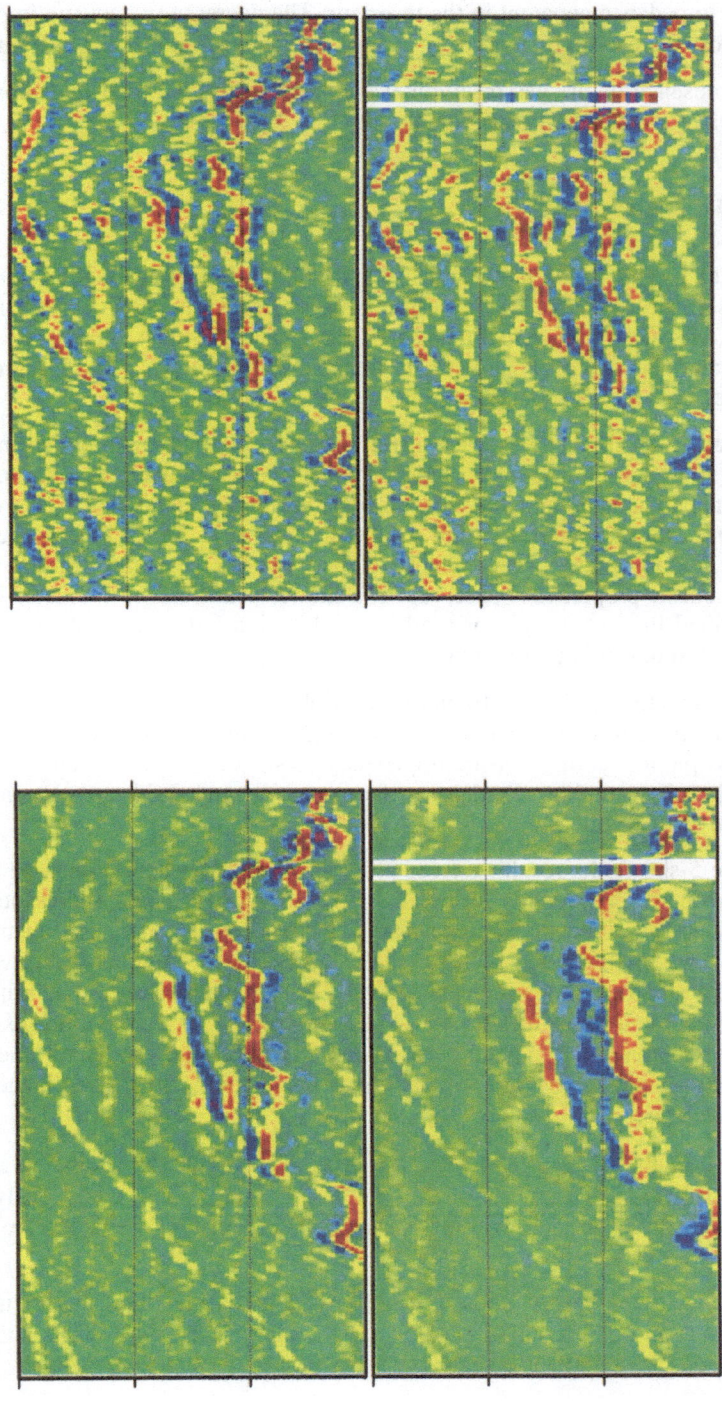

Figure 3. Inversion of intercept and calculated shear seismic. Upper left: The intercept data. Upper right: Calculated shear data. Lower left: Inverted acoustic impedance. Lower right: Inverted shear impedance. A decrease in relative acoustic impedance correlated with no relative variation of the shear impedance indicates gas. The size of the shown section is 0.3 s TWT and 3125 m.

By examining the shear impedance section (see Figure 3; right-hand side) the cause of the acoustic impedance anomaly can be determined. There is no decrease in shear impedance apparent in the reservoir. The decrease in acoustic impedance is therefore not caused by a porosity change, but by a variation in pore fluid.

The presence of the gas reservoir was confirmed by the well log information held back by Texaco Britain Limited.

4. Case Study 2

The second case study presented is based on three seismic angle stack data volumes made available by Fina Exploration Minh Hai B.V., Vietnam. The seismic data was processed to preserve the true amplitudes, and three sets of seismic angle stack data for the angle intervals 0-20 degrees, 20-35 degrees and 35-50 degrees were calculated.

The angle stack volumes were inverted separately for angle impedances, as given in Equation (4). Acoustic impedance, shear wave impedance and Poisson's ratio were calculated on the basis of the angle stack inversion results. The acoustic impedance and the Poisson's ratio results are shown in Figure 4 with the actual well log data inserted. A very good match between the measured well log and the inverted elastic properties is observed. The shown results can be evaluated directly as described in case study 1 or used in combination with the available well log data for lithological interpretation.

In co-operation with Fina Exploration Minh Hai B.V. a well was selected for use in lithological interpretation. The well log was divided into lithological units and cross-plotted as the acoustic impedance versus the Poisson's ratio. In theory the sand-shale variation, the degree of compaction and fluid content should be separated as in Figure 2.

Figure 5 shows the resulting cross-plot. The individual data points are colour-coated with the gamma-ray values and plotted with symbols representing the lithology. Large circles indicate gas sands, medium circles indicate oil sands and water sands, small circles indicate shaly sands, large triangles indicate sandy shales and small triangles indicate shales. The cross-plot shows that lithologies with high gamma-ray values lie above the empirical shale line, while lithologies with low gamma-ray values lie between the empirical shale and sand lines. This corresponds with the overall separation of the symbols in circles (sands) and triangles (shales) on the plot.

In general, the shales (small triangles) and the gas sands (large circles) are well defined. However, the cross-plot shows that oil sands, water sands, shaly sands and sandy shales, all fall within the same ranges making separation of these lithologies uncertain.

Based on these observations it was decided to define three major lithology units: a wet sand unit (consisting of oil sands, water sands, shaly sands and sandy shales), a gas sand unit and a shale unit.

37

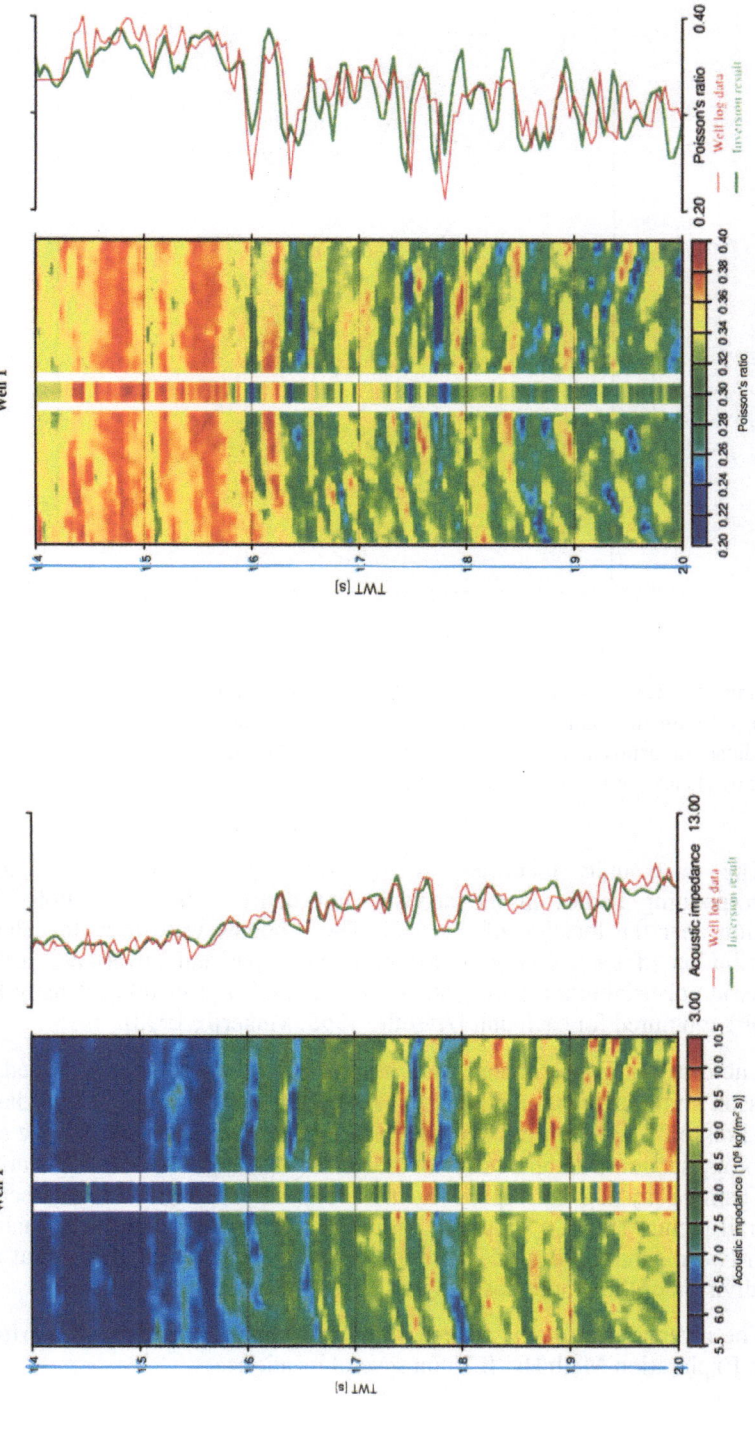

Figure 4. Left: Acoustic impedance with well log inserted. Right: Poisson's ratio with well log inserted. The sections were calculated from the angle stack inversion results. A very good match is observed between the inverted values and the measured values from the well log.

38

Well 1

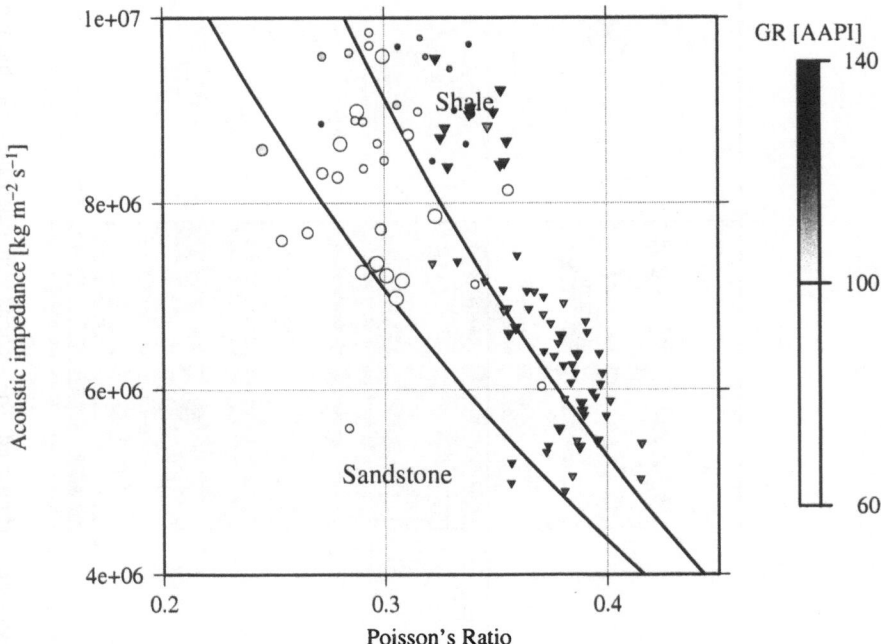

Figure 5. Cross-plot of acoustic impedance versus Poisson's ratio from Well 1 colour-coated with the gamma-ray values. Large circles indicate gas sands, medium circles indicate oil sands and water sands, small circles indicate shaly sands, triangles indicate sandy shales (large) and shales (small).

The three units identified in the well log data were used to isolate corresponding bodies in the acoustic impedance section and Poisson's ratio section near the location of the well. These bodies were used to calculate the distribution of each unit in terms of mean value and covariance assuming a Gaussian distribution. Figure 6 shows the calculated probability density functions (PDF) contoured for each unit [Avseth , 1998; Mukerji , 1998].

The distributions estimated were used to evaluate the acoustic impedance and Poisson's ratio results for lithology. Figure 7 shows the acoustic impedance (top) and Poisson's ratio sections (middle) with the well logs inserted, and the estimated lithology section with an overlay of the gamma-ray log (bottom). The variations of the gamma-ray log correspond to the predicted lithologies in the section. The overall picture shows a sequence of interbedded sands (white) and shales (grey) with two distinct bodies of gas sands (black). The results suggest that a further subdivision of the lithologies into units would be possible.

The inversion results have been verified by subsequent drillings performed by Fina Exploration Minh Hai B.V. on several locations.

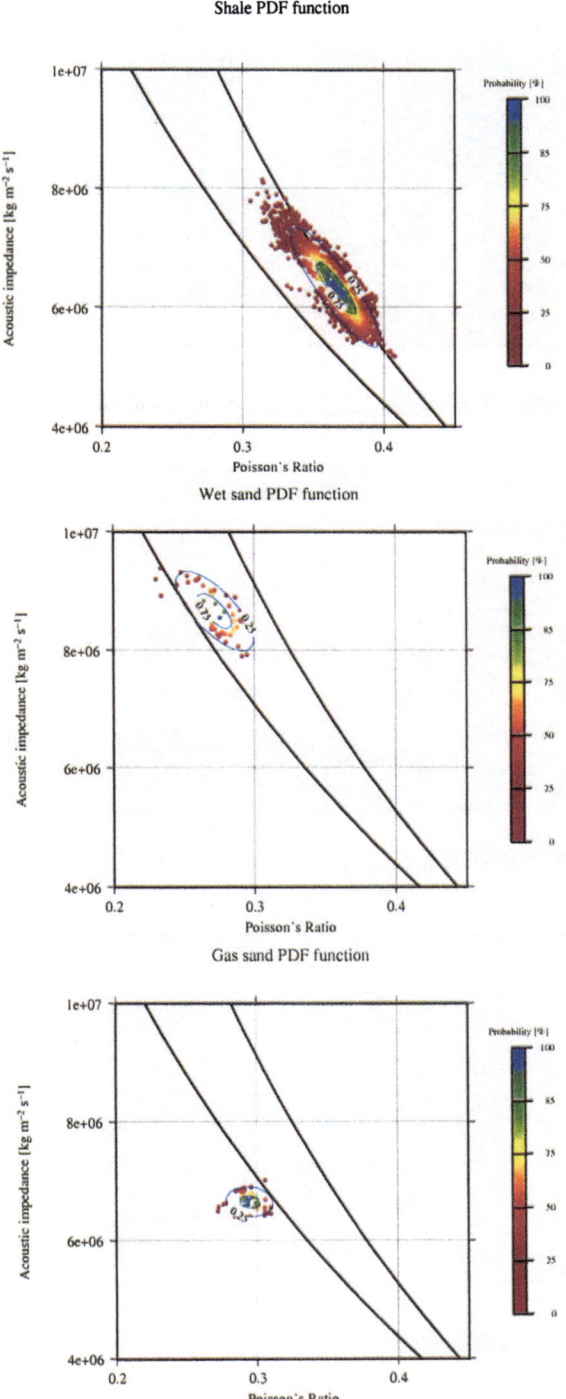

Figure 6. The contour lines for the PDF functions for each body (wet sand, shale and gas sand) identified in the acoustic impedance and Poisson's ratio sections.

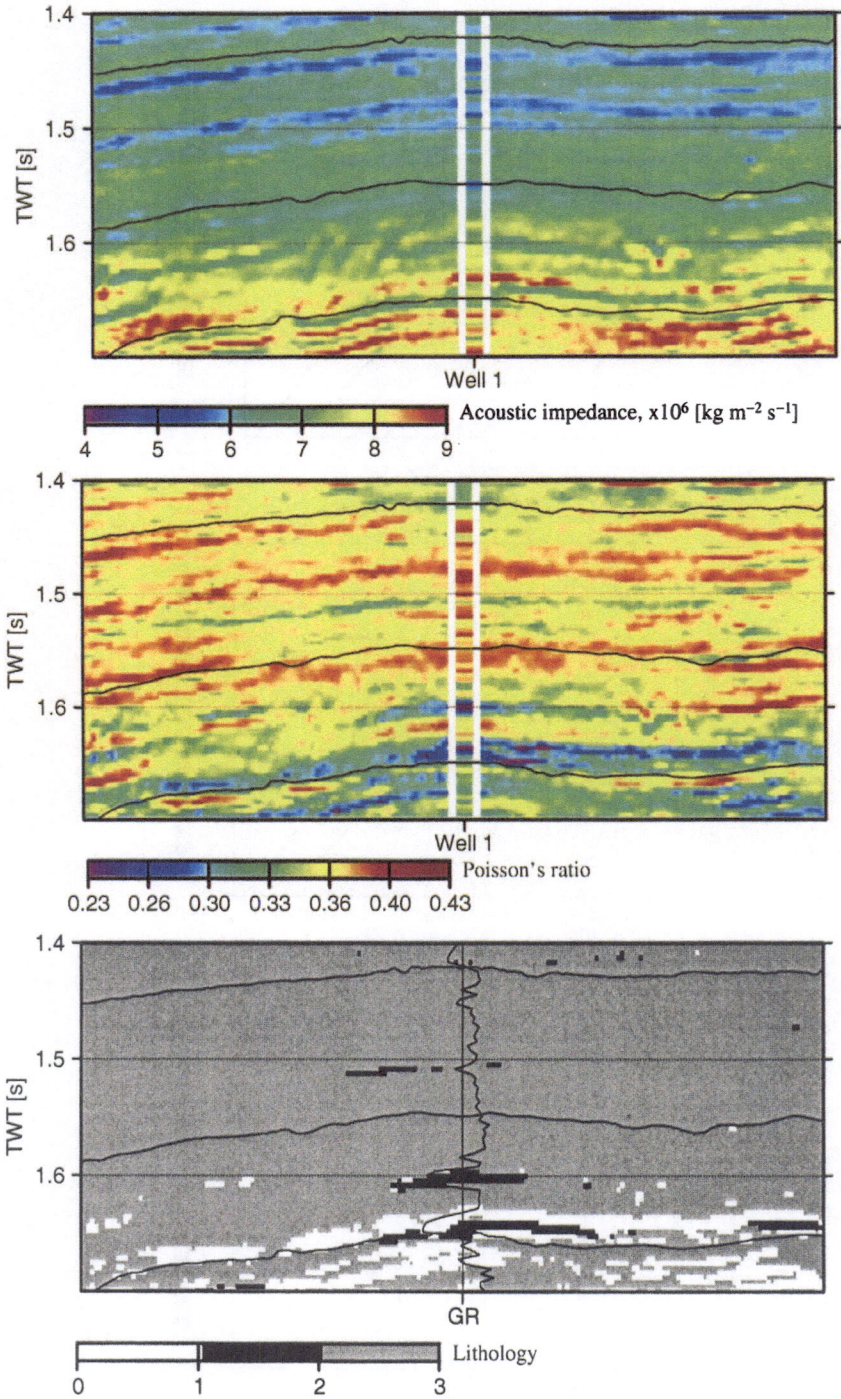

Figure 7. The acoustic impedance and Poisson's ratio sections with well log inserted, and the lithology section estimated from the PDF functions with the gamma ray log overlaid. Wet sand is white, shale is grey and gas sand is black.

5. Conclusion

It has been demonstrated that the elastic quantities of the subsurface can be obtained by implementing inversion of conventional seismic AVO data. The seismic inversion brings the seismic AVO attributes over to physical properties that can be compared directly with the well log data. The case studies show that the extra effort put into making the AVO inversion provides a more detailed and reliable prediction of the subsurface and optimises hydrocarbon detection.

Combining the obtained inversion results with lithology and fluid information from the well log using cross-plotting or statistical techniques enables evaluation of lithology and pore fluid variations on both a gross regional scale and a detailed reservoir scale.

The implementation requires that the seismic data are processed in a true amplitude manner and that good log data is available.

References

Avseth, P. et al., 1998, Statistical discrimination of lithofacies from pre-stack seismic data constrained by well log rock physics: Application to a North Sea turbidite system, Society of Exploration Geophysicists, Expanded Abstracts.

Castagna; J. et al., 1995, Rock Physics - The link between Rock Properties and AVO response: Investigations in Geophysics NO. 8, Society of Exploration Geophysicists, 135-171.

Connoly, P. 1999, Elastic Impedance, The Leading Edge, April 1999, 438-452.

Gardner, G.H.F. et al., 1974, Formation velocity and density - The diagnostic basics for stratigraphic traps, Geophysics 39, 770-780

Mavko, G. et al., 1998, Rock Physics Handbook, Tools for seismic analysis in Porous Media, Cambridge University Press.

Mukerji, T. et al., 1998, Applying statistical rock physics and seismic inversions to map lithofacies and pore fluid probabilities in a North Sea reservoir, Society of Exploration Geophysicists, Expanded Abstracts.

Resnick, J.R., 1993, Seismic Data Processing for AVO and AVA Analysis, in Castagna, J.P. and Backus, M.M. Offset-Dependent Reflectivity - Theory and Practice off AVO Analysis, Society of Exploration Geophysicists, 175.

Shuey, R. T., 1985, A simplification of the Zoeppritz equations: Geophysics, 50, 609-614.

Smith, G. C., and Gidlow, P. M., 1987, Weighted stacking for rock property estimation and detection of gas: Geophysical Prospecting, 35, 993-1014.

Generalized Fourier Transforms and Inverse Scattering

Svend Berntsen

Department of Mathematical Sciences, Aalborg University
Fredrik Bajers Vej 7, DK-9000 Aalborg, Denmark
Email: sb@math.auc.dk
Home page: http://www.math.auc.dk/~sb

Abstract. A general theory of Fourier like transformations is developed. The theory allows the construction of the explicit solution of integral equations with kernels which are products of a Bessel and a Hankel function. Using the established transformation theory the time domain inverse medium acoustic scattering problem is represented by a coupled set of integral equations with a Volterra kernel. The iterated solution of the refractive index is expressed in terms of the time domain far-field of the iterated scattered field. The lowest order refractive index is explicitly found.

1 Introduction

We are going to find a formulation of the inverse problem of an acoustic wave in the time domain. The differential equation is:

$$\nabla^2 u - m(\overline{x})c^{-2}\partial_t^2 u = 0, \tag{1}$$

with m the refractive index, and c the wave velocity outside the scatter. The function $(1-m)$ has compact support inside a sphere of radius a. We are going to discuss a diffraction problem with known incident and scattered far-field. The integral equations of this problem will be formulated and solved by iteration. The explicit solution of the iterative equations for the refractive index is only possible if we are able to solve integral equations with kernels which are products of a spherical Bessel and a spherical Hankel function. The transformation theory needed for the solution of this problem will be developed.

2 Integral Equations of Inverse Scattering

It is assumed that a medium is inside a sphere of radius a. The problem will be to formulate a coupled set of integral equations for the two functions m and u. We will also represent the problem in the frequency space as

$$[\nabla^2 + k^2 m]U = 0, \qquad k^2 = \frac{\omega^2}{c^2}, \qquad U = \mathcal{F}u. \tag{2}$$

It is well known that we may formulate the diffraction problem in the frequency space by the Lippmann- Schwinger integral equation,

$$U^{sc} = k^2 \Phi * [(m-1)U], \tag{3}$$

where $*$ is the convolution operator, and Φ the fundamental solution of the Helmholtz equation given by:

$$\Phi(\overline{x}) = \frac{\exp\{ik \mid \overline{x} \mid\}}{4\pi \mid \overline{x} \mid}. \tag{4}$$

Alternatively we may use the Lipmann-Schwinger equation in the time domain:

$$u^{sc}(\overline{x}, t) = -c^{-2}(\partial_t)^2 \Phi_{k=0} * \left[(m-1)u\left(\overline{y}, t - \frac{\mid \overline{x} - \overline{y} \mid}{c}\right)\right], \tag{5}$$

where $\Phi_{k=0}$ denotes the Greens function for the Laplace equation.

We assume that the incident field is a spherical wave:

$$u^{in}(r, t) = \frac{\sqrt{2\pi}}{r}\delta(ct + r) \qquad r = \mid \overline{x} \mid, \tag{6}$$

The Fourier Transform of the incident wave is:

$$U^{in}(\overline{x}, \omega) = -ikh_0^2(kr) \qquad r = \mid \overline{x} \mid. \tag{7}$$

Let j_n be the spherical Bessel function, $h_n^{1,2}$ the spherical Hankel functions and let Y_n^q be spherical harmonics. Then the fundamental solution has an expansion in terms of these functions for $\mid \overline{x} \mid = x \geq \mid \overline{y} \mid = y$:

$$\Phi(\overline{x} - \overline{y}) = ik \sum_{n=0}^{\infty} \sum_{q=-n}^{n} h_n^{(1)}(kx)j_n(ky)Y_n^q(\hat{x})\overline{Y_n^q(\hat{y})}. \tag{8}$$

Using this and the asymptotic expansion of the Hankel function we obtain from (3) that the far-field pattern for the scattered field has the form:

$$(U_\infty, Y_n^q) = L_n(m-1, Y_n^q) + k^2(-i)^n \langle (m-1)U^{sc}, j_n Y_n^q \rangle. \tag{9}$$

where the brackets () denote inner product on the unit sphere, the brackets $\langle \rangle$ denote the usual inner product in the three dimensional space, and the operator L_n is defined by

$$L_n(m-1, Y_n^q) = k^2(-i)^n \langle (m-1)U^{in}, j_n Y_n^q \rangle. \tag{10}$$

We are going to transform equation (9) to a second kind Volterra integral equation. This will be possible if we can show that L_n has a bounded inverse. We note that the kernel of the operator L_n is proportional to a product of spherical Bessel and Hankel functions j_n and h_0^2. Integral equations with such kernels are analyzed in Sections 3 and 4. In theorem 2 and 3 is proved that the inverse of the operator L_n does indeed exist and L_n^{-1} is a bounded operator for any value of n. Then (9) is equivalent to the integral equation:

$$(m - 1, Y_n^q) + L_n^{-1} k^2 (-i)^n \langle (m - 1) U^{sc}, j_n Y_n^q \rangle = L_n^{-1} (U_\infty, Y_n^q). \tag{11}$$

This integral equation (11) may be transformed to the time domain. It is proved in Section 4 that the operator $L_n^{-1} \mathcal{F}$ is a Volterra integral operator. In a similar way the operator $L_n^{-1} j_n \mathcal{F}$ may explicitly be constructed and again the kernel will be of Volterra type. That is the time domain form of (11) is a Volterra second kind integral equation:

$$(m - 1, Y_n^q) + K_n ([m - 1] u^{sc}, Y_n^q) = L_n^{-1} (U_\infty, Y_n^q). \tag{12}$$

Thus the inverse problem may be formulated as the coupled non linear integral equations (5) and (12). Given m, the first equation (5) has a unique solution U. And for given u equation (12) has a unique solution m. Convergence of the iterated solution of the integral equations (3) and (12) is a difficult problem. And so far the convergence of the iteration has not been proved. The reason is that whereas the inverse integral equation (12) has the attractive form of a second kind Volterra integral equation, the structure of direct problem in the time domain has not the structure of a second kind Fredholm equation (∂_t^2 is an unbounded operator).

In this paper is proved that the iterated refractive index may be found and expressed in terms of the iterated far field. The lowest order of m is the right hand side of (11). We may express this in terms of the time domain far-field:

$$(m^{(1)} - 1, Y_n^q) = L_n^{-1} \mathcal{F} \lim_{R \to \infty} R u^{sc} \left(R, \theta, \phi, t + \frac{R}{c}, Y_n^q \right). \tag{13}$$

The iteration is will now be defined. For a given value of the s'th order iterated $m^{(s)}$ we may solve the forward problem by the Lippmann -Schwinger equation in the time domain:

$$u^{sc(s)}(\overline{x}, t) = -c^{-2} (\partial_t)^2 \Phi_{k=0} * \left[(m^{(s)} - 1) u^{(s)} \left(\overline{y}, t - \frac{|\overline{x} - \overline{y}|}{c} \right) \right]. \tag{14}$$

The s'th iterated of the index m is calculated from (11) or (9) using the $(s - 1)$ iterates for $m - 1$ and U^{sc} in $\langle (m - 1) U^{sc}, j_n Y_n^q \rangle$:

$$(U_\infty, Y_n^q) = L_n (m^{(s)} - 1, Y_n^q) + $$
$$k^2 (-i)^n \langle (m^{(s-1)} - 1) U^{sc(s-1)}, j_n Y_n^q \rangle. \tag{15}$$

The $(s - 1)$ iterated field satisfies (9) that is:

$$(U_\infty^{(s-1)}, Y_n^q) = L_n (m^{(s-1)} - 1, Y_n^q) + $$
$$k^2 (-i)^n \langle (m^{(s-1)} - 1) U^{sc(s-1)}, j_n Y_n^q \rangle. \tag{16}$$

We apply the operator L_n^{-1} on the difference of the two equations (16) and (15). Finally we express the far-field pattern in terms of the far-field in the time domain and find the simple expression for the iterates of the refractive index in spherical coordinates:

$$m_{nq}^{(s)}(r) - m_{nq}^{(s-1)}(r) = \tag{17}$$

$$L_n^{-1} \mathcal{F} \lim_{R \to \infty} \left\{ R[u_{nq}^{sc} - u_{nq}^{sc(s-1)}] \left(R, t + \frac{R}{c} \right) \right\}(r),$$

with m_{nq}, u_{nq} denoting the Fourier coefficients, $m_{nq}^{(s)} = (m^{(s)}, Y_n^q)$ and $u_{nq}^{sc} = (u^{sc}, Y_n^q)$. In Section 4 the operator $L_n^{-1} \mathcal{F}$ is explicitly found see (39). The conclusion of this section is that if the operator $L_n^{-1} \mathcal{F}$ can be constructed the iterates of the refractive index can be expressed in terms of the scattered time domain far-field and the iterates of the scattered far-field.

A very similar formalism for the iterative refractive index expressed in terms of the field at the boundary of a sphere containing the scatterer may be formulated. The only difference will be that the inhomogeneous terms of the integral equations will differ. Thus the left hand side of the basic integral equation (9) for that case will be an integral on the boundary of the sphere, depending on the field and the normal derivative of the boundary field.

3 Generalized Fourier Transforms

The general problem to be treated in the transformation theory here is referred to as "The Generalized Fourier Transform pair." The transform pairs have the form

$$F(\omega) = \mathcal{B}f(\omega) = \int_0^\infty f(r)B(\omega r)dr \tag{18}$$

with the inverse given by

$$f(r) = [\mathcal{B}^{-1}F](r) = P \int_{-\infty}^\infty B(-\omega r)QF(\omega)d\omega. \tag{19}$$

The operators P and Q will to be chosen such that there exists a class of kernels B for which the transform pair (18) and (19) hold. As an example $P = Q = I$ will give a class of transformations including the Fourier Transform. The class of transform pairs which are important from the point of view of inverse scattering, is obtained if we choose $P = ir\partial_r$ and $Q = r\omega$:

$$f(r) = \mathcal{B}^{-1}F = ir\partial_r \int_{-\infty}^\infty r\omega F(\omega)B(-\omega r)d\omega. \tag{20}$$

The canonical kernel which satisfies this transform is:

$$B_0 = \frac{e^{ix}}{ix\sqrt{2\pi}}. \tag{21}$$

The integral equation (18) may explicitly be solved by the generalized distributional Mellin Transform. Let $\mathcal{M}$ denote the Mellin Transform, then equation (18) is equivalent to:

$$\mathcal{M}F(s) = \mathcal{M}B(s)\mathcal{M}f(1-s). \tag{22}$$

An analysis of the Mellin transform of (20) shows that the transformation pair (19) and (20) hold if and only if the kernel B satisfies the equation:

$$[1-s]\sin \pi s \mathcal{M}B(s)\mathcal{M}B(2-s) = \frac{1}{2}. \tag{23}$$

This non-linear equation is solved by:

$$\mathcal{M}B(s) = \psi(s)\mathcal{M}B_0(s), \tag{24}$$

where ψ is any function which is bounded at infinity by a polynomial and satisfies

$$\psi(s)\psi(2-s) = 1. \tag{25}$$

Theorem 1. *The generalized Fourier Transform pair (19) and (20) hold if and only if $\mathcal{M}B$ has the form (24) with ψ any function which satisfies the equation (25).*

An important class of operators for which this theorem applies is those with a kernel consisting of the product between a Bessel and a Hankel function:

$$B_{nm}(x) = i\sqrt{\frac{\pi}{2}}e^{i\frac{\pi}{2}(m-n)}H_m^1(x)J_n(x), \tag{26}$$

where n and m are real and $n \geq m \geq 0$. The Mellin transformation of B is found in standard mathematical tables of integrals, $\mathcal{M}B$ has the form (24) with:

$$\psi_{nm}(s) = \frac{1}{2^{1-s}}\frac{\Gamma(1-s)}{\Gamma(s-1)}\frac{\Gamma[\frac{n+m+s}{2}][\Gamma[\frac{n+s-m}{2}]}{\Gamma[1+\frac{n-s-m}{2}]\Gamma[1+\frac{n+m-s}{2}]}. \tag{27}$$

It is easily proved that the function ψ_{nm} satisfies (25), and the following result is obtained

Theorem 2. *Let $n \geq m \geq 0$ be real. Then any of the functions B_{nm} given by (26) are kernels of a generalized Fourier transformations and (19), (20) hold.*

4 Generalized Fourier Transform and Integral Equations

In this section we are going to apply the transformation theory for the construction of the inverse operator of B defined by (10). We will discuss integral equations:

$$\mathcal{B}_n f = \int_0^\infty dr f(r)B_n(\omega r) = \omega^{-2}F(\omega) = \omega^{-2}\mathcal{F}\phi, \tag{28}$$

with the kernels:

$$B_n(x) = \frac{1}{2\sqrt{2\pi}}(-i)^{n-1}xh_0^2(x)j_n(x), \tag{29}$$

where j_n and h_n denote the spherical Bessel and Hankel functions. That is, the transformation has the form (26), with $m = \frac{1}{2}$ and $n := n + \frac{1}{2}$. Then according to theorem 2 the inverse is given by (20). We assume that the function ϕ in (28) is in the space

$$X = \{\phi \in C^0([-\infty, 0]) \mid \|\phi\| = \|\phi\|_\infty\}, \tag{30}$$

and look for solutions of (28) in the space

$$Y = \{f \in C^0([0, \infty]) \mid \|f\| = \|f\|_\infty\}. \tag{31}$$

The solution of (28) is obtained by the inverse transform (20). However it is proved, that the mapping of the function ϕ of (28) onto f can directly be found:

Theorem 3. *Assume that $\phi = \mathcal{F}^{-1}F$ has compact support on $[-\infty, 0]$, and $\phi \in X$. Then the integral equation (28) has a unique solution in the space Y given by (32). The mapping (32) is continuous on the space X into Y.*

$$f(r) = \mathcal{B}_n^{-1}\omega^{-2}\mathcal{F}\phi(r) = \mathcal{B}_n^{-1}\phi(-2r), \tag{32}$$

with

$$\mathcal{B}_n^{-1}\phi(r) = 2\{rI\phi(r) - n(n+1)I^2\phi(r) + I^3[G_n(t,r)\phi(t)](r)\}, \tag{33}$$

I is the integral operator:

$$I\phi(r) = \int_{-\infty}^r \phi(v)dv \tag{34}$$

and the function G_n is given by:

$$G_n(t,r) = -r\partial_r \sum_{p=2}^n \frac{(-i)^p \overline{a_{pn}}}{r^p(p-1)!}(t-2r)^{p-1}. \tag{35}$$

Outline of proof: The spherical Hankel and Bessel functions may be expressed by finite series of the form:

$$h_n^{(1)}(r) = (-i)^n \frac{e^{ir}}{ir}\left\{1 + \sum_{p=1}^n \frac{a_{pn}}{r^p}\right\} \tag{36}$$

with a_{pn} known complex coefficients, and a corresponding expression for the Bessel function. Using this the kernel of the integral operator $\mathcal{B}_n^{-1}\omega^{-2}\mathcal{F}$, with

$\mathcal{B}_n^{-1}$ given by (20) may explicitly be found. A simple calculation leads to the result (33).

We may apply theorem 3 for a discussion of the operator problem:

$$L_n(m-1, Y_n^q)(\omega) = \mathcal{F}[\phi(t)](\omega). \tag{37}$$

The operator L_n is related to $\mathcal{B}_n$ defined by (28):

$$L_n(m-1, Y_n^q)(\omega) = -2\sqrt{2\pi}\omega^2 \mathcal{B}_n[r((m-1), Y_n^q)(cr)](\omega). \tag{38}$$

Theorem 3 show that the operator L_n has a bounded inverse, and $L_n^{-1}\mathcal{F}$ may explicitly be constructed from (32) and (33). The result is:

$$L_n^{-1}\mathcal{F}[\phi(t)](r) = \frac{-c}{2r\sqrt{2\pi}} B_n^{-1} \phi\left(-\frac{2r}{c}\right), \tag{39}$$

with B_n^{-1} defined by (33). Thus we will be able to find explicit expressions for the iterated refractive index in terms of the time domain scattered field, and the calculated iterated scattered field.

Preliminary numerical calculations indicate that there will be restrictions to the validity of this method. However for the cases mentioned in the conclusion the low order iterate field is very close to the exact field, and the construction of the refractive index is expected to hold for these cases.

This concludes the construction of the solution of the integral equation (28) in terms of the inverse Fourier Transform of the function F.

5 Conclusion

The method of this paper gives explicit expressions for the refractive index in terms of the iterated far-field in the time domain. The validity of the reconstruction is restricted to the cases:

- The scatterer is small.
- The refractive index is close to one.
- The points where we reconstruct the refractive index are sufficiently close to the boundary.

The method has generalizations to point source excitations and to media for which $m = m_0 + \delta m$, where m_0 is a constant and δm is sufficiently close to zero.

References

1. Angel, T.S., Kleinman, R.E., and Roach, G.F.: An inverse transmission problem for the Helmholtz equation. Inverse Problems 6, 299–300 (1990).
2. Colton, D., Kirsch, A., and Päivärinta, L.: Far field patterns for acoustic waves in an inhomogeneous medium. SIAM J. Math. Anal. 20, 1472–1483 (1989).

3. Colton, D., Kress, R.: Time harmonic electromagnetic waves in an inhomogeneous medium. Proc. Royal Soc. Edinburg 116 A, 279–293 (1990).
4. Colton, D., Kress, R.: Inverse Acoustic and Electromagnetic Scattering Theory. Springer-Verlag (1992).
5. He, S., Weston, V.H.: Determination of the permittivity and conductivity in R3 using wave splitting of Maxwell equations. J. Math. Phys. 36, 1776–1789, (1995).
6. Kirsch, A.: An Introduction to the Mathematical Theory of Inverse Problems. Springer-Verlag (1996).
7. Kristensson, G., and Krueger, R.J,: Direct and inverse scattering in the time domain for a dissipative wave equation. Scattering operators. J. Math. Phys. 27, 1667–1682, (1986).
8. Kristensson, G., and Vogel, C.R.,: Inverse problems for acoustic waves using the penalised likelihood method. Inverse Problems 2, 461–479 (1986).
9. Weston, V.H.: Invariant imbedding for the wave equation in three dimensions and the application to the direct and inverse problems. Inverse Problems 6, 1075–1105 (1990).

Multi-Step Samplers for Improving Efficiency in Probabilistic Geophysical Inference

Miguel Bosch[1], Christophe Barnes[2], and Klaus Mosegaard[3]

[1] Facultad de Ingeniería, Universidad Central de Venezuela,
Los Chaguaramos, Caracas, VENEZUELA.
mbosch@reacciun.ve

[2] Institut de Physique du Globe de Paris, 4, place Jussieu,
F-75252, Paris, Cedex 05, FRANCE.
barnes@ipgp.jussieu.fr

[3] Niels Bohr Institute, University of Copenhagen, Julianes Maries Vej 30,
DK-2100 Copenhagen Ø, DENMARK.
klaus@gfy.ku.dk

Abstract. Geophysical inference is characterized by non-linear relationships between model and data parameters, large model spaces describing spatial distributions of media properties, and intensive computations related to the numerical resolution of the forward problem. Although sampling approaches are convenient to solve such inverse problems, sometimes the involved computations are demanding. We consider here sampling techniques directed to improve the efficiency of sampling procedures in large real geophysical applications. We propose a sampling algorithm incorporating classical importance sampling within a two-step (or multistep) Markov chain sampler set-up. The first step of the algorithm is a Metropolis sampler ergodic to an importance density function and the second step is a Metropolis sampler correcting from the bias introduced by the importance density function; the combined algorithm samples the posterior probability density function asymptotically. The importance density is a combination of the prior density and an importance likelihood function, obtained as an approximation of the data likelihood function. Hence, the importance density is an approximation of the posterior density. Although rejection rates of the combined algorithm are larger than the rejection rates of the simple step Metropolis algorithm, the computational sampling efficiency is improved when the calculation of the importance density is much easier than the calculation of the actual posterior density. The algorithm is characterized by high acceptance rates in its second step because the first step serves as a barrier rejecting unlikely samples. Data likelihood approximations can be obtained in several ways : using simplified geophysical simulation, using the likelihood with partial (smaller) sets of the observed data, or using information obtained by preliminary analysis of the data. The technique has been successfully used for the inversion of phase arrival times from an offset vertical seismic profile (OVSP) data.

1 Introduction

Inverse problems in geophysics have been commonly treated using optimization approaches like least squares regression or simulated annealing. The ultimate goal has been to find an optimum (in some sense) model that explains the observations and satisfies some prior information. An alternative approach is sampling models according to a probability density function (*pdf*) combining prior information and the likelihood of the model with the observations. The goal, in this case, is to find a large number of models that explain the data and obey the prior information within uncertainties (Press, [1]; Mosegaard and Tarantola, [2]; Mosegaard, [3]; Bosch, [4]). Sampling solves situations that are difficult to treat by searching an optimal model: multiple solutions located in separate regions of the model space, strongly non-linear relationship between data and model, or presence of categorical parameters in the model.

Nowadays, practical applications demand detailed models, large data sets describing the observations and realistic prior information to include in the inverse problem. Sampling posterior *pdf*s in geophysical applications implies the forward calculation of the geophysical data. In most cases the likelihood function has an open form expression and its evaluation for a particular model represents a significant computation effort. This poses the problem of defining techniques to sample in an efficient way posterior densities defined over very large model spaces (from thousands to millions of parameters).

Extensive work can be found in statistical literature about the Markov-chain samplers (e.g. Smith and Roberts, [5]; Tierney, [6]), commonly used to sample posterior *pdf*s. In particular, the Metropolis sampler is recommended for geophysical inference due to its ability to handle complicated posteriors (Mosegaard and Tarantola, [2]).However, the choice of the sampling technique and its actual performance is highly dependent on the particular type of application.

Several techniques have been used in statistical work to improve the accuracy of the posterior estimates calculated by Monte Carlo methods; they are called *variance reduction techniques* (Wolpert, [7]). One of the better know techniques is *importance sampling* (Hastings, [8]; Kloek and van Dijk, [9]; Evans, [10]). In the present work we describe how this technique can be combined with the Metropolis algorithm, within a multiple step set-up, to improve the sampling efficiency.

Multi-step samplers have been previously used in statistical work to improve sampling efficiency (e.g., Geyer, [11]; Tierney, [6]). Commonly, the first step of the algorithm is a Gibbs sampler or other simpler sampler. This particular formulation is not convenient for geophysical inverse problems because the Gibbs sampler is adequate for simple analytic densities with known and easy to sample conditionals, which is not in general the case in geophysical work. We propose in this work a different technique combining several Metropolis samplers in steps. This algorithm is useful in non-linear geophysical inverse problems, demanding large model spaces and heavy computations.

Sections 2 reviews importance sampling and the Metropolis sampling methods, focusing on their application to statistical inference problems. Section 3 proposes a two-step Metropolis algorithm incorporating an importance density in its first stage. Section 4 discusses different ways to construct importance likelihood functions in geophysical inverse problems, and Section 5 summarizes a practical application of the algorithm to the inversion of real bore-hole travel time seismic data.

2 Statistical Inference

In statistical inference problems, the structure of the posterior *pdf* $\sigma(\mathbf{m})$ is given by

$$\sigma(\mathbf{m}) \propto \rho(\mathbf{m})\, L(\mathbf{m}), \tag{1}$$

with $L(\mathbf{m})$ being the likelihood function, $\rho(\mathbf{m})$ being the prior *pdf*, and $\mathbf{m}$ being a point in the model parameter space $\mathcal{M}$. Inference from the *pdf* involves the calculation of posterior probabilities for model parameters and model structure. The probability that a sample $\mathbf{m}_0$ from the posterior density belongs to a subset $\mathcal{A}$ of the model space is given by the integral,

$$p(\mathbf{m}_0 \in \mathcal{A}) = \int_{\mathcal{A}} \sigma(\mathbf{m}) d\mathbf{m}, \tag{2}$$

With more generality, it is required the calculation of the expected value of model functionals $f(\mathbf{m})$ with an integral of the type,

$$E[f] = \int_{\mathcal{M}} f(\mathbf{m})\, \sigma(\mathbf{m})\, d\mathbf{m}. \tag{3}$$

Notice that the above formula reduces to (2) with $p(\mathbf{m}_0 \in \mathcal{A}) = E(f)$, and $f(\mathbf{m})$ being an indicator function ($f(\mathbf{m}) = 1$ if $\mathbf{m} \in \mathcal{A}$; $f(\mathbf{m}) = 0$ if $\mathbf{m} \notin \mathcal{A}$).

The integral (3) is also useful to calculate the moments of an arbitrary integrable functional $y(\mathbf{m})$. The second moment, for example, would be obtained with $f(\mathbf{m}) = y^2(\mathbf{m})$,

$$E[y^2] = \int_{\mathcal{M}} y^2(\mathbf{m})\, \sigma(\mathbf{m})\, d\mathbf{m}. \tag{4}$$

Marginals or conditionals of the posterior density can be also formulated by (3) using Dirac's $\delta(\mathbf{m})$ distributions.

As already discussed, in our inference problems the posterior density is a complicated function, and integrals (2) and (3) can not be solved analytically. Instead, we construct unbiased estimates of $E[f]$ by using Monte Carlo integration methods. A naive estimate can be obtained using N samples $\mathbf{m}_i$ from a uniform distribution in $\mathcal{M}$

$$E^0 = \frac{\sum_{i=1}^{N} f(\mathbf{m}_i)\, \sigma(\mathbf{m}_i)}{\sum_{i=1}^{N} \sigma(\mathbf{m}_i)}. \tag{5}$$

However, the variance of this estimate is very large. The regions of the model space with large values of the integrand are usually poorly represented in the sample.

2.1 Importance Sampling

Instead of using a uniform distribution as in (5) one can draw samples from an arbitrary distribution with density $\tilde{\sigma}(\mathbf{m}) > 0$. The integral (3) can then be estimated by

$$E^1 = \frac{\sum_{i=1}^{N} f(\mathbf{m}_i)\, \sigma(\mathbf{m}_i)/\tilde{\sigma}(\mathbf{m}_i)}{\sum_{i=1}^{N} \sigma(\mathbf{m}_i)/\tilde{\sigma}(\mathbf{m}_i)}. \tag{6}$$

The idea of importance sampling consists of making an adequate selection of the sampling density $\tilde{\sigma}(\mathbf{m})$, currently called the importance density function. Statisticians established (see Kloek and van Dijk [9] or Evans [10] for a review) that the variance of estimate E^1 is significantly reduced when $\tilde{\sigma}(\mathbf{m})$ is taken to be similar to $\sigma(\mathbf{m}_i)$.

The reason behind the variance reduction is that more samples are taken from the regions where the density $\sigma(\mathbf{m}_i)$ has an important contribution to the integral value $E(f)$. Indeed, the variance would be optimal if the samples are taken from a distribution with density proportional to $|f(\mathbf{m})|\sigma(\mathbf{m})$. Nevertheless, as often ones needs to estimate many different functions f in the same problem, it is usually the convenient choice to produce samples from a density as close as possible to $\sigma(\mathbf{m})$. The computational effort needed for the generation of the samples is another practical aspect to take into account in the choice of the importance density function; easy to sample importance densities are naturally preferred.

An important variant of the method is called adaptive importance sampling (e.g., Man-Suk and Berger, [12]). It consists of periodically optimizing the parameters of the importance density, in order to better approximate the posterior density. In this way the data likelihood evaluations that are progressively made along the sampling procedure can be used to improve the importance density.

A straightforward application of importance sampling to inference problems consists in drawing samples from the prior density (i.e. $\tilde{\sigma}(\mathbf{m}) = \rho(\mathbf{m})$) to calculate $E(f)$ by

$$E^2 = \frac{\sum_{i=1}^{N} f(\mathbf{m}_i)\, L(\mathbf{m}_i)}{\sum_{i=1}^{N} L(\mathbf{m}_i)}. \tag{7}$$

Possibly useful for preliminary diagnostics, E^2 is in most cases a very rough estimate of integral $E(f)$ because the likelihood function can be wildly different from the prior density.

A better estimation is obtained by sampling from the importance density $\tilde{\sigma}(\mathbf{m}) \propto \rho(\mathbf{m})I(\mathbf{m})$, with $I(\mathbf{m})$ being an approximation of the likelihood function. In this case, the importance density is closer than the posterior density, than in the precedent case. The integral $E(f)$ is estimated by

$$E^3 = \frac{\sum_{i=1}^{N} f(\mathbf{m}_i)\, L(\mathbf{m}_i)/I(\mathbf{m}_i)}{\sum_{i=1}^{N} L(\mathbf{m}_i)/I(\mathbf{m}_i)}. \tag{8}$$

Of course, the most convenient situation for inference purposes would be established by drawing samples from the posterior density itself (i.e. $\tilde{\sigma}(\mathbf{m}) = $

$\sigma(\mathbf{m})$), yielding to the integral estimate

$$E^4 = \frac{1}{N} \sum_{i=1}^{N} f(\mathbf{m}_i). \tag{9}$$

2.2 The Metropolis Sampler

The Metropolis algorithm (Metropolis et al., [13]; Hastings, [8]) is a Markov-chain Monte Carlo method that is used, in long runs, to produce samples from posterior densities, needed in (9). Its convenience to geophysical inference has been shown in the works of Mosegaard and Tarantola [2], Mosegaard [3] and Bosch [4]; these works provide thorough descriptions of the way the Metropolis algorithm is applied to solution of geophysical inverse problems.

Briefly, the Metropolis algorithm operates as follows. Suppose that we have a process, a proposal Markov-chain, that is able to generate samples from an arbitrary proposal density $\phi(\mathbf{m})$ in sufficiently long runs. This property is called ergodicity in statistical work; so consider that we have already constructed such a chain ergodic to the proposal density. The Metropolis algorithm defines a Markov-chain that is ergodic to a target density $\pi(\mathbf{m})$, that we call here the posterior chain. Denoting by $\mathbf{m}_t$ the sample drawn from the posterior chain at the step t, the next sample to be drawn from the posterior chain, $\mathbf{m}_{t+1}$, is obtained by the following rules.

1. Consider $\mathbf{m}_t$ to be the current sample in the proposal chain and generate, conditionally to $\mathbf{m}_t$, a next sample $\mathbf{m}_{t+1}^{\text{can}}$ form the proposal chain. This will be the candidate sample for the posterior chain.
2. Accept the candidate sample, making $\mathbf{m}_{t+1} = \mathbf{m}_{t+1}^{\text{can}}$ with probability

$$p_1 = \text{Min}\left[1, \frac{\pi(\mathbf{m}_{t+1}^{\text{can}})\phi(\mathbf{m}_t)}{\pi(\mathbf{m}_t)\phi(\mathbf{m}_{t+1}^{\text{can}})}\right]; \tag{10}$$

otherwise, make $\mathbf{m}_{t+1} = \mathbf{m}_t$.
3. Return to label (1).

A large amount of work has been centered on practical and theoretical subjects of the Metropolis sampler performance (Tierney, [6]; Robert, [14]; Cowles and Carling [15]) and particularly on the problem of convergence.

Notice that for the case in which the proposal density is the prior density, and the target density is the posterior density in (1), the Metropolis acceptance probability is

$$p_1 = \text{Min}\left[1, \frac{L(\mathbf{m}_{t+1}^{\text{can}})}{L(\mathbf{m}_t)}\right], \tag{11}$$

with $L(\mathbf{m})$ being the likelihood function. Hence, the Metropolis rule only depends on the ratio of the likelihood function evaluated at the candidate and at the current model. This is the setting in which we use the Metropolis rule in our inference problems. We construct the prior chain (i.e. ergodic to the prior density) depending on the particular prior information available. A Gibbs sampler, a Metropolis sampler, sampling from multivariate gaussian densities, or combinations of several sampling methods can be used, depending on the problem.

In geophysical inverse problems the likelihood function is difficult to compute as it involves the solution of a geophysical forward problem. Hence, one important factor related to the sampling efficiency is the acceptance rate p_1. If the Metropolis rule yields too low acceptance rates, samples would be often repeated in consecutive steps of the chain and several likelihood evaluations would be needed in order to produce a transition to a different point of the model space. For this reason, sometimes, the acceptance rate has been used as one criterion of efficiency.

However, the acceptance rate is only a rough criterion of efficiency. For instance, high acceptance rates can be obtained with a proposal chain providing very small changes in model parameters at consecutive chain steps, i.e. small step lengths in the model space. In this case the high acceptance rate would not indicate and efficient sampling of the posterior density, because the variability of the chain (the movement of the chain path in the model space) would still be very slow. A chain is usually said to be fast mixing if it can move reasonably well in the model space in order to explore the important areas of the posterior density support in the given number of steps.

3   The Combined Metropolis Algorithm

In this section we present a two-step Metropolis algorithm that is designed to provide good candidate models for the computer intensive likelihood evaluation, that is, models that have a high probability of being accepted in the second step of the Metropolis algorithm. The algorithm is inspired by the classical idea of importance sampling, but it operates by incorporating the importance density function in the two-step Metropolis setup.

We are going to consider a function $I(\mathbf{m}) \geq 0$, here called the importance likelihood function, to transform expression (1) into the equivalent,

$$\sigma(\mathbf{m}) \propto \rho(\mathbf{m})\, I(\mathbf{m}) \left[\frac{L(\mathbf{m})}{I(\mathbf{m})} \right]. \tag{12}$$

The above equation can be regarded as the product of an importance density function defined as $\tilde{\sigma}(\mathbf{m}) \propto \rho(\mathbf{m}) I(\mathbf{m})$ and a likelihood correction term $L(\mathbf{m})/I(\mathbf{m})$. Notice that the data likelihood is divided by the importance likelihood to correct for the bias introduced by the importance density in the first factor.

Applying recursively the conventional Metropolis rule (11) to the structure of the above expression, we can define a posterior Markov-chain (i.e. ergodic to the posterior density) in the following way. As for the single step Metropolis algorithm, assume that a prior Markov-chain (i.e. ergodic to the prior density) is available. Consider that the model $\mathbf{m}_t$ has been obtained from the posterior Markov-chain. The next model $\mathbf{m}_{t+1}$ in the posterior chain is obtained as follows:

1. Consider $\mathbf{m}_t$ to be the current sample in the prior chain, and draw the next sample, $\mathbf{m}_{t+1}^{\text{precan}}$, from the prior chain. This sample is considered a pre-candidate sample of the posterior.

2. Make $\mathbf{m}_{t+1}^{\text{can}} = \mathbf{m}_{t+1}^{\text{precan}}$ with probability

$$p_2 = \text{Min}[1, I(\mathbf{m}_{t+1}^{\text{precan}})/I(\mathbf{m}_t)];$$

otherwise, make $\mathbf{m}_{t+1} = \mathbf{m}_t$ and return to label (1).

3. Make $\mathbf{m}_{t+1} = \mathbf{m}_{t+1}^{\text{can}}$ with probability

$$p_3 = \text{Min}\left[1, \frac{L(\mathbf{m}_{t+1}^{\text{can}})\, I(\mathbf{m}_t)}{I(\mathbf{m}_{t+1}^{\text{can}})\, L(\mathbf{m}_t)}\right];$$

otherwise, make $\mathbf{m}_{t+1} = \mathbf{m}_t$.

4. Return to label (1).

In the above algorithm, steps (1) and (2) alone define a chain ergodic to the importance density. Then, steps (3) and (4) transform this chain to be ergodic to the posterior density. The importance likelihood $I(\mathbf{m})$ can be conceived as a smooth approximation of the likelihood function, with longer tails than the likelihood function (see figure 1). Its role consists of focusing the first sampling stage on some relevant regions of the model space. The evaluation of the importance likelihood does not involve the evaluation of the data likelihood function, which is commonly in geophysics a heavy calculation. Instead, the importance likelihood should be easier to evaluate and may follow some simple multivariate parametric model or a mixture of them. Hence, a high computational cost is only associated with the last step of the sampling.

If $I(\mathbf{m}) \approx a\, L(\mathbf{m})$, with a being a proportionality constant, the acceptance probability of the second sampling step would be $p \approx 1$, and we could expect a high acceptance rate in this last step. Even if we use conveniently large step lengths in the prior walk, the first sampling step of the combined algorithm would serve as a "filter", rejecting unlikely samples before they are presented to the second step. This justifies the use of the importance density function as a way to improve the efficiency of the overall sampling procedure, if the importance density is easier to calculate than the posterior density.

However, the overall rejection rate of the combined algorithm is equal or larger to the rejection rate of the single step Metropolis algorithm. In fact, $p_2 p_3 \leq p_1$ because

$$\text{Min}\left[1, \frac{I(\mathbf{m}_{t+1}^{\text{precan}})}{I(\mathbf{m}_t)}\right] \text{Min}\left[1, \frac{L(\mathbf{m}_{t+1}^{\text{precan}})\, I(\mathbf{m}_t)}{L(\mathbf{m}_t)\, I(\mathbf{m}_{t+1}^{\text{precan}})}\right] \leq \text{Min}\left[1, \frac{L(\mathbf{m}_{t+1}^{\text{precan}})}{L(\mathbf{m}_t)}\right]. \quad (13)$$

Although the two-step algorithm has the same or more overall rejections than the single algorithm, the rejections are concentrated in the first step, which only demands evaluation of the importance likelihood. The overall sampling efficiency

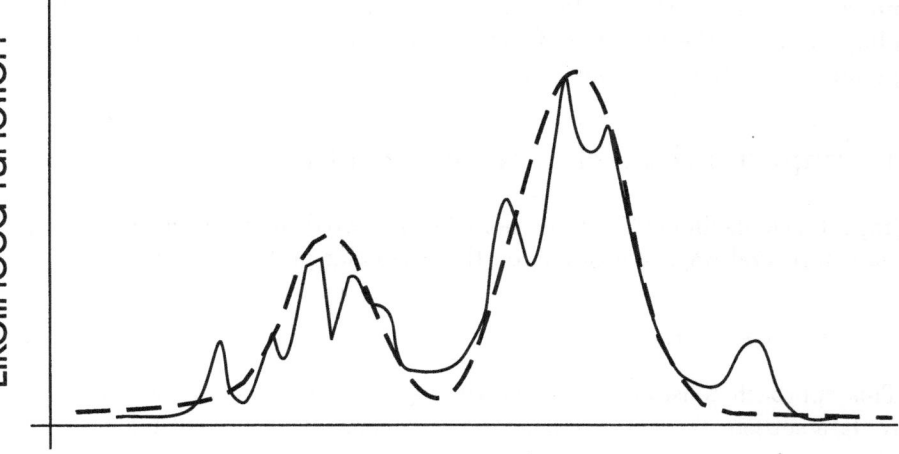

Model parameter space

Fig. 1. Simplified illustration of (solid) a data likelihood function defined over a one dimensional parameter space and (dashed) an adequate importance sampling likelihood.

is improved if one defines the function $I(\mathbf{m})$ to be an acceptable approximation to the data likelihood function and one that is inexpensive to calculate. Then, the actual computational cost of pre-candidate rejections is smaller compared to the computational cost of candidate rejections in the single step Metropolis sampler. Consequently, for the same total computational cost, more iterations are produced by the combined algorithm than by the single step Metropolis sampler.

Also, fast mixing prior chains could be incorporated into the algorithm. Because of large rejection rates the simple Metropolis algorithm is not performant with fast mixing prior chains. The combined algorithm here described tolerates large rejection rates, as rejection is concentrated in the first step while keeping a large acceptance rate in the second step. For the same reason, the combined algorithm is able to speed-up the initial *burn-in period*, in which the chain is still influenced by its starting point in the model space.

As for the classical applications of importance sampling, we have here that: (1) The efficiency of the sampling process would be improved when the importance density is approximately similar to the posterior density and easier to calculate. (2) The selection of the importance density preserves the process unbiased affecting only its speed of convergence to the posterior distribution. It is important to remark that the effect of the importance density in the first sampling stage is removed in the second sampling stage by dividing the likelihood function by $I(\mathbf{m})$.

These results on the two-step algorithm are valid for multiple steps by recursivity. Further improvement of the algorithm could incorporate an adaptation of $I(\mathbf{m})$ in order to better approximate the likelihood function as the iterations progresses. This could be done, as for conventional adaptive importance sampling, by using the likelihood function evaluations to periodically optimize the parameters of the importance likelihood.

4 Importance Likelihoods in Geophysical Problems

Importance likelihood functions can be constructed in geophysical inverse problems in several ways, depending on the particular problem at hand.

4.1 Data Partition

This approach consists in dividing the data set D in a convenient partition of N data subsets D_n and defining a series of consecutively inclusive data subsets $\tilde{D}_n$,

$$\tilde{D}_n = \bigcup_{k=1}^{k=n} D_k \Longrightarrow \tilde{D}_1 \subset \tilde{D}_2 \subset ... \subset D. \tag{14}$$

Let us denote by $L(\mathbf{m})$ the likelihood function of a model $\mathbf{m}$ derived from the complete data set D, and similarly denote $\tilde{L}_n(\mathbf{m})$ the likelihood function of the model derived from the data set $\tilde{D}_n$. The likelihood function $\tilde{L}_n(\mathbf{m})$ can be regarded as an approximation of the likelihood function $\tilde{L}_{n+1}(\mathbf{m})$ and can be used as an importance sampling likelihood. In this way we can transform expression (1) for the posterior density into,

$$\sigma(\mathbf{m}) \propto \rho(\mathbf{m}) \, \tilde{L}_1(\mathbf{m}) \frac{\tilde{L}_2(\mathbf{m})}{\tilde{L}_1(\mathbf{m})} ... \frac{\tilde{L}_n(\mathbf{m})}{\tilde{L}_{n-1}(\mathbf{m})} ... \frac{L(\mathbf{m})}{\tilde{L}_{N-1}(\mathbf{m})}. \tag{15}$$

A multi-step Metropolis sampling algorithm can be adapted to this structure. If one considers a partition of the data set in only two subsets, then a two-step sampler would be implemented as described in the preceding section.

If we consider that uncertainties in data are independent across data subsets, the likelihood function can be factorized and the above equation is simpler,

$$\sigma(\mathbf{m}) \propto \rho(\mathbf{m}) \prod_n L_n(\mathbf{m}), \tag{16}$$

with $L_n(\mathbf{m})$ being the likelihood function of the model derived from the data set D_n. An example of an importance likelihood function derived from the data subset $\tilde{D}_n$ considering a generalized Gaussian model for data uncertainties is,

$$I(\mathbf{m}) = L_n(\mathbf{m}) = \exp\left[-\frac{1}{p} \sum_{d_i^{\mathrm{obs}} \in \tilde{D}_n} \frac{|g_i(\mathbf{m}) - d_i^{\mathrm{obs}}|^p}{\tilde{\sigma}_i^p} \right], \tag{17}$$

with d_i^{obs} being the ith observed data, $\tilde{\sigma}_i$ the data uncertainty and $g_i(\mathbf{m})$ the simulated data.

Different criteria could be used to setup convenient data partitions. When possible they should be composed by a representative sample of all kinds of data. But differences in efficiency between different kinds of geophysical data simulation should also count, leaving heavy simulations for the last steps of the algorithm.

As in other importance sampling techniques, the efficiency improvement is not warrant in all circumstances. It depends on the choice of data partition and on the problem. We recommend that the partition separates redundant or complementary data. In this way, consecutive likelihood functions in (16) have approximate shapes, yielding to low rejection rates between steps. If there is contradictory data, it should be better grouped into the same data subset. Otherwise, at some level of the algorithm, the proposal models would be contradictory (i.e. located at a different region of the model space) with the likelihood function, producing high rejection rates.

4.2 Fast Geophysical Modeling

The likelihood function based on fast but inaccurate data simulation can be used as importance likelihood function for preliminary rejection of unlikely models before using better quality data simulation. In this way the better quality simulation is made for models that have passed this preliminary selection. An example of importance likelihood function having an approximate forward problem solution $\tilde{g}(\mathbf{m}) \approx g(\mathbf{m})$ and following a generalized Gaussian model is

$$I(\mathbf{m}) = \exp\left[-\frac{1}{p}\sum \frac{|\tilde{g}_i(\mathbf{m}) - d_i^{\text{obs}}|^p}{\tilde{\sigma}_i^p}\right]. \tag{18}$$

The dispersion used in the importance likelihood function should be made greater than the dispersion used in the data likelihood function, $\sigma < \tilde{\sigma}$, to make the first a smoother function than the latter.

4.3 Analysis of Data

Before our inversion, the data may have lead to preliminary characterization of the model parameters through conventional data processing, conventional inversion or qualitative data interpretation. Those results, which may be preliminary images of the study zone (e.g. migrated seismic sections, tomographic images), can well be used to construct an importance sampling likelihood function. As an example, consider that we have a model $\tilde{m}$ which is the result of a preliminary inversion of the data and consider that we associate to each model parameter independent generalized Gaussian uncertainties $\tilde{\sigma}$. Then an importance sampling likelihood function can be

$$I(\mathbf{m}) = \exp\left[-\frac{1}{p}\sum \frac{|m_i - \tilde{m}_i|^p}{\tilde{\sigma}_i^p}\right]. \tag{19}$$

The dispersion $\tilde{\sigma}$ should be sufficiently large to account for the inaccuracy of the preliminary result and to account also for the possible multi-modality of the data likelihood function. If the preliminary result is obtained by optimization (e.g. least squares) it is important to note that conventional dispersion estimates obtained only measure the local dispersion around the solution and much larger dispersions should be used to define the importance function. If we have preliminary model solutions representing several local maxima of the likelihood function a multi-modal importance function can be well defined as a summation of importance functions for each local maxima. Multi-modal importance density functions are discussed in the work of Man-Suk and Berger [12].

4.4 External Information

In inference problems we often have information that we have not included in the prior density or in the likelihood function for practical implementation reasons or because we do not consider this information reliable enough. This information may come, for instance, from additional data sets or geological hypothesis. This kind of external information, can be represented in $I(\mathbf{m})$ and used to construct the importance likelihood function. If we decide to represent this information by indicating a central model and its parameter variances, a representation of $I(\mathbf{m})$ as in (19) could be used.

It is important to conceptually distinguish importance sampling information represented in $I(\mathbf{m})$ from prior information. Importance sampling information is used as a sampling guide and it is removed in the final step of the sampling algorithm. Prior information, on the contrary, is considered as true information directly contributing to the posterior probability density function. This is another practical advantage of introducing importance sampling. Loose intuitive information about the solution may be used to focus the sampling in some areas of the model space, expecting that this would increase the sampling efficiency. Of course the sampling efficiency would be improved, in fact, only if this external information is nicely consistent with the information represented by the posterior density.

5 An Application to Seismic Data Inversion

The combined Metropolis sampler has been applied with success to the estimation of P-wave and S-wave velocities from bore-hole seismic arrival times provided by an offset vertical seismic profile real experiment (OVSP). To illustrate the method we present a brief description of this application. Further information is presented in the work of Barnes et al. [16].

The seismic data was recorded by 155 three-component geophones (seismographs measuring the velocity vector) located inside a vertical well from 1250 to 4000 m depth. The instruments received the seismic signal from an explosive (air-gun) source located at the surface with 2000 m offset from the top of the well; figure 2 describes the geometry of the offset vertical seismic profile (OVSP) experiment. The obtained seismograms show complicated signals including refracted, reflected and diffracted wave phases due to discontinuities in the medium physical properties. The traveltimes from the source to geophones for 28 significant phases were picked from the seismogram and used as data in the inverse problem.

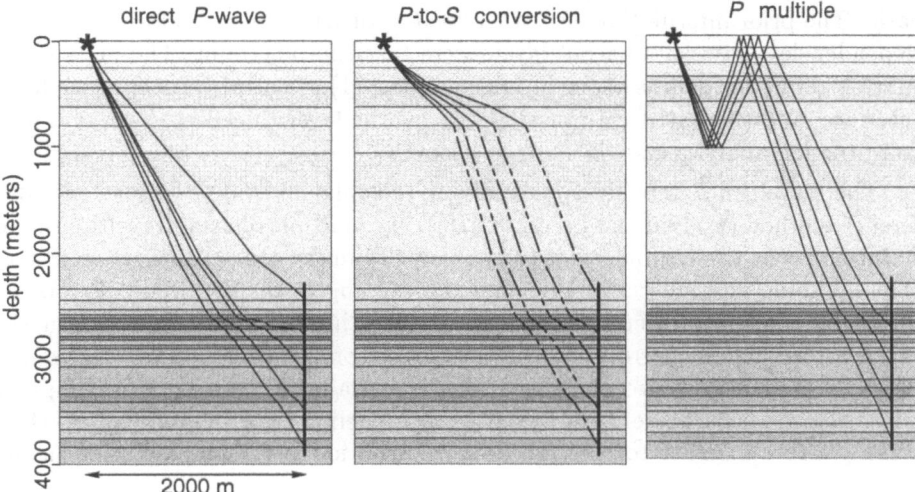

Fig. 2. Layered representation of the media where the geometry of acquisition for the seismic data is shown. The symbol in the surface indicates the source position, and the vertical bold line indicates the position of the array of geophones (inside the well). Raypaths have been plotted to illustrate three different wave phases; solid lines represent P-wave raypaths and dashed lines represent S-wave raypaths.

Prior information was incorporated into the inverse problem:

– structural information from migrated images (low spatial resolution) obtained from seismic data recorded at the surface,
– information on part of the physical parameters from direct physical measurements in the bore-hole (acoustic and density logs), and
– information from regional geological surveys (e.g. information about statistical relationship between P-wave velocity and S-wave velocity).

Structural information showed that the medium was approximately horizontally layered. As regional studies confirmed this observation, we used it in order to justify the horizontally layered assumption made to simplify the parameterization (physical parameters depended only on the depth variable) and thus decrease the number of parameters describing the medium. According to migrated images and the analysis of the log data we considered that the medium presented 34 main layers. Moreover, as the log data showed weak variability of the seismic velocities inside a layer, we considered these parameters homogeneous (constant) for each layer. Hence, we had $34 \times 3 = 102$ model parameters, defining for each layer: (1) the layer bottom depth, (2) the P-wave layer velocity, and (3) the S-wave layer velocity.

Information from log data was used to infer the prior *pdf* for the P-wave velocities and the depths of interfaces. As direct information on S-wave velocity was not available, we used petrophysical studies on samples about P versus S-wave velocity relationship (Castagna et al., [17]), for rocks in the same region, in order to infer statistical information on S-wave velocity over P-wave velocity ratio. The prior information on both velocities distributions was log-normal and dependencies between adjacent layers were taken into account. The prior information about the depths of the layers combined (1) a central value for each layer interface, and (2) statistical constrains about the layer thickness, both obtained from the log analysis and the surface seismic.

The traveltimes were simulated solving the Eikonal equation by a finite difference method (Podvin and Lecomte, [18]). For multiple phases, traveltime computation needs three simulations (down-going from the source then reflected and thus up-going and reflected again thus down-going to the receivers). Errors induced by the space discretization have been estimated and introduced in the inversion as modeling errors. As the traveltime computation is time consuming, we have used a grid size of 30 m which is a trade-off between the computer time cost and accuracy of time calculation (modeling errors should be less than picking errors in order to preserve data information and then resolution in the solution).

To solve the problem it was used a Markov chain Monte Carlo method: the Metropolis sampler. We recall that the Gibbs sampler can not be used here because we do not have the conditionals of the posterior density. The single step Metropolis algorithm was implemented at a first glance, and it turned out to be impossible to achieve equilibration in misfit and posterior parameter statistics within the available computation time. Data misfit reduction was not satisfactorily reduced (large misfit compared with data uncertainties), using several trials for the prior chain step length and step directions. This showed that the algorithm was not completing its initial phase of misfit reduction (burn-in period).

The difficulty for treating this inverse problem with the conventional Metropolis algorithm, suggested the use of the multi-step metropolis sampling scheme. The approximation of the likelihood function with the data partition approach (16-17) was natural here, because the calculation of the arrival times could be separated for the different phases. The arrival time data set was divided into six subsets (see figure 3) and the Metropolis algorithm was decomposed in six steps, correspondingly. Candidate models were confronted with each data subset in sequence, allowing to reject unlikely models using the simulation of a few wave phases. The simulation of additional phases was only performed if the model was accepted in the previous test.

The six subsets were ordered following two criteria: the data precision (energy, picking uncertainty), and the sequence of calculation in the forward problem solution. Group 1 in the sequence, for instance, included the direct P-wave arrival and the first P-wave multiple (water layer multiple). These were the phases with smallest uncertainties and their calculation was necessary for simulating the other phases travel times (saving computing time). Groups 5 and 6

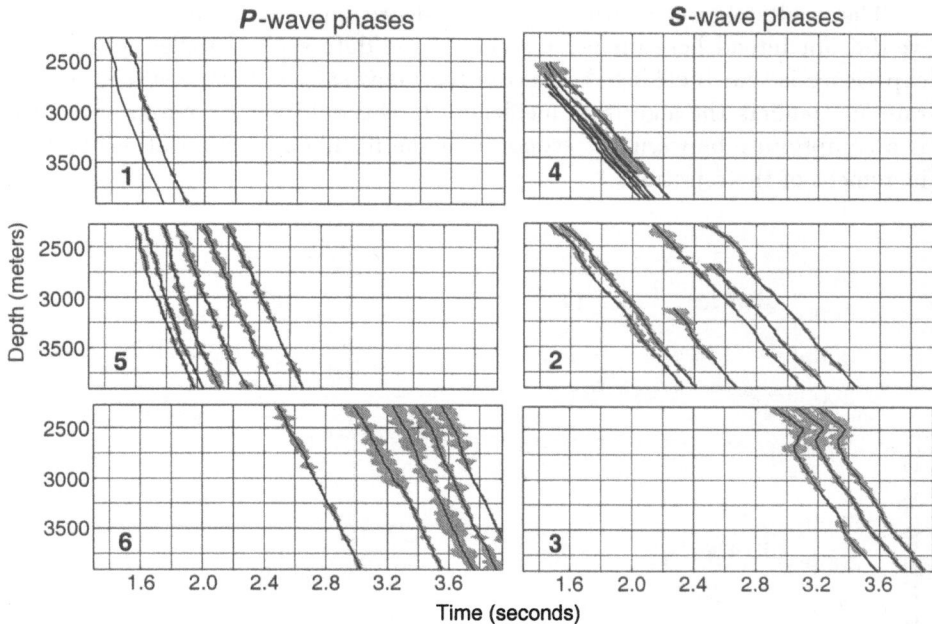

Fig. 3. The arrival times for 28 selected observed wave phases grouped in six data subsets. The solid black lines indicate the arrival times picked from the seismogram and the light gray regions show the uncertainty (mean deviation) associated with the picking operation. The numbers represent the sequence in which the data subsets are used by the multi-step sampling algorithm.

in the sequence were P-multiples and the travel time calculation was more time consuming than for the P-to-S converted phases (groups 2, 3 and 4). Groups 2 and 3 provided accurate information about the S-velocity model (which was not well constrained by the prior information).

The multiple step Metropolis algorithm resulted in a major improvement in sampling efficiency. Choosing an adequate preliminary random walk (found empirically after several trials), it has allowed a fast misfit reduction (burn-in-period) and an acceptable convergence of posterior statistics. To check the convergence of posterior statistics several univariate marginal plots and bivariate marginal cross-plots were monitored, and four independent chains were run. We verified that at least three of the four independent chains converge to similar plots in each monitored variable, or pair of variables. As usual in Monte Carlo, the burn-in period outcome samples were left out from posterior statistics.

Our comparison of the performance of the two algorithms is here qualitative. The multi-step Metropolis algorithm (1) passed the burn-in period by reducing satisfactorily the misfit between calculated and observed data, and (2) fulfilled convergence criteria. Whereas, the single-step algorithm could not pass the prior stage of reducing the data misfit.

The results obtained with the six-step algorithm were satisfactory. Although we did not intend here further division of the data set in smaller subsets, our experience is that breaking the data in too many subsets increases the rejection rate and renders the algorithm inefficient. In this data partition approach there is a compromise between the economy in the likelihood computations and the increment of the rejections.

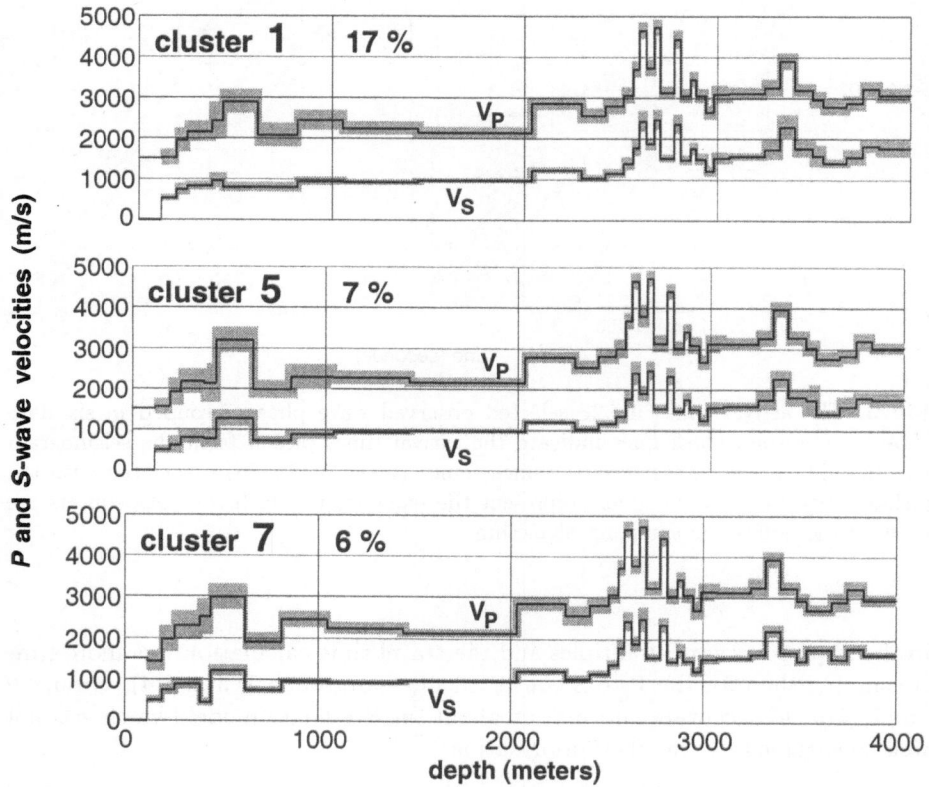

Fig. 4. Three of the most representative model clusters obtained after the analysis of the posterior model set generated by the sampling algorithm. Each cluster being not so far from a multivariate Gaussian statistical model, we represent its centroid (mean model) and the standard deviation of the model parameters. The mean model is indicated by a black bold line, and the standard deviation is given by the gray region around the mean model. The global probability for each mode is given in percent.

The results shown here were obtained using the four independent Markov chains running in parallel (during five days in four work-stations) to produce an overall posterior sample set of 600.000 models (joint P-wave velocity, S-wave velocity models, and depth interface parameters). This posterior model sample set was used to estimate posterior statistics on model parameters. The posterior *pdf* being clearly multi-modal, we have used a clustering algorithm which has allowed to identify 15 separated and significant clusters. Figure 4 shows three of these clusters.

The multi-modality of the posterior *pdf* was clearly seen in marginal histograms on model parameters. It was expected because an observed phase could be explained by several different synthetic phases; the set of simulated phases was larger than the actual data phases. The differences among the clusters shown in figure 4 are particularly clear at the uppermost and lower-most zones of the model. This differences are enough to change (shift) the matching between some observed and calculated phases. It was verified, for instance, that the first observed phase of data group 3 was explained in a different way for the clusters one and seven. For the first cluster it was fitted by a conversion on the 5th interface while for the seventh cluster it was fitted by a conversion on the 4th interface.

6 Conclusions

In the first sections of the work we reviewed sampling techniques that are useful in complex inference problems where the evaluation of the likelihood function represents a major computational effort: importance sampling and the Metropolis sampler. Classical importance sampling is a variance reduction technique that allows to focus the sampling in some regions of the model space, without introducing a bias in posterior estimates. If the importance density is suitable selected to mimic the posterior density the technique is able to improve the sampling efficiency.

The Metropolis sampler is an algorithm that provides, asymptotically, a sample of the posterior *pdf*. It is a convenient method in problems with complex posterior information, as it is the case for geophysical inverse problems. However, the actual performance of the Metropolis sampler is known to be affected by the computational limitations in evaluating the likelihood function (and thus a limitation in the total number of collected samples). The speed of convergence, the sample mixing and the accuracy of estimates are highly dependent on the problem at hand and on the particular implementation of the algorithm.

The contribution of the present work is to combine the two techniques portrayed above to design a multi-step Metropolis algorithm been able to significantly improve the sampling performance. The first step of the algorithm samples from an importance density, $\tilde{\sigma}(\mathbf{m}) \propto \rho(\mathbf{m})I(\mathbf{m})$, and the second step samples from the posterior density. The importance density is here defined as the product of the prior density and an importance likelihood function that should be much easier to calculate than the actual data likelihood function. Although the overall acceptance rate of the combined algorithm is smaller than the acceptance rate of the single Metropolis algorithm, the acceptance rate of the second step of the combined algorithm is significantly increased. As the second step is involving the major computational effort, the overall computational efficiency is improved.

Different ways to construct importance likelihood functions for geophysical inverse problems were discussed. This functions can be defined with basis on

(1) the information from the preliminary analysis of the data, (2) the misfit of the observed data with simplified simulations of the geophysical response of the models, (3) the misfit between calculated and observed data for suitably chosen subsets of the total available data set, or (4) additional information that is not incorporated to the posterior density.

The algorithm has been tested on a seismic real inverse problem with success. The design of the importance likelihood function followed here a partition of the travel time data in six subsets according to different groups of wave phases; a six-step Metropolis sampler was implemented correspondingly. The candidate models were confronted progressively with the six subsets of the observed data allowing fast rejection of unlikely models.

Acknowledgements. We thank the Institut de Physique du Globe de Paris, the Central University of Venezuela, the University of Copenhagen and the Geophysic Center of the Ecole des Mines de Paris. Part of this work has been funded by the UMR 7580 of the CNRS (France) and the program GeoFrance-3D. We also gratefully thank ELF company as they provided us the OVSP data. We thank the editor Per Christian Hansen, and the comments of an anonymous referee that helped to improve the manuscript.

References

1. Press, F., Earth models obtained by Monte Carlo inversion, *Journal of Geophysical Research, 73*, 5223-5234, 1968.
2. Mosegaard, K., and A. Tarantola, Monte Carlo sampling of solutions to inverse problems, *Journal of Geophysical Research, 100* 12,431-12,447, 1995.
3. Mosegaard, K., Resolution Analysis of General Inverse Problems through Inverse Monte Carlo Sampling: *Inverse Problems 14*, 405-426, 1998.
4. Bosch, M., Lithologic tomography: from plural geophysical data to lithology estimation, *Journal of Geophysical Research, 104*, 749-766, 1999.
5. Smith, A. F., and G. O. Roberts, Bayesian computation via the Gibbs sampler And related Markov chain Monte Carlo methods, *J. R. Statist. Soc. B, 55*, 3-23, 1993.
6. Tierney, L., Markov-chains for exploring posterior distributions, *Annals of Statistics, 22*, 1702-1762, 1994.
7. Wolpert, R., Monte Carlo integration in Bayesian statistical analysis, *in* Statistical Multiple Integration, Flournoy N. and Tsutakawa R. K. Eds., Contemporary Mathematics Series, American Mathematical Society, v.115, 101-115, 1989.
8. Hastings, W. K., Monte Carlo sampling method using Markov chains and their applications, *Biometrika, 57*, 97-109, 1970.
9. Kloek, T., and H. K. van Dijk, Bayesian estimates of equation system parameters: an application of integration by Monte Carlo, *Econometrica, 46*, 1-19, 1978.
10. Evans, M., Adaptive importance sampling and chaining, *in* Statistical Multiple Integration, Flournoy N. and Tsutakawa R. K. Eds., Contemporary Mathematics Series, American Mathematical Society, v.115, 137-143, 1989.

11. Geyer, Ch. J., Practical Markov Chain Monte Carlo, Statistical Science, 7, 473-511, 1992.

12. Man-Suk, O., and J. O. Berger, Integration of multimodal functions by Monte Carlo importance sampling, *Journal of the American Statistical Association*, 88, 450-456, 1993.

13. Metropolis, N. , A. W. Rosenbluth, M. N Rosenbluth, A. H. Teller, and E. Teller, Equation of state calculations by fast computing machines, *The Journal of Chemical Physics, 21*, 1087-1092, 1953.

14. Robert, Ch. P., Convergence control methods for Markov Chain Monte Carlo Algorithms, *Statistical Science, 10*, 231-253, 1995.

15. Cowles, M. K., and B. P. Carling, Markov Chain Monte Carlo convergence diagnostics: A comparative review, *Journal of the American Statistical Association, 91*, 883-904, 1996.

16. Barnes, Ch., M. Charara, and A. Tarantola, Traveltime inversion of bore-hole seismic data using Monte Carlo Methods, submitted to the Geophysical Journal International, 1999.

17. Castagna, J. .P, M. L Batzle, T. K. and Kan, Rock Physics — The Link Between Rock Properties and AVO Response *Offset-Dependant Reflectivity — Theory and Practice of AVO Analysis*, (Society of Exploration Geophysics: Investigations in Geophysics), p 135, 1993

18. Podvin, P., and I. Lecomte, Finite difference computations of travel times in very contrasted velocity models: massively parallel approach and its associated tools, *Geophysical Journal International*, 105, 271-284, 1991.

Full Waveform Inversion of Seismic Data for a Viscoelastic Medium

Marwan Charara[1,2], Christophe Barnes[2], and Albert Tarantola[2]

[1] Schlumberger
[2] Institut Physique du Globe de Paris

Abstract. A linear viscoelastic media with a constant quality factor provide a reasonably realistic model of Earth for many seismic application. A linear viscoelastic relation, based on the superposition of relaxation mechanisms, produced formulations that allow the computation of synthetic seismograms in time-domain for arbitrary spatial distributions of quality factors. By using this constitutive relation and the adjoint state technique, we propose a new formulation for the viscoelastic seismic inverse problem. This approach reduces the computer memory requirements to the point where computations are practical. For illustration, this full viscoelastic waveform inversion has been applied for a numerical experiment.

1   Introduction

Full waveform seismic inversion is a non-linear problem. This inverse problem can be solved by iteratively improving an initial model of the subsurface by matching the measured seismic data with the modeled data, using for instance gradient algorithms.

When the Earth model can be described by a simple model with a limited number of parameters (e.g., a vertical succession of horizontal homogeneous layers), the computation of the gradient of the misfit function with respect to the model parameters can be computed by the finite difference method at the cost of at least one modeling per parameter. Such algorithms have been used for example by Pan et al. (1988), Amundsen and Ursin (1991), Martinez and Mac Mechan (1991), Helgensen and Landrø (1993), and Zhao et al. (1994). Numerical tests on synthetic data show that the parameters of a plane-layer model can be recovered successfully.

When the earth model is more complicated (e.g., generally inhomogeneous), the number of parameters is usually so large that the numerical computation of the misfit function gradient by finite difference is prohibitive. For that particular reason, analytical expression of the gradient is to be preferred by using for instance the adjoint state technique Lions (1971). For the seismic case formulations have been developed by Lailly (1984) and Tarantola (1984) in the acoustic approximation, then by Tarantola (1986) for the elastic approximation. The feasibility and limitation of such inversion have been investigated on seismic synthetic data by Canadas (1986) and Gauthier et al. (1986) for the acoustic case and for the elastic case by Mora (1987). Application on real data have been reported by Pica et al. (1990) for the acoustic case and for the elastic case by Crase et al. (1990).

The elastic approximation of the Earth model is in many cases not realistic as the waves undergoes attenuation and dispersion. Fitting the recorded data with synthetic data generated with an elastic modeling method that neglects absorption obviously results in a biased estimation of the earth model or in a distorted images of the subsurface. Most viscoelastic inversions are performed for a simple 1D viscoelastic earth model as it is possible to compute the misfit function gradient by finite difference method or one can use pseudo-analytical forward problem in the frequency domain that allows to get a simple anelastic strain-stress relation (Dietrich and Bouchon, 1985). When the medium is heterogeneous, most efficient method for modeling wave propagation are in time domain (Emmerich and Korn, 1987), and it would be a great advantage to use a formulation of the misfit function gradient expressed in the time domain. Tarantola (1988) proposed a general expression of the gradient for an attenuating medium in the time domain; however, his formulation is based on a constitutive law that has the form of a time convolution integral which is intractable in a numerical computation. Formulations in linear viscoelasticity based on the superposition of relaxation mechanisms (Day and Minster, 1984; Emmerich and Korn, 1987), allows the computation of synthetic seismograms for arbitrary spatial distributions of quality factors. The key concept of this approach is the replacement of a time convolution between stress and strain by "strain memory variables" (Carcione, 1990) through a set of first order temporal partial differential equations to solve. Using the same formulation for the constitutive law, and making the realistic approximation of a constant quality factor model for an attenuating medium, we propose a new analytical expression for the viscoelastic parameter components of the gradient. We illustrate the feasibility of the proposed method by a numerical experiment of inverting an OVSP data (offset vertical seismic profiling).

2 Constitutive Law for an Attenuating Medium

For an anisotropic linear viscoelastic material, the most general relation between the components of the stress tensor σ^{ij} and ε^{kl} is (Christensen, 1982)

$$
\sigma^{ij}(\mathbf{x},t) = \int_{-\infty}^{+\infty} d\tau \, \Psi^{ijkl}(\mathbf{x}, t-\tau) \dot{\varepsilon}^{kl}(\mathbf{x}, \tau)
$$

$$
= \int_{-\infty}^{+\infty} d\tau \dot{\Psi}^{ijkl}(\mathbf{x}, t-\tau) \, \varepsilon^{kl}(\mathbf{x}, \tau), \tag{1}
$$

where t is the time, $\mathbf{x}$ is the position vector, and Ψ^{ijkl} is a fourth-order tensorial relaxation function. The dot above a variable denotes a time derivative. This expression is not suited for solving initial value problem as it is required to know the whole history of the strain fields. By restating the relaxation function in term of relaxed modulus associated with a function of time taking into account the history of interactions

$$\Psi^{ijkl}(\mathbf{x}, t) \equiv \Psi^{ijkl}(\mathbf{x}, +\infty) \chi_\nu(\mathbf{x}, t), \qquad (2)$$

with $\Psi^{ijkl}(\mathbf{x}, +\infty)$ the relaxed state, and the function $\chi_\nu(\mathbf{x}, t)$

$$\chi_\nu(\mathbf{x}, t) = \left[1 - \frac{1}{L_\nu} \sum_{\ell=1}^{L_\nu} \left(1 - \frac{\tau_{\varepsilon\ell}^{(\nu)}(\mathbf{x})}{\tau_{\sigma\ell}^{(\nu)}(\mathbf{x})} \right) \exp\left(-\frac{t}{\tau_{\sigma\ell}^{(\nu)}(\mathbf{x})} \right) \right] H(t), \qquad (3)$$

describing the behavior of the material with respect to time. τ_σ et τ_ε are respectively, for a mechanism ℓ, the relaxation time of the stress σ under a constant strain and the strain relaxation time ε under constant stress. The Heaviside function $H(t)$ ensures causality. The index ν is a function of $ijkl$. For invariance-under-rotation reasons and also for physical considerations (principal mode of deformation of the medium) (Helbig, 1993; Carcione and Cavallini, 1994) an appropriate association for the temporal function $\chi_\nu(\mathbf{x}, t)$ are the six non-degenerated eigenvalues of the stiffness matrix $\Psi^{ijkl}(\mathbf{x}, +\infty)$ which in the isotropic case degenerates to the incompressibility modulus κ and twice the shear modulus μ.

In terms of rate of relaxation, needed if we prefer to use the second equality of Equation (1), the relation (2) becomes

$$\dot{\Psi}^{ijkl}(\mathbf{x}, t) = \Psi^{ijkl}(\mathbf{x}, 0^+) + \Psi^{ijkl}(\mathbf{x}, +\infty) \times \sum_{\ell=1}^{L_\nu} \left(1 - \frac{\tau_{\varepsilon\ell}^{(\nu)}(\mathbf{x})}{\tau_{\sigma\ell}^{(\nu)}(\mathbf{x})} \right) \phi_\ell^{(\nu)}(\mathbf{x}, t), \quad (4)$$

with

$$\phi_\ell^{(\nu)}(\mathbf{x}, t) = \frac{1}{L_\nu \, \tau_{\sigma\ell}^{(\nu)}(\mathbf{x})} \exp\left(-\frac{t}{\tau_{\sigma\ell}^{(\nu)}(\mathbf{x})} \right) H(t). \qquad (5)$$

By using Equations (4) and (5) in Equation(1), the new constitutive law becomes

$$\sigma^{ij}(\mathbf{x}, t) = \Psi^{ijkl}(\mathbf{x}, 0^+) \varepsilon^{kl}(\mathbf{x}, t)$$
$$+ \Psi^{ijkl}(\mathbf{x}, +\infty) \times \sum_{\ell=1}^{L_\nu} \left(1 - \frac{\tau_{\varepsilon\ell}^{(\nu)}(\mathbf{x})}{\tau_{\sigma\ell}^{(\nu)}(\mathbf{x})} \right) \epsilon_\ell^{kl}(\mathbf{x}, t), \qquad (6)$$

where we define the memory strain variables $\epsilon_\ell^{kl}(\mathbf{x}, t)$ as

$$\epsilon_\ell^{kl}(\mathbf{x}, t) = \int_{-\infty}^{+\infty} dt' \phi_\ell^{(\nu)}(\mathbf{x}, t - t') \varepsilon^{kl}(\mathbf{x}, t'). \qquad (7)$$

The temporal convolution of the constitutive relation can be replaced now by solving the following set of linear first order time differential equations:

$$\dot{\epsilon}_\ell^{kl}(\mathbf{x}, t) = \phi_\ell^{(\nu)}(\mathbf{x}, 0^+) \varepsilon^{kl}(\mathbf{x}, t) - \frac{1}{\tau_{\sigma\ell}^{(\nu)}(\mathbf{x})} \epsilon_\ell^{kl}(\mathbf{x}, t). \qquad (8)$$

3 The Case of a Constant Quality Factor

Attenuation and dispersive effects are often quantified by the quality factor Q. This quantity is defined as the number of wavelengths a wave can propagate through the medium before its amplitude has decreased by a factor $e^{-\pi}$. The quality factor has been found to be nearly constant as a function of frequency (McDonal et al., 1958; Murphy, 1982; and Spencer, 1981) for the seismic exploration bandwidth (approximately 1–200 Hz).

By choosing a constant ratio of the relaxation times for the different mechanisms:

$$r^{(\nu)}(\mathbf{x}) = \frac{\tau_{\varepsilon\ell}^{(\nu)}(\mathbf{x})}{\tau_{\sigma\ell}^{(\nu)}(\mathbf{x})} = \chi_\nu(\mathbf{x},0^+) = \frac{\Psi^{ijkl}(\mathbf{x},0^+)}{\Psi^{ijkl}(\mathbf{x},+\infty)} = cst, \tag{9}$$

it is possible to model a constant quality factor[1] over a frequency bandwidth by determining the optimal coefficient a and b in the relation:

$$Q_\nu(\mathbf{x}) = a + b\frac{\Psi^{ijkl}(\mathbf{x},+\infty)}{\Delta\Psi^{ijkl}(\mathbf{x})}, \tag{10}$$

with $\Delta\Psi^{ijkl}(\mathbf{x})$ representing the difference between the relaxed and unrelaxed modulus:

$$\Delta\Psi^{ijkl}(\mathbf{x}) = \Psi^{ijkl}(\mathbf{x},+\infty) - \Psi^{ijkl}(\mathbf{x},0^+). \tag{11}$$

Now we can simplify the constitutive law:

$$\sigma^{ij}(\mathbf{x},t) = \Psi^{ijkl}(\mathbf{x},0^+)\varepsilon^{kl}(\mathbf{x},t) + \Delta\Psi^{ijkl}(\mathbf{x})\sum_{\ell=1}^{L_\nu}\epsilon_\ell^{kl}(\mathbf{x},t). \tag{12}$$

4 The Viscoelastic Wave Equation and its Transposed

The aim of least-squares inversion is to obtain the model $\mathbf{m}$ minimizing the misfit function (Tarantola, 1988)

$$S[\mathbf{m}] = \frac{1}{2}\left[\langle C_{\mathbf{D}}^{-1}(\mathbf{u}[\mathbf{m}] - \mathbf{u_{obs}}), (\mathbf{u}[\mathbf{m}] - \mathbf{u_{obs}})\rangle \right. \tag{13}$$
$$\left. + \langle C_{\mathbf{M}}^{-1}(\mathbf{m} - \mathbf{m_{prior}}), \mathbf{m} - \mathbf{m_{prior}}\rangle\right]$$

where $\mathbf{u_{obs}} = \{u^i(\mathbf{x}_r,t)_{obs}\}$ are the observed displacement data and $\mathbf{u_{cal}} = \mathbf{u}[\mathbf{m}] = \{u^i(\mathbf{x}_r,t)_{cal}\}$ are the computed data at receiver location $\mathbf{x}_r$. $C_{\mathbf{D}}$ is the covariance describing data uncertainties. The data vector $\mathbf{u}$ belongs to the 'data space' $\mathbf{D}$. The model $\mathbf{m}$ belongs to the 'model space' $\mathbf{M}$, $\mathbf{m_{prior}}$ is some a priori model and $C_{\mathbf{M}}$ is the covariance operator describing uncertainties in

---

[1] Two mechanisms $L_\nu = 2$ are usually enough to model the constant quality factor (Emmerich and Korn, 1987). For instance, by choosing $a = 1.86$ and $b = 3.248$, we are able to model a constant quality factor in the range of 10 and 40 Hz (Charara, 1996).

$\mathbf{m}_{\text{prior}}$. Let $\mathbf{g}$ be the wave propagation operator linking the vector displacement $\mathbf{u}$ to the vector parameter $\mathbf{m}$. For simplicity, if we drop the second term in the misfit function (13) associated with $\mathbf{m}_{\text{prior}}$, the gradient of the misfit function will be

$$\widehat{\gamma}\left(\mathbf{m}\right) = \mathbf{C}_{\mathbf{M}}^{-1}\gamma = \frac{\partial S}{\partial \mathbf{m}}\left(\mathbf{m}\right) = \mathbf{H}^{\mathsf{T}}\left(\mathbf{m}\right)\,\delta\widehat{\mathbf{u}}\left(\mathbf{m}\right) \tag{14}$$

where

$$\mathbf{H}\left(\mathbf{m}\right) = \frac{\partial \mathbf{g}}{\partial \mathbf{m}}\left(\mathbf{m}\right) \tag{15}$$

and

$$\delta\widehat{\mathbf{u}}\left(\mathbf{m}\right) = \mathbf{C}_{\mathbf{D}}^{-1}\delta\mathbf{u}\left(\mathbf{m}\right) = \mathbf{C}_{\mathbf{D}}^{-1}\left(\mathbf{g}\left(\mathbf{m}\right) - \mathbf{u}_{\mathbf{obs}}\right). \tag{16}$$

The hat is used to denote an entity of the dual space. In equations (14) and (16), $\delta\widehat{\mathbf{u}}\left(\mathbf{m}\right)$ may be identified as the weighted residuals. The matrix $\mathbf{H}\left(\mathbf{m}\right)$ will in the continuous case be an operator, known as the linear Born operator. In what follows, the meaning of the Born operator and its transpose $\mathbf{H}^{\mathsf{T}}\left(\mathbf{m}\right)$ will be explicitly derived for the viscoelastic wave equation.

For a heterogeneous linear anisotropic viscoelastic medium, displacement $u^i(\mathbf{x},t)$ satisfies, inside a volume V surrounded by a surface S, the relationship

$$\begin{cases} \rho(\mathbf{x})\frac{\partial^2 u^i}{\partial t^2}(\mathbf{x},t) - \frac{\partial \sigma^{ij}}{\partial x^j}(\mathbf{x},t) = \phi^i(\mathbf{x},t) \\ \\ \sigma^{ij}(\mathbf{x},t) - \Psi^{ijkl}(\mathbf{x},0^+)\varepsilon^{kl}(\mathbf{x},t) - \Delta\Psi^{ijkl}(\mathbf{x})\displaystyle\sum_{\ell=1}^{L_\nu}\epsilon_\ell^{kl}(\mathbf{x},t) = M^{ij}(\mathbf{x},t) \end{cases} \tag{17}$$

where $\rho(\mathbf{x})$ is the mass density, $M^{ij}(\mathbf{x},t)$ and $\phi^i(\mathbf{x},t)$ are moment density and force density terms, respectively.

For the surface S of the medium, the stress satisfies

$$n^j\left(\xi\right)\sigma^{ij}\left(\xi,t\right) = \tau^i\left(\xi,t\right) \quad \text{for } \xi \in S, \tag{18}$$

$n^j\left(\xi\right)$ being the outgoing unit normal to the surface and $\tau^i\left(\xi,t\right)$ the surface traction.

Let $u^i\left(\mathbf{x},t\right)$ be the solution of the wave equation and let

$$\mathbf{m} = \left(\phi^i, \rho, \tau^i, M^{ij}, \Psi^{ijkl}, \Delta\Psi^{ijkl}\right)^{\mathsf{T}} \tag{19}$$

be the model parameters. A perturbation on the model parameters $\mathbf{m} \to \mathbf{m} + \delta\mathbf{m}$ will give rise to a perturbation of the displacement field $\mathbf{u} \to \mathbf{u} + \delta\mathbf{u}$, i.e.,

$$\begin{cases} (\rho + \delta\rho)(\mathbf{x})\frac{\partial^2\left(u^i + \delta u^i\right)}{\partial t^2}(\mathbf{x},t) - \frac{\partial\left(\sigma^{ij} + \delta\sigma^{ij}\right)}{\partial x^j}(\mathbf{x},t) = (\phi^i + \delta\phi^i)(\mathbf{x},t) \\ \\ \left(\sigma^{ij} + \delta\sigma^{ij}\right)(\mathbf{x},t) - \left(\Psi^{ijkl} + \delta\Psi^{ijkl}\right)(\mathbf{x},0^+)\varepsilon^{kl}(\mathbf{x},t) \\ \qquad - \left(\Delta\Psi^{ijkl} + \delta\Delta\Psi^{ijkl}\right)(\mathbf{x})\displaystyle\sum_{\ell=1}^{L_\nu}\epsilon_\ell^{kl}(\mathbf{x},t) = \left(M^{ij} + \delta M^{ij}\right)(\mathbf{x},t) \end{cases} \tag{20}$$

and

$$n^j \left(\sigma^{ij} + \delta\sigma^{ij}\right)(\xi, t) = \left(\tau^i + \delta\tau^i\right)(\xi, t) \text{ for } \xi \in S. \tag{21}$$

Substracting equations (17) to (20) and neglecting higher order terms give

$$\begin{cases} \rho(\mathbf{x})\frac{\partial^2 \delta u^i}{\partial t^2}(\mathbf{x}, t) - \frac{\partial \delta\sigma^{ij}}{\partial x^j}(\mathbf{x}, t) = -\delta\rho(\mathbf{x})\frac{\partial^2 u^i}{\partial t^2}(\mathbf{x}, t) + \delta\phi^i(\mathbf{x}, t) \\ \\ \delta\sigma^{ij}(\mathbf{x}, t) - \Psi^{ijkl}(\mathbf{x}, 0^+)\delta\varepsilon^{kl}(\mathbf{x}, t) - \Delta\Psi^{ijkl}(\mathbf{x})\sum_{\ell=1}^{L\kappa}\delta\epsilon_\ell^{kl}(\mathbf{x}, t) \\ \\ \quad = \delta M^{ij}(\mathbf{x}, t) + \delta\Psi^{ijkl}(\mathbf{x}, 0^+)\varepsilon^{kl}(\mathbf{x}, t) + \delta\Delta\Psi^{ijkl}(\mathbf{x})\sum_{\ell=1}^{L\kappa}\epsilon_\ell^{kl}(\mathbf{x}, t) \end{cases} \tag{22}$$

the perturbation on $\epsilon_\ell^{kl}(\mathbf{x}, t)$ being

$$\delta\epsilon_\ell^{kl}(\mathbf{x}, t) = \int_{-\infty}^{+\infty} dt' \phi_\ell^{(\nu)}(\mathbf{x}, t - t')\delta\varepsilon^{kl}(\mathbf{x}, t'),$$

and also by substracting Equations (18) to (21), we get

$$n^j \delta\sigma^{ij}(\xi, t) = \delta\tau^i(\xi, t) \text{ for } \xi \in S. \tag{23}$$

We notice that, to a first approximation, the perturbation of the field $\delta u^i(\mathbf{x}, t)$ obeys the same wave equation as the original field, except for a different source term. The result may be stated in terms of a Green's function G^{ij} by using the representation theorems (Tarantola, 1988):

$$\begin{aligned} \delta u^i(\mathbf{x}, t) = &\int_V dV(\mathbf{x}') \int_{t_0}^{t_1} dt' G^{ij}(\mathbf{x}, t; \mathbf{x}', t')\delta\phi^j(\mathbf{x}', t') \\ &+ \int_S dS(\xi') \int_{t_0}^{t_1} dt' G^{ij}(\mathbf{x}, t; \xi', t')\delta\tau^j(\xi', t') \\ &- \int_V dV(\mathbf{x}') \int_{t_0}^{t_1} dt' G^{ij}(\mathbf{x}, t; \mathbf{x}', t')\delta\rho(\mathbf{x}')\frac{\partial^2 u^j}{\partial t'^2}(\mathbf{x}', t') \\ &- \int_V dV(\mathbf{x}') \int_{t_0}^{t_1} dt' \frac{\partial G^{ij}}{\partial x'^k}(\mathbf{x}, t; \mathbf{x}', t')\delta M^{jk}(\mathbf{x}', t') \\ &- \int_V dV(\mathbf{x}') \int_{t_0}^{t_1} dt' \frac{\partial G^{ij}}{\partial x'^k}(\mathbf{x}, t; \mathbf{x}', t')\delta\Psi^{jklm}(\mathbf{x}', 0^+)\epsilon^{lm}(\mathbf{x}', t') \\ &- \int_V dV(\mathbf{x}') \int_{t_0}^{t_1} dt' \frac{\partial G^{ij}}{\partial x'^k}(\mathbf{x}, t; \mathbf{x}', t')\delta\Delta\Psi^{jklm}(\mathbf{x}')\sum_{\ell=1}^{L\nu}\epsilon_\ell^{lm}(\mathbf{x}', t) \end{aligned} \tag{24}$$

The linear relationships in Equation (24) define the application of the linear Born operator, which can be written as

$$\delta\mathbf{u} = \mathbf{H}\delta\mathbf{m} = \left(\mathbf{A}\ \mathbf{B}\ \mathbf{C}\ \mathbf{D}\ \mathbf{E}\ \mathbf{F}\right) \begin{pmatrix} \delta\phi^j \\ \delta\tau^j \\ \delta\rho \\ \delta M^{jk} \\ \delta\Psi^{jklm} \\ \delta\Delta\Psi^{jklm} \end{pmatrix}. \tag{25}$$

Equation (24) gives the Fréchet derivatives, or the kernel of the Born operator, for the different model parameters,

$$\mathbf{A}\left(\mathbf{x},t,\mathbf{x}'\right) = +\int_{t_0}^{t_1} dt'\, \mathbf{G}^{ij}\left(\mathbf{x},t;\mathbf{x}',t'\right)$$

$$\mathbf{B}\left(\mathbf{x},t,\xi'\right) = +\int_{t_0}^{t_1} dt'\, \mathbf{G}^{ij}\left(\mathbf{x},t;\xi',t'\right)$$

$$\mathbf{C}\left(\mathbf{x},t,\mathbf{x}'\right) = -\int_{t_0}^{t_1} dt'\, \mathbf{G}^{ij}\left(\mathbf{x},t;\mathbf{x}',t'\right)\frac{\partial^2 \boldsymbol{u}^j}{\partial t'^2}\left(\mathbf{x}',t'\right)$$

$$\mathbf{D}\left(\mathbf{x},t,\mathbf{x}'\right) = -\int_{t_0}^{t_1} dt'\, \frac{\partial \mathbf{G}^{ij}}{\partial x'^k}\left(\mathbf{x},t;\mathbf{x}',t'\right)$$

$$\mathbf{E}\left(\mathbf{x},t,\mathbf{x}'\right) = -\int_{t_0}^{t_1} dt'\, \frac{\partial \mathbf{G}^{ij}}{\partial x'^k}\left(\mathbf{x},t;\mathbf{x}',t'\right)\varepsilon^{lm}\left(\mathbf{x}',t'\right)$$

$$\mathbf{F}\left(\mathbf{x},t,\mathbf{x}'\right) = -\int_{t_0}^{t_1} dt'\, \frac{\partial \mathbf{G}^{ij}}{\partial x'^k}\left(\mathbf{x},t;\mathbf{x}',t'\right)\sum_{\ell=1}^{L_\nu}\epsilon_\ell^{lm}\left(\mathbf{x}',t'\right)$$

If we rewrite the equation (25) in the dual domain, we obtain the relation

$$\delta\hat{\mathbf{m}} = \mathbf{H}^{\mathsf{T}}\delta\hat{\mathbf{u}}(\mathbf{x},t). \tag{26}$$

The above equation can be equivalently be stated as

$$\begin{aligned}
\delta\widehat{\phi}^j(\mathbf{x}',t') &= \int_V dV(\mathbf{x})\int_{t_0}^{t_1} dt\, \mathbf{A}^{\mathsf{T}}\left(\mathbf{x},t,\mathbf{x}'\right)\delta\hat{u}^i\left(\mathbf{x},t\right) \\[4pt]
\delta\widehat{\tau}^j(\xi',t') &= \int_S dS(\xi)\int_{t_0}^{t_1} dt\, \mathbf{B}^{\mathsf{T}}\left(\mathbf{x},t,\mathbf{x}'\right)\delta\hat{u}^i\left(\xi,t\right) \\[4pt]
\delta\widehat{\rho}(\mathbf{x}') &= \int_V dV(\mathbf{x})\int_{t_0}^{t_1} dt\, \mathbf{C}^{\mathsf{T}}\left(\mathbf{x},t,\mathbf{x}'\right)\delta\hat{u}^i\left(\mathbf{x},t\right) \\[4pt]
\delta\widehat{M}^{jk}(\mathbf{x}',t') &= \int_V dV(\mathbf{x})\int_{t_0}^{t_1} dt\, \mathbf{D}^{\mathsf{T}}\left(\mathbf{x},t,\mathbf{x}'\right)\delta\hat{u}^i\left(\mathbf{x},t\right) \\[4pt]
\delta\widehat{\Psi}^{jklm}(\mathbf{x}',0^+) &= \int_V dV(\mathbf{x})\int_{t_0}^{t_1} dt\, \mathbf{E}^{\mathsf{T}}\left(\mathbf{x},t,\mathbf{x}'\right)\delta\hat{u}^i\left(\mathbf{x},t\right) \\[4pt]
\delta\widehat{\Delta\Psi}^{jklm}(\mathbf{x}') &= \int_V dV(\mathbf{x})\int_{t_0}^{t_1} dt\, \mathbf{F}^{\mathsf{T}}\left(\mathbf{x},t,\mathbf{x}'\right)\delta\hat{u}^i\left(\mathbf{x},t\right)
\end{aligned} \tag{27}$$

It can be shown (Tarantola, 1988) that the operator $\mathbf{H}$ and its transpose have identical kernels. The terms $\delta\hat{\phi}^j$, $\delta\hat{\tau}^j$, $\delta\hat{\rho}$, $\delta\widehat{M}^{jk}$, $\delta\widehat{\Psi}^{jklm}$, $\delta\widehat{\Delta\Psi}^{jklm}$ may be identified as being the components of the gradient $\hat{\gamma}(\mathbf{m})$. A new field $u^i(\mathbf{x}', t')$ may now be defined, given by

$$
\begin{aligned}
\overleftarrow{u}^i(\mathbf{x}', t') &= \int_V dV(\mathbf{x}) \int_{t_0}^{t_1} dt' \, \overleftarrow{G}^{ji}(\mathbf{x}', t'; \mathbf{x}, t) \delta\hat{u}^i(\mathbf{x}, t) \\
&= \int_V dV(\mathbf{x}) \int_{t_0}^{t_1} dt' \, \overrightarrow{G}^{ij}(\mathbf{x}, t; \mathbf{x}', t') \delta\hat{u}^i(\mathbf{x}, t)
\end{aligned}
\tag{28}
$$

The field introduced in equation (28) can be interpreted as the field resulting from propagation backwards in time of the data residuals, acting as sources at the receiver locations. This field must satisfy null final conditions,

$$
\begin{aligned}
\overleftarrow{u}^i(\mathbf{x}, T) &= 0 \\
\frac{\partial \overleftarrow{u}^i}{\partial t}(\mathbf{x}, T) &= 0 \;.
\end{aligned}
\tag{29}
$$

In Equation (28), use has been made of the fact that the anti-causal Green's function $\overleftarrow{G}^{ji}$ needed for the calculation of the reverse field can be shown to be related to the original (causal) Green's function $\overrightarrow{G}^{ij}$ through

$$
\overrightarrow{G}^{ij}(\mathbf{x}, t; \mathbf{x}', t') = \overleftarrow{G}^{ji}(\mathbf{x}', t'; \mathbf{x}, t).
\tag{30}
$$

Combining Equations (24) and (29), exchanging primed and unprimed variables yields

$$
\begin{aligned}
\delta\hat{\phi}^i(\mathbf{x}, t) &= u^i(\mathbf{x}, t) \\
\delta\hat{\tau}^i(\xi, t) &= u^i(\xi, t) \\
\delta\hat{\rho}(\mathbf{x}) &= -\int_{t_0}^{t_1} dt \frac{\partial u^i}{\partial t}(\mathbf{x}, t) \frac{\partial u^i}{\partial t}(\mathbf{x}, t) \\
\delta\hat{M}^{ij} &= -\varepsilon^{ij}(\mathbf{x}, t) \\
\delta\widehat{\Psi}^{ijkl}(\mathbf{x}, 0^+) &= -\int_{t_0}^{t_1} dt \varepsilon^{ij}(\mathbf{x}, t) \epsilon^{kl}(\mathbf{x}, t) \\
\delta\widehat{\Delta\Psi}^{ijkl}(\mathbf{x}) &= -\sum_{\ell=1}^{L_\nu} \int_{t_0}^{t_1} dt \varepsilon^{ij}(\mathbf{x}, t) \epsilon_\ell^{kl}(\mathbf{x}, t).
\end{aligned}
\tag{31}
$$

For the isotropic case, Ψ^{ijkl} is a just a linear combination of the eigenstiffness moduli κ and μ:

$$
\Psi^{ijkl}(\mathbf{x}) = \delta^{ij}\delta^{kl}\kappa(\mathbf{x}) + \left(\delta^{ik}\delta^{jl} + \delta^{il}\delta^{jk} - \frac{2}{3}\delta^{ij}\delta^{kl}\right)\mu(\mathbf{x}) \;.
\tag{32}
$$

Thus, the isotropic viscoelastic components of the misfit gradient are

$$\delta\widehat{\kappa}(\mathbf{x}, 0^+) = -\int_{t_0}^{t_1} dt\varepsilon^{ii}(\mathbf{x}, t)\varepsilon^{jj}(\mathbf{x}, t)$$

$$\delta\widehat{\Delta\kappa}(\mathbf{x}) = -\sum_{\ell=1}^{L_\nu}\int_{t_0}^{t_1} dt\varepsilon^{ii}(\mathbf{x}, t)\epsilon_\ell^{jj}(\mathbf{x}, t)$$

$$\delta\widehat{\mu}(\mathbf{x}, 0^+) = -\left(\delta^{ik}\delta^{jl} + \delta^{il}\delta^{jk} - \tfrac{2}{3}\delta^{ij}\delta^{kl}\right)\int_{t_0}^{t_1} dt\varepsilon^{ij}(\mathbf{x}, t)\varepsilon^{kl}(\mathbf{x}, t) \qquad (33)$$

$$\delta\widehat{\Delta\mu}(\mathbf{x}) = -\left(\delta^{ik}\delta^{jl} + \delta^{il}\delta^{jk} - \tfrac{2}{3}\delta^{ij}\delta^{kl}\right)\sum_{\ell=1}^{L_\nu}\int_{t_0}^{t_1} dt\varepsilon^{ij}(\mathbf{x}, t)\epsilon_\ell^{kl}(\mathbf{x}, t).$$

5 The Numerical Experiment

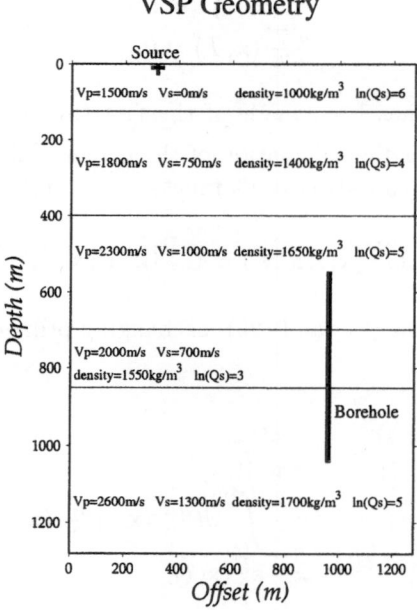

Fig. 1. Synthetic experiment configuration where the cross near the surface denotes the source position, whereas the solid line denotes the location of the two component receivers.

For our numerical experiment, we have chosen to invert OVSP data because these data are often used for quantifying the attenuation of shear and compressional waves in the Earth. The geometry of the experiment is detailed on the (Figure1) and the generated seismograms by finite difference approximation of the wave

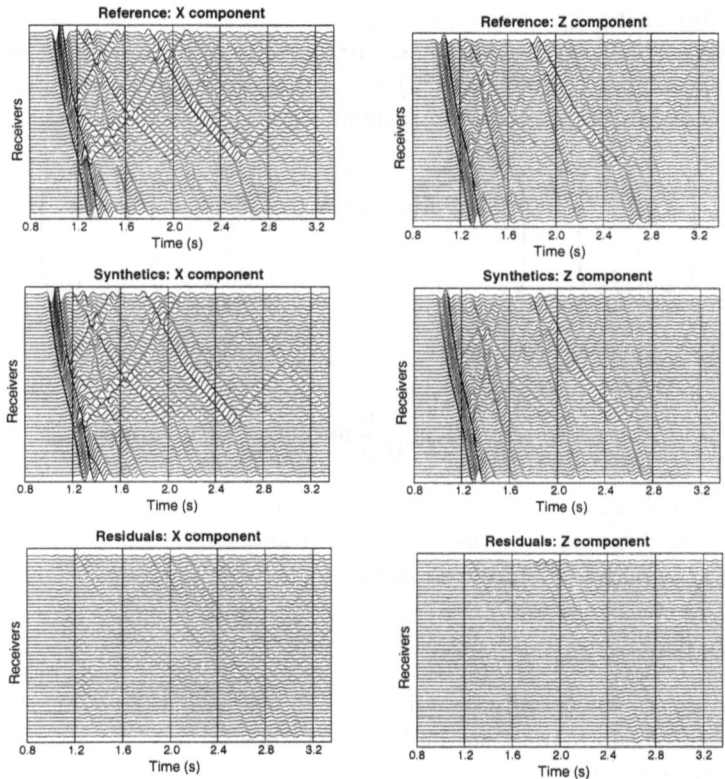

Fig. 2. On the top: the reference displacement component seismograms obtained for this experiment using the finite difference approximation to the wave equation. On the middle the two component seismograms obtained with our best inverted parameter model. On the bottom the difference between the reference seismograms and seismograms with the parameter model obtained from inversion, showing that only 6% of the energy of the reference seismograms have not been explained.

equation are on the top of the Figure (2). As the shear modulus controls the propagation for both type of waves (see relations in 35), but also for physical reasons (shear waves are more attenuated than compressional waves), the shear modulus μ is often chosen as the parameter controlling the attenuation. For the same reason and for minimizing the time computation, we will restrict ourselves to that parameter for describing the attenuation.

5.1 Parametrization

The choice of the parameters to invert is crucial as shown by Debski and Tarantola (1995) for surface seismic data. Following the same approach, we have found that a good set of parameters for OVSP data to invert are the mass density ρ, the compressional and shear wave velocity V_P and V_S, and the logarithm of the shear quality factor Q_s (Charara, 1996). As for each iteration, we are dealing with small perturbations of the model parameters, we can link the derived expression of gradient components (33) and (31) to the desired parameter gradient components to invert through the Jacobian matrix :

$$
\begin{pmatrix} \delta \widehat{V}_P \\ \delta \widehat{V}_S \\ \delta \widetilde{\rho}^* \\ \delta \ln \widehat{Q}_S \end{pmatrix} = \begin{pmatrix} 2\rho V_P & 0 & 0 & 0 \\ -\frac{8}{3}\rho V_S & 2\rho V_S & 0 & \frac{2b\rho V_S}{b-a+Q_s} \\ V_P^2 - \frac{4}{3}V_S^2 & V_S^2 & 1 & 0 \\ 0 & 0 & 0 & \frac{-b\rho V_S^2}{(b-a+Q_s)^2} \end{pmatrix} \begin{pmatrix} \delta \widehat{\kappa} \\ \delta \widehat{\mu} \\ \delta \widehat{\rho} \\ \delta \widehat{\Delta\mu} \end{pmatrix}, \tag{34}
$$

knowing the following relations

$$
V_P = \sqrt{\frac{\kappa + \frac{4}{3}\mu}{\rho}}, \; V_S = \sqrt{\frac{\mu}{\rho}} \text{ and } Q_s = Q_\mu = a + b\frac{\mu}{\Delta\mu}. \tag{35}
$$

For fast computation, we restricted ourselves to a 1D inversion. The number of parameters are the same as the vertical grid nodes used for our forward modelling, i.e., 125 nodes per physical parameter.

5.2 The Inversion Algorithm

The iterative inversion algorithm (in our case we used the conjugate gradient method) can be summarized in the following steps:
for a given iteration k

1. Solve the forward problem using as parameters $\mathbf{m}_k$ to obtain seismograms $\mathbf{u}_k$.
2. Calculate the weighted residuals $\delta\widehat{\mathbf{u}}_k$. Exit if converged.
3. Propagate the weighted residuals as acting sources backward in time and compute the gradient components $\widehat{\gamma}_k$ by using the Equations (33) and (31).
4. Convert the gradient with the desired parametrization $\widehat{\gamma}_k^*$ through the relation (34).
5. Calculate the conjugate gradient $\widehat{\phi}_k$ and the step length α_k
6. Calculate new model $\mathbf{m}_{k+1} = \mathbf{m}_k - \alpha_k \phi_k$

5.3 Initial Model and Inversion Results

To be realistic, we have generated data that have a frequency bandwidth of 10 to 40 Hz. Local optimization methods, based on gradient algorithms, cannot recover the low spatial frequencies of the model due to the missing low frequency bandwidth in the seismic data (Gauthier et al., 1986; Mora, 1987). For that reason, the initial should contain the low frequency of the model, i.e., a smoothed model (see the doted curves in figure 3) not so far from the true model parameters (the thin solid line in the same figure).

After 60 iterations, our inversion algorithm was able to find an Earth model (thick solid lines in figure 3) able to generate seismograms (middle of figure

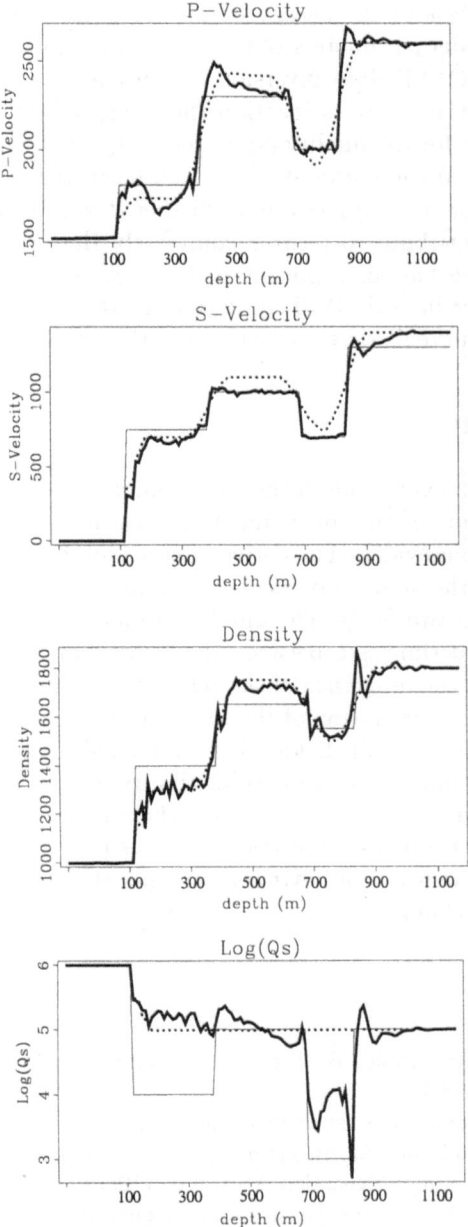

Fig. 3. 1D isotropic viscoelastic inversion on synthetic OVSP data. Doted lines for the initial model, thick solid line for the parameter model at the current iteration (here after 60 iterations) and the thin solid line for the true model. Above the receiver level, as expected for an OVSP geometry, the model is not recovered for all parameters. At the level of the receivers (550-1050m), the model is recovered. The small layer is well recovered for V_P, V_S and ρ parameters. For the $\ln Q_S$ parameter, the small layer is more or less recovered. This shows that this parameter is not as much resolved as the others.

2) that explains 94% of the energy of the reference seismograms (top of the figure 2). The model parameters obtained by inversion have the same features as for a classical OVSP data inversion for an elastic medium (Charara et al., 1996), i.e., above the receivers location the model is not well resolved and this is particularly true for the quality factor model Q_s; however at the receiver level (550 to 1050 meters) the parameters are well recovered and the anomaly is well positioned and more or less recovered. Further iterations to explain the left 6% of energy may contribute to recover completely the anomaly; however, as the rate of convergence has an exponential decay, we decided to stop at the 60th iteration. Moreover in real OVSP data, due to the noise level contained in the data, it would be unrealistic to recover more than that.

6 Conclusion

We have derived an expression in the time domain for the viscoelastic parameter gradient components of the misfit function for a nearly constant quality factor over a frequency bandwidth. The derived expression for the viscoelastic gradient components have the same expression as for the elastic case, i.e., a correlation at zero lag between two fields. The unrelaxed modulus parameter gradient component correlates at time zero between the forward and backward in time strain fields. However for the difference of the relaxed and unrelaxed modulus gradient component, it is the correlation of the forward in time memory strain fields and the backward in time strain fields. The numerical example of 1D viscoelastic inversion of OVSP data illustrates the feasibility of the proposed viscoelastic inversion. For sonic data, for instance, where the quality factor is known to be not constant, we could think not to restrict ourselves to a constant ratio for the time relaxations, at the expense of having an unrelaxed modulus parameter gradient component per mechanism.

References

1. Amundsen, L., and Ursin, B.: Frequency-wavenumber inversion data. Geophysics **56** (1991)1027–1039
2. Canadas, G.: Nonlinear seismic inversion using the 2-D full wave equation: First numerical results in heterogeneous media and strategy for inversion. 48 Mtg,. Eur. Assn. Geoscient. Eng., Expanded Abstracts, (1986) C033
3. Carcione, J. M.: Wave propagation in anisotropic linear viscoelastic media: theory and simulated wavefields. Geophys. J. Int. **101** (1990) 739–750
4. Carcione, J. M., and Cavallini, F.: A rheological model for anelastic anisotropic media with applications to seismic wave propagation. Geophys. J. Int. **119** (1994) 338–348
5. Charara, M.: Tomographie d'un milieu visco-élastique par ajustement de la forme d'onde. Phd Thesis, Université Paris 7 (1996)
6. Charara, M., Barnes, C., Tarantola, A.: Constrained waveform inversion of seismic well data. Inverse Methods. Lecture Notes in Earth Sciences **63** (1996)
7. Christensen, R. M.: Theory of viscoelasticity. Academic Press (1982)

8. Crase, E., Pica, A., Noble, M., McDonald, J., and Tarantola, A.: Robust elastic nonlinear waveform inversion: Application to real data. Geophysics 55 (1990) 527–538

9. Day, S. M., and Minster, J. B.: Numerical simulation of wavefields using a Padé approximant method. Geophys. J. Roy. Astr. Soc. 78 (1984) 105–118

10. Debski, W. and Tarantola A.: Information on elastic parametersobtained from the amplitudes of reflected waves. Geophysics 60 (1995) 1426–1436

11. Dietrich, M., and Bouchon, M.: Measurements of attenuation from vertical seismic profiles by iterative modeling. Geophysics 50 (1985) 931–949

12. Emmerich, H., and Korn, M.: Incoporation of attenuation into time-domain computations of seismic wave fields. Geophysics 52 (1987) 1252–1264

13. Gauthier, O., Virieux, J., and Tarantola, A.: Two-dimensional non-linear inversion of seismic waveforms: Numerical results. Geophysics 51 (1986) 1387–1403

14. Helbig, K., 1993: Foundation of Anisotropy for Exploration Seismics. Handbook for Geophysical Exploration, Pergamon, Oxford (1993)

15. Helgensen, J., and Landrø, M.: Estimation of elastic parameters from AVO effects in the $\tau - p$ domain. Geophys. Prosp. 41 (1993) 341–366

16. Lailly, P.: The seismic inverse problem as a sequence of before stack migrations. In Santosa, F., Pao, Y., Symes, W., and Holland, C., Eds., Inverse problems of acoustic and elastic waves: Soc. Ind. Appl. Math. (1984) 206–220

17. Lions, J. L.: Optimal control of systems governed by partial differential equations. Springer-Verlag (1971)

18. McDonal, F. J., Angona, F. A., Mills, R. L., Sengbush, R. L., van Nostrand, R. G., White, J. E.: Attenuation of shear and compressional waves in Pierre shale. Geophysics 23 (1958) 421–439

19. Martinez, R.D., and McMechan, G.A.: $\tau - p$ seismic data for viscoelastic media–Part 2: Linearized inversion. Geophy. Prosp. 39 (1991) 157–181

20. Mora, P.: Nonlinear 2-D elastic inversion of multioffset seismic data. Geophysics 52 (1987) 1211–1228

21. Murphy, W. F. III: Effects of partial stauration on attenuation in Massilon sandstone and Vycor porous glass. J. Acoust. Soc. Am. 71 (1982) 1458–1468

22. Pan, G.S., Phinney, R.A., and Odom, R.I.: Full-waveform inversion of planewave seismograms in stratified acoustic media: Theory and feasibility. Geophysics 53 (1988) 21–31

23. Pica, A., Diet, J. P., and Tarantola, A.: Nonlinear inversion of seismic reflection data in a laterally invariant medium. Geophysics 55 (1990) 284–292

24. Spencer, J. W. Jr.: Stress relaxations at low frequencies in fluid-saturated rocks: Attenuation and modulus dispersion. J. Geophys. Res. 86 (1981) 1803–1812

25. Tarantola, A.: Inversion of seismic reflection data in the acoustic approximation. Geophysics 49 (1984) 1259–1266

26. Tarantola, A.: A strategy for nonlinear elastic inversion of seismic reflection data. Geophysics 51 (1986) 1893–1903

27. Tarantola, A.: Theoratical background for the inversion of seismic waveforms including elasticity and attenuation. Pure Appl. Geophys. 128 (1988) 365–399

28. Zhao, H., Ursin, B., and Amundsen, L.: Frequency-wavenumber elastic inversion of marine seismic data. Geophysics 59 (1994) 1868–1881

Integrating Surface Reflection Data Inversion and Offset-VSP Modeling for Imaging of Converted S-Waves in Gulf Coast Hydrocarbon Reservoirs

Hugues Djikpéssé[1,2] and Christophe Barnes[1]

[1] Institut de Physique du Globe; 4, place Jussieu; 75252 Paris Cedex 05; France
[2] Formally, Jason Geosystems b.v., P.O. Box 1573, 3000 BN Rotterdam, The Netherlands. hdjikpesse@jasongeo.com

Abstract. Hydrocarbon reservoir characterization (lithological identification, porosity estimation, etc.) based on seismic reflection waveform inversion is to be known difficult to achieve. But, at least in principle, a multiparameter waveform inversion even for marine reflection seismic data should provide the necessary informations.

Djikpéssé and Tarantola (1999) retrieved physical fields for two independent parameters (for P and S-waves, as density is not resolved), from marine reflection data collected in the Gulf of Mexico. Considering a linear elastic medium and adopting a least-absolute-values fitting criterion, their data included phenomena difficult to model such as events related to unconsolidated sediments located at the sea floor and which have a great influence upon the amplitudes of converted S-waves. As phases in observed data are too complicated to be interpreted, they did not see any converted PSP or PSSP waves on the seismograms recorded at the surface.

In this paper, once a physical solution from the inverse problem has been obtained, we performed numerical offset-VSP experiments allowing to collect both horizontal and vertical components of the displacement fields inside the medium. Comparing seismograms obtained with and without the second parameter (related to S-waves) allows to view the generation of converted shear waves and to quantify the effect of P to S conversions on the amplitudes of the data. When introducing Poisson's ratio contrasts, wavefields inside the medium are mainly dominated by converted shear waves. Our conclusion is that these conversions cannot be neglected. Indeed, when their effects are not or badly taken into account, processed P-waves can suffer from large amplitude errors. Therefore, elastic modeling and inversion are necessary when extracting information from the signal amplitudes in order to characterize reservoirs.

1 Introduction

A linear elastic earth model can be described with a set of three parameters like density, compressional wave velocity and shear wave velocity. The long wavelengths of the compressional wave velocity control the travel time of P-waves. This information has to be used to define a velocity model used as starting model in the waveform inversion. This is a difficult task and is outside the scope of this article (see, e.g., Barnes, 1997). Detailed information about the reflectivity, lithology, porosity or petrophysical properties of the subsurface is contained in the short wavelengths. Dębski and Tarantola (1995) showed that two parameters (impedance of P-waves and Poisson' s ratio) are the most independent (in term of a posteriori probability density) and then the better couple of parameters when we seek to obtain uncorrelated images that can be recovered from seismic reflection data. The acoustic impedance contrasts are well resolved (with small offset reflections), but provide incomplete information about petrophysical properties of the subsurface. One needs to extract a second parameter related to shear wave properties such as the Poisson's ratio. This second parameter is essential because it gives further and discriminant information on lithology and hydrocarbon (oil, gas, coal, etc.) accumulations in geophysical exploration and on water in geothermal exploration. Unfortunately, in marine reflection experiments only P-waves are recorded. Information related to Poisson's ratio is contained in the multi-offset reflection seismograms. Much of reported work of extracting a second parameter from conventional marine data set, has been of qualitative nature, and have required the use of signal processing techniques in order to reduce complexity in observed data before interpretation.

Those processes (migration followed by AVO analysis) are based on amplitudes that are not always significant since they neglect both local sea bottom conditions and converted energy between P- and S-waves.

Kim and Serif (1992) showed that the PSSP reflection coefficient at the sea bottom is proportional to V_s^3, where V_s is the S-wave velocity at the sea floor. Consequently, in an unconsolidated marine area with recent sedimentation, (like the Gulf of Mexico) where shear velocity at the sea bottom is approximately of a few tens m/s, the unconsolidated sea bottom sediments have a great attenuative influence on the amplitude of converted waves. Could the water bottom and unconsolidated sediment layers, explain the non observation of any converted waves at the surface? Those physical events are difficult to model and lead into large residuals that tend to bias inversion results when using a l_2 norm of fit. An alternative is to measure the fit between observed and synthetic seismograms, with a robust criterion such as the least-absolute-values criterion (e.g., Gill and Murray, 1974).

Using the same Gulf of Mexico data set, Djikpéssé and Tarantola (1999) showed that the minimization of the least-absolute-values of the difference between observed and synthetic seismograms can be done using gradient techniques, even for large size problem (about 10^6 model parameters). The gradient based algorithm for large size ℓ_1 norm optimization is described in the Appendix. The next section is a synopsis of their paper, demonstrating the feasibility of extracting perturbations of acoustic impedance and Poisson's ratio from a conventional set of marine reflection seismograms. Adopting a ℓ_1 fitting norm has allowed to interpret the complete recorded wavefileds including reflections coming from sea floor and those coming from unconsolidated sediments which are widely muted. Interested readers may refer to the original 1999 Djikpéssé and Tarantola's paper for further details.

In order to illustrate the effect of the P-S waves conversion - which are usually neglected by traditional AVO processes - on the signal amplitudes, we have performed in the third section, several numerical source-offset VSP experiments, using the resulting models of surface reflection data inversion. When introducing Poisson's ratio contrasts, wavefields inside the medium are mainly dominated by converted shear waves which are highly sensitive to porous layers of unconsolidated sediments. These VSP experiments have allowed a better understanding of the weakness of ocean bottom converted PSSP waves, that are recorded at the sea surface in Gulf coast area of recent sedimentation.

Table 1: Acquisition parameters of the 1990 EXXON Gulf of Mexico data.

No. shot records:	512	Source type:	airgun
No. traces:	301	No. source subarrays:	4
Min. offset:	147 m	No. guns / subarray:	6
Max. offset:	4650 m	Source depth:	6 m
Receiver array length:	15 m	Source array length:	18.5 m
No. hydrophones in array:	17	Total source volume:	3836 cuin
Receiver array depth:	5 m		
High cut filter:	110 Hz		

2 Field Example

2.1 Data Contents and Starting Model

During a surface seismic reflection experiment, Exxon collected a data set of 512 shot records in the Gulf of Mexico. We wish to obtain a model that generates synthetic seismograms that are close to the observed ones. Table 1 summarizes source and acquisition parameters. Figure 1 shows one of the 512 shot records which correspond to shot point 270. Data have been bandpass filtered with a maximum frequency of 35 Hz, to save computer time. The seismograms are dominated by unconverted P-waves and no converted shear waves (PSP) are observed. Conventional seismic data tend to contain two distinct types of information (Claerbout 1985): information on long wavelengths (typically larger than 200 m) related to the velocities of the medium, and information on acoustic impedance and Poisson's ratio contrasts, with a wavelength of approximately 10 m. Since the seismic inverse problem of surface data is nonlinear, it can be solved using gradient methods in a series of iterations. Each iteration of our inversion can be viewed as a linearized inversion in which the residuals are treated as diffraction points (Tarantola, 1987). The following steps must be carried out prior to short wavelengths inversion.

Because the long wavelengths of the velocity do not create a backscattered field, it is difficult to recover them using gradient methods in a reasonable number of iterations. For our field example, a velocity model is built using a prestack migration-based velocity analysis (Migpack from Dataid). A comparison of travel time for different offsets between observed and synthetic data generated with this velocity model shows that the arrival times of the direct wave and the principal reflectors (including the reflection at 2.3 s) are explained approximately within a quarter of the fundamental period of the estimated wavelet. This gives us confidence that a smoothed version of this 2-D model can be used as an initial velocity model.

Forward modeling requires, in addition to P-wave velocity, models of density and Poisson's ratio. An initial density model (ρ) was derived from empirical relationships (Gardner et al., 1974) for clastic sequences, using the smoothed P-wave velocity (V_P):

$$\rho = \rho_0 \, (V_p/V_0)^{k_0} \,, \tag{1}$$

with $k_0 = 0.25$, $\rho_0 = 340$ kg m^{-3} and $V_0 = 1$ m/s. The initial Poisson's ratio is set to 0.48 at the water bottom, and decreases with depth, reaching a value of 0.35 approximately at 1.4 km depth. Below this depth, the initial Poisson's ratio is kept constant. Note that, since the initial model of acoustic impedance and Poisson's ratio is smooth (except at the water bottom), the reflected waves – generated with this initial model – are of very weak amplitudes compared to the direct wave.

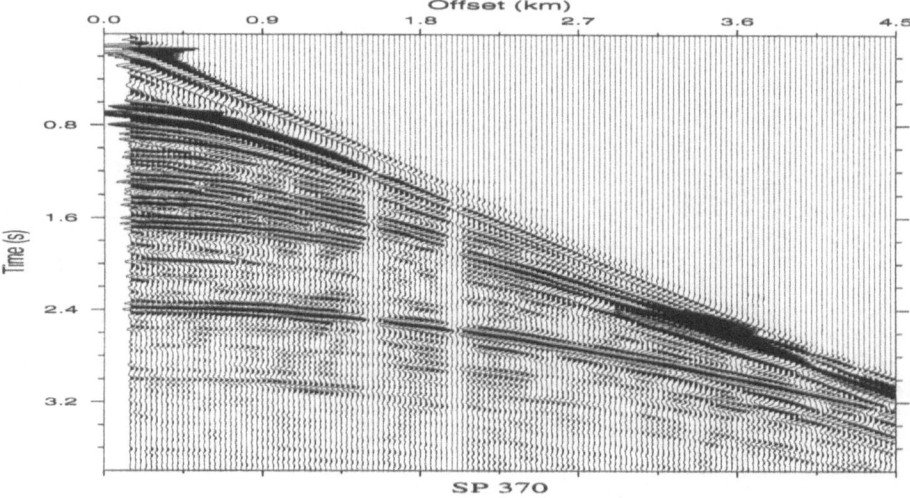

Fig. 1. One of the 512 shot gathers after low-pass filtering in order to save computer time. The maximum frequency is 35 Hz. The weak amplitude zones at 1.5 and 2.0 km offset in this and the next figure are due to missing recorded traces.

Since the waveform of the reflected energy depends both on the shape of the reflectors and on the source, any uncertainty about the source will lead to an equivalent uncertainty in the reflector characteristics. Therefore, its waveform and radiation pattern have been estimated from the direct wave observable on near traces, with the help of an airgun modelling package (Nucleus, from Seres).

For a given model, synthetic seismograms are computed by solving the wave equations using a finite difference approximation with convolutional operators of length 8 in space on a staggered grid and the time extrapolation is performed by a Taylor expansion (see, e.g., Djikpéssé, 1996).

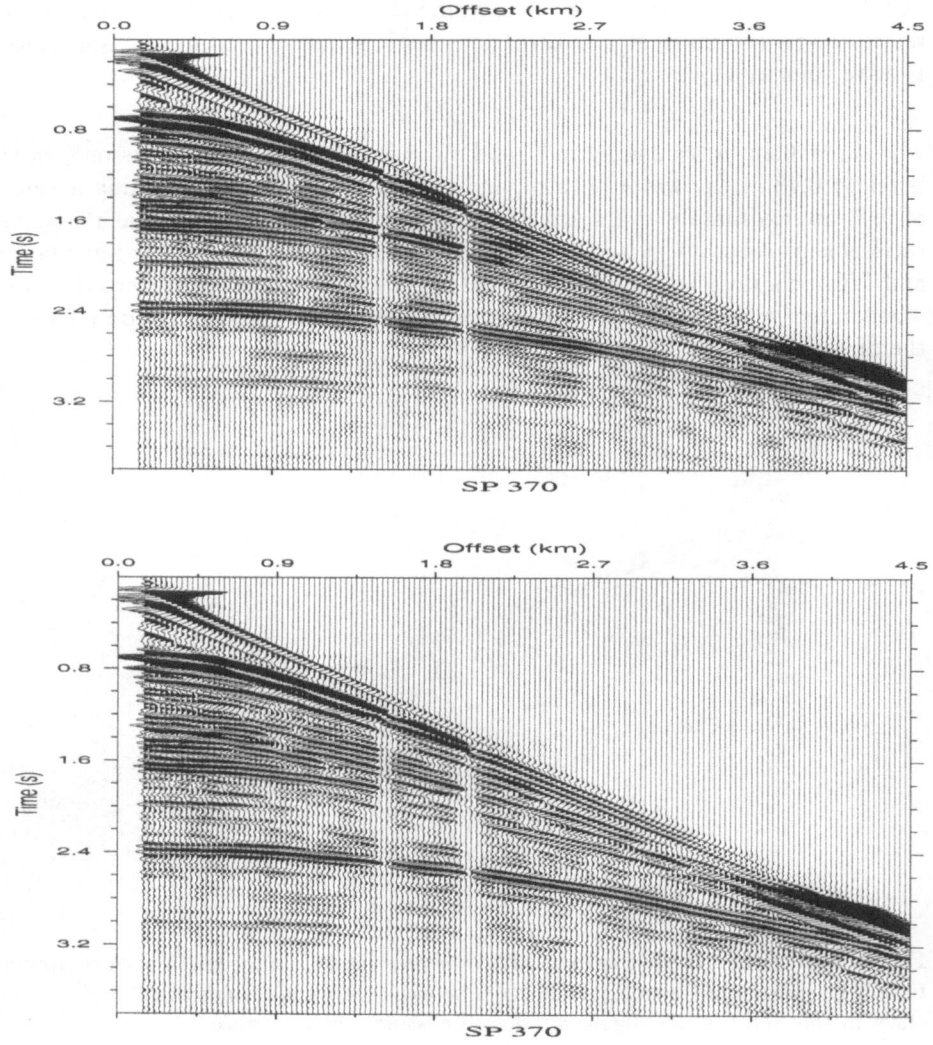

Fig. 2. Synthetic seismograms computed for all offsets using (Top) the final model obtained after the inversion of acoustic impedance contrasts. (Bottom) the final model obtained after the inversion of acoustic impedance and Poisson's ratio contrasts.

2.2 Short Wavelengths Inversion

Small offset data are in principle sufficient to retrieve P-reflectivity of the subsurface. The first 750 m offset were used to obtain the acoustic impedance contrasts. Calculating synthetic seismograms for all offsets with the final P-impedance model and initial Poisson's ratio leads to the seismograms shown in Figure 2, which are compared to the observed data, displayed in Figure 1. This comparison suggests that, apart from the water bottom reflection, the near offsets are well explained. Most of the residual errors increase with offset. Whereas the amplitude in the synthetic data for the 2.3 s reflection decreases considerably with offset, it remains nearly constant for the complete offset in the real data. Consequently, all offsets are included, and an inversion is carried out for Poisson's ratio, leading to the synthetic seismograms shown in Figures 2. Compared to the fit obtained after the inversion for the P-impedance only it can be noticed that, for instance, the amplitude behavior of the reflections at 2.0 s and 2.3 s are better explained after this additional inversion for Poisson's ratio. Now the variation of the amplitude as a function of the offset has been well explained, which was not possible with the acoustic impedance model alone.

2.3 Final Model Interpretation

The final models of acoustic impedance and Poisson's ratio are shown in Figure 3. The perturbations that have been added to the initial model to obtain the final model are displayed in Figure 4. Since the initial model is smooth, the acoustic impedance contrasts give the P-reflectivity of the medium. Apart from the water bottom reflector, the first coherent and continuous reflector is located around 1 km depth and marks the transition from less consolidated sediments to slightly more compact rocks. A normal fault goes from shot 160 to shot 41. The main reflectors are mostly flat-lying. A large drop in acoustic impedance appears around 2.2 km depth. It corresponds to a reflection at 2.3 s on the seismograms. However, the single image of acoustic impedance is not sufficient to characterize the presence of fluids in the subsurface. When looking at the Poisson's ratio model, the layer of unconsolidated sediments appears to have variations distinctly more important and heterogeneous than the acoustic impedance contrasts. A physical explanation is that the unconsolidated rocks from recent sedimentation located at the sea bottom, are porous. This porosity is very variable and results in some strong heterogeneities in the macroscopic Poisson's ratio model. We cannot exclude that this strong heterogeneity in the Poisson's ratio is an artefact due to the lack of convergence of the inversion for acoustic impedance for the reflections coming from unconsolidated sediments. Below those first layers, the contrasts of Poisson's ratio are weaker, except for the strong reflector at 2.2 km depth. The large drop in both acoustic impedance and Poisson's ratio indicates the presence of an hydrocarbon reservoir. The resulting two-dimensional models of P-impedance and Poisson' s ratio contrasts are anti-correlated almost everywhere in depth, except where hydrocarbons are present.

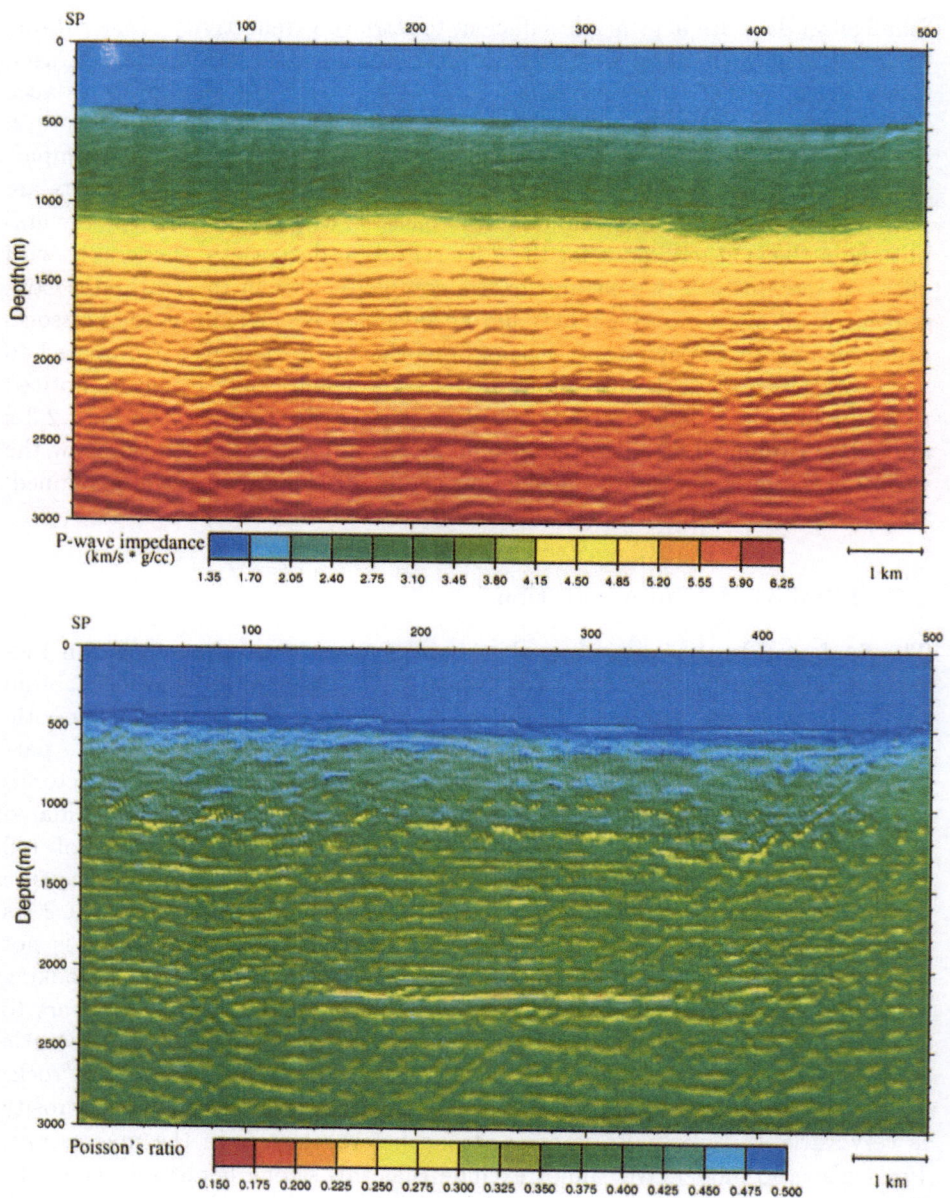

Fig. 3. Final model obtained after adding the pertubations (Figure 4) to the initial model. Top: P-wave impedance. Bottom: Poisson's ratio. An attenuative rock of high Poisson's ratio similar to unconsolidated sediments located between 0.4 and 1 km depth, appears below the gas sand reflector located at 2.2 km.

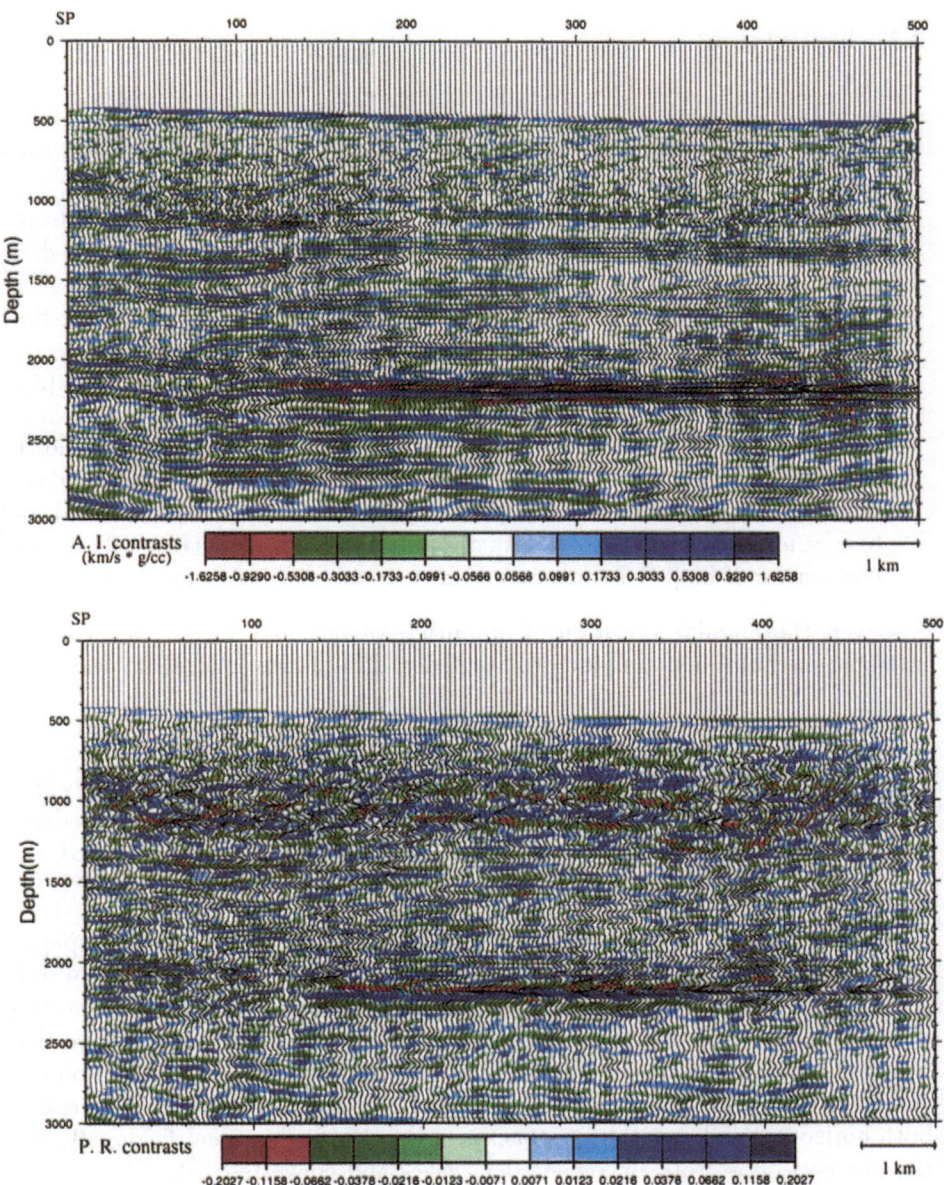

Fig. 4. Top: Final acoustic impedance contrasts. Note the gas sand reflector at about 2.2 km. Bottom: Final Poisson's ratio contrasts. When compared to the impedance contrasts, we see that Poisson's ratio is much more heterogeneous at the top (unconsolidated sediments).

3 Converted Shear Waves Analysis

3.1 Motivation

The inversion of Gulf of Mexico data set suggest the following conclusions:

- The angular variation of the amplitude of reflected seismograms could not be explained by the inversion of acoustic impedance contrasts only.
- An additional optimization of Poisson's ratio contrasts was necessary to obtain a good fit between observed and synthetic seismograms at intermediate and far angles.
- Converted shear waves were not visible neither on observed, nor on the final synthetic seismograms.
- Adopting a ℓ_1 criterion, the physical events that are difficult to model (e.g., unconsolidated sediment reflections, energy reflected at critical angles or refracted waves) are not muted. The layer of unconsolidated sediments and the gas-sand reservoir have higher Poisson's ratio contrasts. The unconsolidated sediments' layer is delimited respectively the the sea floor (at its top) and by a reflector around 1035 m with strong Poisson ratio contrasts. Those two boundaries of high Poisson's ratio contrasts might trap shear waves.

However, the angular variation of amplitude of reflected waves (in an elastic isotropic medium) is mainly due to the generation of converted shear waves.

- Why are converted shear waves not visible on data?
- How complex is wave propagation in a realistic earth model?
- Are the inverted Poisson's ratio contrasts physically meaningful or are they some artifacts (due to a lack of convergence during the acoustic impedance inversion) allowing the minimization of the misfit between observed and synthetic seismograms?

Answering those issues requires the ability to view and analyze waves propagating inside the final model found after the interpretation of Gulf of Mexico data set. Using a source located at 7.5 m depth in the water layer (i.e., generating only P-waves), we made two comparative numerical experiments of offset vertical seismic profiles (OVSP). OVSP surveys differ from surface seismic reflection surveys in that multi-component receivers located along a vertical line can collect both horizontal and vertical components of the the displacement fields, allowing the observation of both unconverted P-waves and converted (PS, PSP) waves, when they exist.

3.2 Acquisition Geometry and Numerical Settings

For each numerical experiment, we used an isotropic explosive source located at shot point 354 and at 7.5 m depth, and having a Gaussian function of 18 Hz central frequency as signature. The simulations were performed using 7.5 m grid spacing and a time sample interval of 0.0012 s. The same finite difference scheme, previously described in section 3.1.1, was used with a grid of 600 x 512, simulating a medium of 4500 m of lateral extend and 3840 m depth. In addition to a free surface condition implemented above the water layer, an absorbing boundary layer (Cerjean et al. 1986) of 600 m width was applied at the borders of the model. Two components based receivers located along a vertical line at shot point 288 (i.e., at 1.5 km offset from the source point) and regularly 7.5 m spaced between 0 and 3 km depth, allowed to collect both horizontal (Ux) and vertical (Uz) component of the displacement field. The vertical component of the displacement field enhances P-waves, while shear waves are highlighted through its horizontal component.

3.3 Use of Both Acoustic Impedance Contrasts and Poisson's Ratio Contrasts

An OVSP simulation is performed with the final model obtained after the complete waveform inversion: using both acoustic impedance contrasts and Poisson's ratio contrasts. The vertical and horizontal components of the displacement field are displayed in Figure 5, showing a strong complexity of waves propagating inside the inverted model.

The direct P-wave is recorded the first receiver located at free surface at 1.025 s, with a quite horizontal polarization angle. It reaches the sea floor in A and generates a reflected wave (AB) and a transmitted wave which reaches the base of unconsolidated sediments layer in D. The reflected wave reaching the free surface in B, is reflected again and becomes a secondary source that will be recorded at the last receiver in C. In point I, the first direct wave reaches the last receiver located at 3 km depth at 1.625 s with a polarization of about 45 degrees, after having being transmitted (in F) by the gas sand reflector. Several other water bottom multiples lead in a series of transmitted waves (e.g., EK, GL, JM) with similar phases as those described by the curve BC. Three other strong reflections (following the first direct transmitted wave) are respectively described when reaching the free surface by phases DE, FG and IJ. These reflections are respectively associated with the interface around 1050 m depth (Figure 4), with the gas-sand reservoir around 2250 m depth and with the latest strong acoustic impedance contrasts around 3100 m depth.

Seismograms are mainly dominated by coherent converted shear waves, in particular by a series of transmitted S-waves with polarization angles between 42 and 30 degrees. Reflected S-waves are less visible compared to transmitted S-waves and are identified as up-going waves that are converted into P-wave at the sea floor. The observation of reflected S-waves is emphasized with the horizontal component of the displacement field. Transmitted S-waves still have higher amplitudes. The propagation of reflected S-waves can be easily followed. For example, two S-waves are respectively reflected on the gas-sand reservoir head and on the high Poisson' ratio contrasts around 3100 m depth. Those converted PS waves reach the floor as described by phases FP and IQ, before being reflected again. Due to the strong Poisson's ratio contrasts at the sea bottom, transmitted PSP (travelling towards the free surface and which suffer of at least

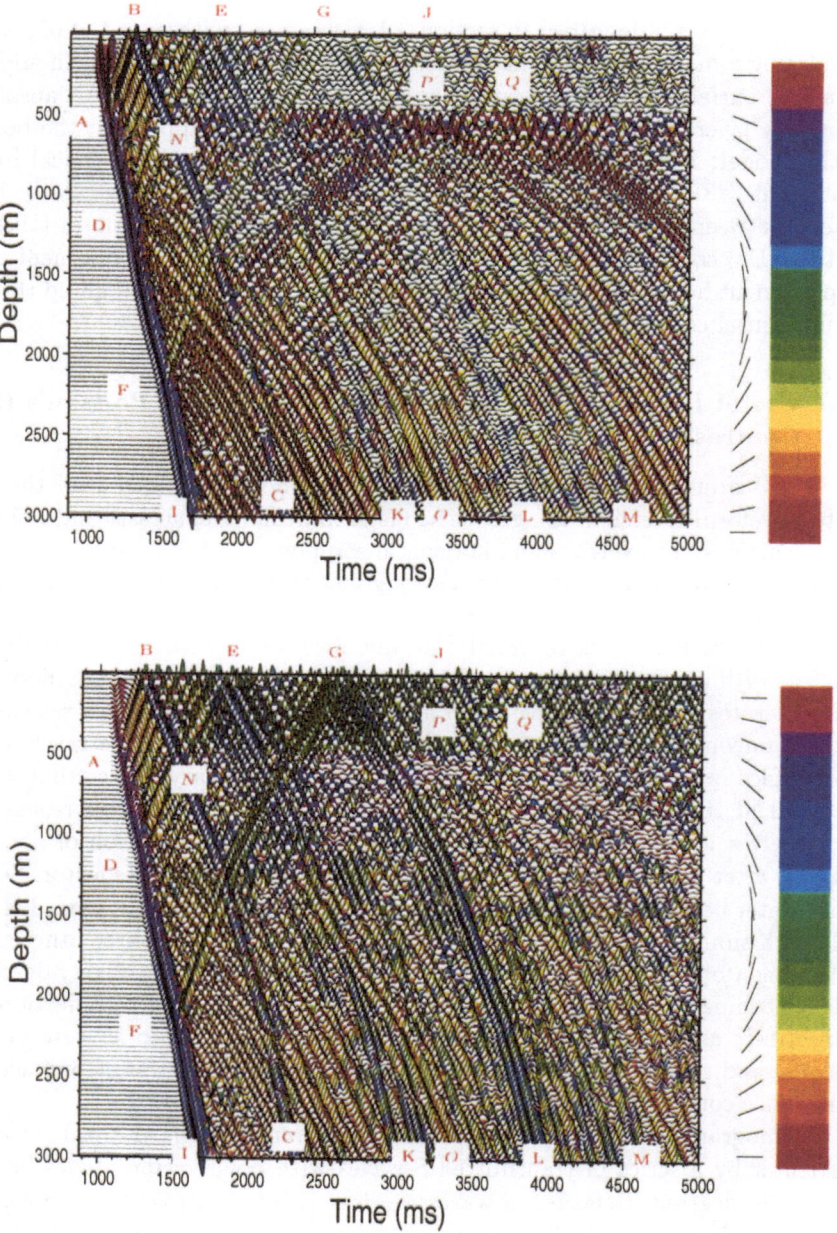

Fig. 5. OVSP modeling using the final model obtained after the waveform inversion of Gulf of Mexico reflection seismograms. The color code indicates the polarization of seismic waves and therefore a better differentiation between P- and S-waves. Top: Horizontal component of the displacement. Bottom: Vertical component. Before Poisson's ration optimization, seismograms were essentially dominated by P-waves (AD, DF, FI, AB, DE, FG, BC, EK, GL, JM). When adding Poisson's ratio contrasts, coherent converted shear waves now appear. In particular reflected converted shear wave between FP and IQ.

a double conversion) are not visible at the surface, although all seismograms have been amplified by an automatic gain control, in order to observe the different conversions independently of their true amplitude.

3.4 Use of Acoustic Impedance Contrasts Only

When simulating an OVSP using acoustic impedance contrasts only (i.e., keeping the initial model of Poisson's ratio and density), seismograms are exclusively dominated by reflected and transmitted unconverted P-waves generated inside the medium. The only visible shear waves, was one that is converted from P to S-wave on the sea floor. Its phase was described by points N and O showing much lower velocity than those related to unconverted P-waves. No obvious other converted PS waves are visible on the seismograms, with the initial Poisson' ratio model.

4   Summary

In this paper, two numerical experiments are presented with and without including Poisson's ratio contrasts estimated from an accurate inversion of marine seismic reflection seismograms. They suggest the following remarks:

1. Poisson's ratio contrasts extracted with a ℓ_1 norm of fit are meaningful as signified by the gas sand reservoir and by the generation inside the medium of coherent and energetic converted shear waves as physically expected.
2. The porous layer of unconsolidated sediments in gulf cost areas of recent sedimentation such as Gulf of Mexico, have great influence upon the amplitude of converted waves and therefore on their visibility on seismograms recorded at the free surface.
   - Indeed, shear waves propagating inside this porous layer are attenuated and therefore have very weak amplitudes, compared to unconverted P-waves.
   - Furthermore, energy of up-going shear waves reaching the top of the unconsolidated layer are mainly reflected down inside the medium: transmitted SP waves that travelling towards the free surface have too weak amplitudes for being visible, although after applying a strong amplification to recorded OVSP seismograms.
3. Whereas unconverted P-waves dominated wavefields when only acoustic impedance contrasts are used, converted S-waves dominate VSP seismograms when taking Poisson's ratio contrasts into account. Neglecting those conversion effects when processing surface seismic data, may lead into erroneous reservoir delineation and therefore into undesirable drilling costs.

A challenge for future research will be to quantify the certainty of results of the type presented here, especially for an integrated inversion of surface reflection and Offset-VSP seismic data.

94

Acknowledgments

We appreciate fruitful discussions with E. Juvé, M. Charara, E. Chaljub, A. Pica and A. Tarantola.

References

1. Barnes, C., 1997, Le problème inverse en tomographie géophysique: introduction d'informations *a priori* et méthodes de Monte Carlo. Application à l'inversion de temps d'arrivée. Thèse de Doctorat, Université Paris 7, France.
2. Cerjean, C., Kosloff, D., Kosloff, R., and Reshef, M., 1985, A nonrefelcting boundary condition for discrete acoustic-wave and elastic-wave equations, Geophysics, vol. 50, 705-7-8.
3. Claerbout, J. F., 1985, Imaging the earth's interior: Blackwell Scientific Publications, Inc.
4. Crase, E., Pica, A., Noble, M., McDonald, J., Tarantola, A., 1990, Nonlinear inversion: Application to real data: Geophysics, **55**, 527-538.
5. Dębski, W., and Tarantola, A., 1995, Information on elastic parameters obtained from amplitude of reflected waves: Geophysics, **60**, 1426-1436.
6. Djikpéssé, H. A., 1996, Caractérisation lithologique du sous-sol: Inversion multiparamètres de données de sismique réflexion marine: Thèse de Doctorat, Université Paris 7, France.
7. Djikpéssé, H. A., and Tarantola, A., 1999, Multiparameter ℓ_1 norm waveform fitting: Interpretation of Gulf of Mexico reflection seismograms: Geophysics, **64**, no. 4, in press.
8. Gardner, G. H., Gardner, L. W., and Gregory, A. R., 1974, Formation velocity and density – the diagnostic basis of stratigraphic traps: Geophysics, **39**, 770-780.
9. Gill, P. E., and Murray, W., Eds, 1974, Numerical methods for constrained optimization: Academic Press.
10. Kim, N. W., and Serif, A. J., 1992, Marine PSSP reflections with a bottom velocity transition zone: Geophysics, **57**, 161-170.
11. Tarantola, A., 1987, Inverse problem theory – methods for data fitting and model parameter estimation: Elsevier Science Publications.

A Appendix: Large Scale ℓ_1 Norm Optimization Through Iterative Gradient-Based Methods

The measure of quality for the retrieved model depends on the norm under which observed seismograms are compared to synthetic ones. Waveform inverse problems are usually solved by assuming that uncertainties in the prior model, measurements and in calculated data are gaussianly distributed. Then, one typically uses an ℓ_2 (least-squares) criterion of fit, which is often chosen because it leads to simple computations. In our case, we cannot use it for the data. It is well known that the least-squares criterion is not robust, and gives solutions that can easily be biased by the presence of large amplitudes in the data if they are difficult to explain. An alternative is to use an ℓ_1 (least-absolute-values) criterion of fit (e.g., Gill and Murray, 1974). This criterion is not often used in geophysics, perhaps because it is widely believed that it necessarily leads to linear programming techniques. It is not the case for our problem. The minimization of the least-absolute-values of the difference between observed and computed seismograms can be done using gradient techniques, even for large model (about 10^6 parameters). Following ideas of Tarantola (1987), the gradient is estimated at each iteration by correlating two particular fields. The first field is obtained by propagating the actual source in the current medium. The missing field results from the propagation of the signs of the residuals, acting as if they were sources, backwards in time.

We wish to infer measurements produced by some physical system S. Let $\mathbf{m}$ represent any particular value of the parameters describing the system. Assuming Laplacian independent uncertainties related to data, leads in the following standard misfit function to minimize (Tarantola, 1987):

$$S(\mathbf{m}) = \sum_{i=1}^{N} \left| d_{obs}^i - d_{calc}^i \right| \qquad (2)$$

where N is the number of observations, $\mathbf{d}_{obs}$ the observed measurements and $\mathbf{d}_{calc}$ are synthetic data. The ℓ_1 norm misfit function is widely solved using linear programming techniques. When the use of linear programming techniques become expensive and if the misfit function $S(\mathbf{m})$ is differentiable either analytically or numerically, it becomes more efficient to use iterative gradient-based algorithms to minimize $S(\mathbf{m})$. The most popular of gradient methods is the well known steepest descent method. Assuming a starting model $\mathbf{m}_n$ at iteration n, we wish obtain a best model at iteration $n+1$, such that

$$\mathbf{m}_{n+1} = \mathbf{m}_n - \mu_n \gamma_\mathbf{n}, \qquad (3)$$

where μ_n represents a positive step length to scale the direction of steepest ascent γ_n with respect to an ad-hoc ℓ_2 norm. The steepest ascent direction is related to the gradient $\hat{\gamma}_n$ of the misfit function $S(\mathbf{m})$, through uncertainties σ_M on the prior model such that:

$$\gamma_n = (\sigma_\mathbf{M}) \, \hat{\gamma}_n, \qquad (4)$$

$$\hat{\gamma}_n = \frac{\partial S}{\partial \mathbf{m}}(\mathbf{m}_n). \qquad (5)$$

At each iteration, synthetic seismograms are calculated and compared to the observed seismograms. The gradient of the misfit function is calculated by correlating the forward propagating wave-field with another wave-field propagating backwards in time with final conditions instead of initial conditions and the weighted residuals acting as sources at the receiver locations (Tarantola, 1987). The weighted residuals are the derivatives of the misfit function $S(\mathbf{m})$ with respect to the ith calculated data sample:

$$\hat{r}_n^i = -\text{sgn}\left(d_{obs}^i - d_{cal}^i\right) \qquad (6)$$

where

$$\text{sgn}(x) = \left\{ \begin{array}{l} +1 \text{ if } x > 0, \\ 0 \text{ if } x = 0, \\ -1 \text{ if } x < 0. \end{array} \right.$$

The gradient consists of the model perturbations in the elastic parameters, with which the model is updated for the next iteration. To optimize the size of this perturbation, in other words the step-length, an additional forward modelling for each parameter and iteration is required (Crase et al., 1990). The iterative gradient algorithm can be summarized in the following steps for all iterations n:

1. Initialize the Earth model $\mathbf{m}_n$ using as parameters ρ, λ, and μ.
2. Solve the forward problem to obtain seismograms $\mathbf{p}$.
3. Calculate the residuals and misfit. Exit if converged.
4. Propagate residual *signs* backwards in time.
5. Correlate wave-fields to obtain the gradient for P-impedance and/or Poisson's ratio.
6. Calculate step-length for desired parameter.
7. Update model.

One aspect of the algorithm is of particular interest as described by the equation 6. Within the context of least-squares, the residual seismograms (observed minus computed seismograms) must be propagated backwards into the medium at each iteration. When adopting an ℓ_1 norm, only the *signs* of the residuals are back propagated. Strong and erroneous residual amplitudes no longer bias the inversion results. It is interesting to note that each iteration appears to act like a migration of a binary field containing only 0's and 1's.

Inversion for the Moment Tensor and Source Location Using Love- and Rayleigh-Type Waves

L. Engell-Sørensen[1] and G. F. Panza[2]

[1] University of Bergen, Parallab, Thormøhlensgate 55, N-5020 Bergen, Norway,
Email: lisbeth@ii.uib.no
[2] University of Trieste, Department of Earth Sciences
Via E. Weiss, 1, 34127 Trieste, Italy

Abstract. The main purpose is to analyse global inversion of wave form data for the moment tensor, source time function, and source location retrieval using three component data from seismic stations. For this purpose we have extended to the SH component of motion the method proposed by Šílený et al. [59]. The main advantage of considering all three components at each recording station is to reduce the number of necessary stations and hence not only the cost of the seismic network, but also the noise introduced by the insufficient knowledge of the physical properties along the source-receiver path.

It is well known that the source depth is poorly resolved by Love waves, whereas Love waves add information to the moment tensor components (see Ben-Menahem and Harkrider [5]). Rayleigh waves include information of both source depth and earthquake mechanism (see Keilis-Borok and Yanovskaya [30]). It has in addition been shown by Ben-Menahem and Harkrider [5] that the radiation patterns of Rayleigh waves (for couple- and double-couple sources) unlike the fundamental Love mode may be sensitive to small variations in frequency. For a given source and frequency the radiation pattern may differ considerably from one mode to another.

In this study the source is assumed to be a point source in space but not in time. The procedure used finds the global solution of moment tensor and source location in depth. Green functions are interpolated between two extreme models of the structure, representative of the considered area. The first step retrieves the source location and the moment rate tensor components as functions of time, and in the second step, these time dependent moment rate tensor components are inverted to obtain an average constant moment tensor and a source time function common to all moment tensor elements. The object function minimised in the first step is the L_1-norm of the difference between observed and synthetic seismograms in time domain. Synthetic seismograms are calculated by normal mode summation (see Panza [46], Florsch et al. [20]). The method has been applied to local three-component strong motion recordings of Friuli (Italy) earthquakes.

1 Introduction

It has been demonstrated theoretically by Keilis-Borok and Yanovskaya [30] and by means of examples by Ben-Menahem and Harkrider [5] (BMH) that the radiation patterns of Rayleigh waves with periods between 50 and 350 s may depend strongly on the depth of the source, and unlike the fundamental Love mode may be rather sensitive to small variations in frequency. In addition, for a given source and frequency the radiation pattern may differ considerably from one mode to another (BMH). This paper will concern complete synthetic seismograms for periods between 0.1 and 20.0 s and we will investigate the resolution in source parameters for real data using three-component recordings. The method, proposed for Rayleigh waves by Šilený et al. [59] is here extended to the SH component of motion.

Theoretical developments of the algorithms needed for moment tensor inversion of Love-type waves has been carried out. The Love wave equations have been compared with what has been obtained by Aki and Richards [1] previously for another co-ordinate system, Hankel function, and Fourier transform convention. The synthetic seismograms obtained have been compared with synthetic calculations of the fundamental mode Love waves for a double-couple source using other computer programs based on Aki and Richards [1], Takeuchi and Saito [62], and spherical geometry (Schwab and Knopoff [56]) (not shown here). In addition published recorded data (Walter [65]) support our theoretical derivations (not shown here). The general expression for Love wave excitation as a function of the moment tensor has been tested by inversion of noise-free synthetic seismograms for the moment rate tensor, the factored constant moment tensor, and the source time function. The "observed" seismograms at receivers with good azimuthal coverage were obtained for double-couple (D-C) and compensated-linear-vector-dipole (CLVD) sources and the seismograms represented vertical component P-SV waves alone and vertical and radial component P-SV waves and SH waves together.

Synthetic seismograms and Green function derivatives (the kernal functions) are calculated by normal mode summation (see [46], [20]). The used structural model (Friuli7W, Mao and Suhadolc [33]) is assumed to simulate the Friuli (Italy) area.

The inversion method has been applied to local three-component recordings obtained by the Friuli Strong Motion Network [9]. The source parameters are obtained for five earthquakes in Friuli Venezia Giulia recorded by at least two three-component accelerometers: (1) Trasaghis, February 21, 1994, at 07 h 31 m (Md = 3.1), (2) Trasaghis, February 22, 1994, at 04 h 14 m (Md = 2.4), (3) Gemona, July 25, 1995, at 11 h 53 m (Md = 3.0), (4) Claut, February 27, 1996, at 11 h 13 m (Md = 4.0), (5) Claut, April 13, 1996, at 13 h 00 m (Md = 4.3). The error estimates and the size of the smallest eigenvalues are used to evaluate the reliability of the source parameters. The epicentres, a priori source depths, and polarity information is obtained from OGS bulletins (see [41], [42], [43], [44], [45]).

The application of the inversion method in this paper differs from the previous applications for Rayleigh-type waves alone (see Šilený et al. [59], [60], Campus et al. [8]): the number of random steps is large, in order to allow a larger search area of the model parameters in which the inversion problem is nonlinear and the initial trial is chosen randomly. It is well known that errors in the attenuation - velocity structure or the presence of anisotropy (vertical or horizontal) will be reflected into the estimate of the moment-rate function (Kravanja et al. [32]). Only one model parameter in which the problem is nonlinear, namely the source depth is allowed to vary in this work. The structural model is assumed to be one-dimensional. By assuming a quadratic form in the minimum of the object function, theoretical standard deviations can be estimated for the source depth, and for that source depth the moment-rate tensor resolution is found from the least-squares solution to the linear inverse problem. However, for the model parameter in which the problem is nonlinear (the source depth) this is not a good method since it is local. Therefore we have included in this work the estimate of the global a posteriori distribution of the source depth. For this purpose we have used the Monte Carlo method of random search, outlined by Mosegaard and Tarantola [38]. The result of the analysis is a table of source depths, the probability of their occurrence, and for each source depth in the table: the related moment-rate functions, their covariance matrices, and the factored moment-tensor and source-time function. Here we show selected estimates of the *a posteriori* depth distribution only.

2 Inversion Method

The moment-tensor has been found by inverting wave forms for large earthquakes the last two decades: (1) Dziewonski and Gilbert [13], Gilbert and Dziewonski [24], Gilbert and Buland [23], and Masters and Gilbert [34] used normal mode data (free oscillation data) and (2) Kanamori and Given [28] and Kanamori and Stewart [29] (similar surface wave excitation formulas as McCowan [35] and Mendiguren [36]) used surface waves. (3) Hartzel and Heaton [25], Ohlson and Apsel [40], Frankel and Venneberg [21] studied local strong motion data for extended sources and (4) Mendiguren [36], Kikuchi and Kanamori [31], Trehu et al. [64], Sipkin [61], Dziewonski and Woodhouse [14], and Nabelek [39] studied teleseismic (body wave) data. The source model obtained in the method applied here is a point source and hence resembles the methods used for teleseismic, normal mode, and surface wave data.

The source is allowed to have finite duration in time, but is assumed to be a point source in space. However, the distribution in space and the distribution in time cannot be separated for small earthquakes. If the source-time function reflects laterally distributed source effects within the spatial distribution of the source, the estimated moment-rate tensor solution might deviate from a pure shear slip (D-C) or crack model (CLVD), although it is not an explosive or implosive source. Therefore the trace of the moment-rate tensor is generally not constrained to be zero in our calculations because this would give a false D-C or CLVD source estimate. Laterally distributed source effects might in addition result in erroneous crack-model sources, whereas they really are double-couple components. In the application to real data we nevertheless impose the non-volumetric constraint because we have few recording stations. This is not a strong constrain because no volumetric components are expected in the North-Eastern Italy region of compression stress with no vulcanic or geothermal activity. The source duration is allowed to be larger than what can be tolerated by the point source approximation (see Bukchin [6], Engell-Sørensen [19]). The reason is that we might expect secondary source effects due to nearby faults and false source effects due to errors in the one-dimensional velocity structure (Kravanja et al. [32]) and because of the error in origin time obtained from kinematic data.

The method simultaneously, in an iterative scheme, inverts linearly for the moment rate tensor as function of time and minimises nonlinearly the L_1- or L_2-norm of the difference between observed and synthetic seismograms in order to obtain the source depth (step 1). Since strong variations with time of the source mechanism is not considered to be physically related but more likely due to structural heterogeneities a zero'th order moment tensor related to the common source-time function is estimated as a secondary result (step 2). In the description of step 1 we include the formulas for Laplacian distributed data-errors only because we here minimise the L_1-norm , whereas the L_2-norm is minimised by Šílený et al. [59]. We choose the L_1-norm since we intend to make agreement between data and synthetic seismograms in all the selected range of time and not just in the range with the highest amplitude signals.

2.1 Nonlinear Inversion for Source Depth and Structure (Step 1)

The inversion procedure applied is as follows (see also Šílený et al. [59]): Synthetic seismograms (base functions) $u_{ki}, k = 1, ..., N_S, i = 1, ..., M$, where N_S is the number of stations/components, are constructed for a set, M, of discrete values (grid points) of source location and structure. The base functions corresponding to intermediate values of source location and structure are computed by linearly interpolating the discrete base functions.

Let $d_k(t)$ be the k'th component of N_S observed seismograms. The inverse problem of minimising the difference between observed and theoretical seismograms can be written in the form of minimising the error $e_k(t)$ in the following equation:

$$d_k(t) = u_k(t) + e_k(t), k = 1, ..., N_S, \tag{1}$$

where

$$u_k(t) = \sum_{i=1}^{M} w_i' u_{ki}(t), k = 1, ..., N_S, \tag{2}$$

and $\mathbf{w}' = (w_1', ..., w_M')$ are the weights of the base functions in the linear interpolation. If the data errors $e_k(t)$ are Laplacian distributed with zero mean then the maximum likelihood estimate is given by the M dimensional vector $\mathbf{w}'$

which maximises the conditional distribution of $\mathbf{d}(t)$ given $\mathbf{w}'$:

$$p(\mathbf{d}|\mathbf{w}') = c_1 \exp(-\sum_{i=1}^{N_S} |d_k(t) - u_k(t)|/\sigma_k), \tag{3}$$

where c_1 is a constant, and σ_k is data uncertainty. In a discretized form equation (3) becomes:

$$p(\mathbf{d}'|\mathbf{w}') = c_1 \exp(-\sum_{k=1}^{N_S}\sum_{i=1}^{N_D} |d_k(t_i) - u_k(t_i)|/\sigma_{ki}), \tag{4}$$

where N_D is the number of discretized points in the time functions $d_k(t)$, $u_k(t)$, and $e_k(t)$, and $\mathbf{d}'$ is the $N_S \times N_D$ -dimensional data vector with single discretized channels sequentially ordered. Inserting equation (2) in equation (4) the maximum likelihood estimate is obtained by the vector $\mathbf{w}'$ which maximises:

$$p(\mathbf{d}'|\mathbf{w}') = c_1 \exp(-1/2\sum_{k=1}^{N_S}\sum_{i=1}^{N_D} |d_k(t_i) - \sum_{j=1}^{M} w'_j u_{kj}(t_i)|/\sigma_{ki}). \tag{5}$$

Laplacian distributed data errors are assumed in our tests, and the L_1-norm given by $-\ln(p(\mathbf{d}'|\mathbf{w}'))$ from equation (5) is minimized. Minimisation proceeds until a certain minimum value is obtained or a certain number of weights is tested. The MINUIT computer program (James and Roos [26]) is used for function minimisation. Local error analysis at the minimum $\mathbf{w}'_{min}$ is studied, assuming that the object function (the normalised L_1-norm with respect to L_1-norm of data) as a function of the model parameters in which the problem is nonlinear has a parabolic form so that the inverse of the second derivative matrix is the covariance matrix (Eadie et al. [17]). In addition the Davidon-Fletcher-Powell algorithm (Press et al. [49]) is used to obtain a local covariance matrix.

2.2 Linear Inversion for the Moment-Rate Tensor (Step 1 Continued)

We use the moment tensor description of a seismic source and formulate the displacement at a point in space due to a seismic source as a convolution in time between the stress glut and the gradient of the propagation effect. We follow the classical moment tensor description by Backus and Mulcahy [2], [3], [1]. We expand Green's tensor, G, in a Taylor series about a reference point, ξ^0 (see Doornbos [16]) and use the zero'th order approximation of the Taylor series expansion. We assume that the moment tensor, $M_{ij}(t)$ is monotone (e.g. increasing) with non-zero time-derivative $\dot{M}_{ij}(t)$. Hence the i'th component of displacement at $\mathbf{x}$ at time t can be expressed as

$$u_i(\mathbf{x}, t) = M_{jk}(t) * G_{ij,k}(\xi^0, \mathbf{x}, t)$$

$$= \dot{M}_{jk}(t) * H_{ij,k}(\xi^0, \mathbf{x}, t), \tag{6}$$

where $H_{ij,k}(\xi^0, \mathbf{x}, t)$ is the i'th component of the medium response to a Heaviside-time-function dipole-source in the j'th direction with the moment in the k'th direction, the tensor G_{ij} is the Green's function which is supposed to satisfy the reciprocity theorem, a comma between subscripts is used in spatial derivatives, and the Einstein summation convention for repeated indices is used.

The source displacement response, $u_i(\mathbf{x}, t)$ for Love and Rayleigh waves in terms of the moment-rate tensor-components $\dot{M}_{jk}(t)$ and the medium response expressed by $H_{ij,k}(\xi^0, \mathbf{x}, t)$ as seen in equation (6) has been derived. The expressions for the synthetic seismograms in terms of the moment tensor components given in Aki and Richards [1] for Love and Rayleigh waves (equation 7.147 for Love waves; equation 7.149 for Rayleigh waves) are equivalent to the derived formulas. As conventionally done for the moment tensor inversion problem (see e.g. Gilbert and Dziewonski [24], Dziewonski et al. [15], Doornbos [16], Sipkin [61], Šílený et al. [59]) we introduce the following notations for the moment rate components and step functions, respectively:

$$F_m(t) = \dot{M}_{jk}(t), j, k = 1, 2, 3, m = 1, ..., 6, \tag{7}$$

$$\Phi_{im}(\xi^0, \mathbf{x}, t) = H_{ij,k}(\xi^0, \mathbf{x}, t), j, k = 1, 2, 3, m = 1, ..., 6, i = 1, ..., N_S. \tag{8}$$

Hence equation (6) becomes

$$u_i(\mathbf{x}, t) = F_m(t) * \Phi_{im}(\xi^0, \mathbf{x}, t), i = 1, ..., N_S, m = 1, ..., 6. \tag{9}$$

We assume that the moment rate tensor components, $F_m(t)$ have source-time functions, which can be parameterized by triangles, $T_{\Delta\tau}(t)$ of half-width, $\Delta\tau$ and unit height (see Nabelek [39], Šílený et al. [59]):

$$F_m(t) = \sum_{n=1}^{N_t} F_{nm} T_{\Delta\tau}[t - (n-1)\Delta\tau], m = 1, ..., 6, \tag{10}$$

where F_{nm} are constants, and N_t is the assumed number of triangles, equal for all moment tensor components. By inserting equation (10) into equation (9) we get

$$u_i(\mathbf{x}, t) = \sum_{n=1}^{N_t} F_{nm} T_{\Delta\tau}[t - (n-1)\Delta\tau]$$

$$*\Phi_{im}(\xi^0, \mathbf{x}, t), i = 1, ..., N_S, m = 1, ..., 6. \tag{11}$$

This expression can be rewritten in a more compact form as

$$u_i(\mathbf{x}, t) = w_p A_{ip}(t), i = 1, ..., N_S, p = 1, ..., 6N_t, \tag{12}$$

where $p = m + 6(n - 1)$, and $w_p = F_{nm}$, $p = 1, ..., 6N_t, n = 1, ..., N_t$, $m = 1, ..., 6$ are the "weights" on the $6N_t$ triangles, i.e. the weights on the N_t triangles for each of the six moment rate tensor components. The functions $A_{ip}(\xi^0, \mathbf{x}, t) = T_{\Delta\tau}[t - (n-1)\Delta\tau] * \Phi_{im}(\xi^0, \mathbf{x}, t), i = 1, ..., N_S, n = 1, ..., N_t, p = 1, ..., 6N_t$ are the kernel functions sifted $(n-1)\Delta\tau$ time units. The system of equations to be solved for all discretized data can be written

$$\mathbf{Aw} = \mathbf{d}, \tag{13}$$

where $\mathbf{A}$ is a matrix of the discretized elements of $A_{kp}(t), k = 1, ..., N_S, p = 1, ..., 6N_t$ with $6N_t$ columns and $N_S \times N_D$ rows, N_D is the number of discretized points in the time functions $d_k(t), u_k(t), e_k(t)$. For non-singular matrices $\mathbf{A}^T\mathbf{A}$ we obtain the least-squares solution

$$\mathbf{w} = (\mathbf{A}^T\mathbf{A})^{-1}\mathbf{A}^T\mathbf{d}. \tag{14}$$

The arranging of the weights $w_p = F_{nm}$ in the order given by p gives to the matrix $\mathbf{A}^T\mathbf{A}$ the special form of a "block-Toeplitz" matrix ($\mathbf{A}^T\mathbf{A}$ is composed of N_t 6×6 sub matrices, where diagonal blocks are equal, and off-diagonal blocks are equal along the diagonal). The system of equations (13) is solved by recursive techniques (see Robinson [52]). The weights $\mathbf{w}$ represented as a function of time give the moment-rate tensor elements $M_{jk}(t)$ as functions of time.

2.3 Linear Inversion with Data Errors (Step 1 Continued).

We add the following to the method by Šilený et al. [59]. Let $d_k(t)$ and $u_k(t)$ be k'th-component of the observed and synthetic seismograms, respectively. The inverse problem of minimising the least-square difference between observed and theoretical seismogram can be written in the form of minimising the error $e_k(t)$ in the following equation:

$$d_k(t) = u_k(t) + e_k(t), k = 1, ..., N_S. \tag{15}$$

If we assume $e_k(t)$ to be Gaussian with zero mean and covariance matrix $\mathbf{C} = \mathbf{V}^{-1}$, $\mathbf{w} = (F_{11}, ..., F_{N_t6})^T$ to be the unknown random variables, and $\mathbf{d} = (d_1, ..., d_{N_S})^T$, then the maximum likelihood estimate is given by the vector $\mathbf{w}$, which maximises the conditional distribution of $d_k(t)$, given $\mathbf{w}$:

$$p(\mathbf{d}(t)|\mathbf{w}) = c_3 \exp(-1/2 \sum_{k=1}^{N}[d_k(t)-u_k(t)]^T \mathbf{V}[d_k(t)-u_k(t)]), k = 1, ..., N_S, \tag{16}$$

where c_3 is a constant. By expanding p in a Taylor series we obtain that $\mathbf{w}$ must fulfil the equation (see Tarantola [63], Engell-Sørensen [18], Šilený et al. [60])

$$\mathbf{A}^T\mathbf{Vd} = \mathbf{A}^T\mathbf{VAw}, \tag{17}$$

where $\mathbf{A}$ is a matrix of the discretized elements of $A_{kp}(t), k = 1, ..., N_S, p = 1, ..., 6N_t$ with $6N_t$ columns and $N_S \times N_D$ rows, N_D is the number of discretized points in the time functions $d_k(t), u_k(t), e_k(t)$, and we assume three components at each station. For non-singular matrices $\mathbf{A}^T\mathbf{VA}$ we obtain

$$\mathbf{w} = (\mathbf{A}^T\mathbf{VA})^{-1}\mathbf{A}^T\mathbf{Vd}. \tag{18}$$

The resolution of the moment-rate tensor is given by the covariance matrix

$$\mathbf{C}_w = (\mathbf{A}^T \mathbf{V} \mathbf{A})^{-1}. \tag{19}$$

It seems inconsistent, that we previously assumed Laplacian distributed data errors with respect to source location and structure and here assume Gaussian distributed data errors with respect to the moment-rate tensor. However, this is done mainly because we do not want to weight too much the large amplitude part of the selected recorded signal.

After outlining the formulas applied, step one can be explained in detail: The weights and moment rate tensor related to each sum of weighted base functions are found simultaneously in an iterative inversion scheme. First, a random search in the location - structure space choose one combination of weights, $\mathbf{w}'$ for all six moment - tensor kernels. Second, a weighted sum of kernels are then calculated for each of the six Green's function derivatives, and by linear inversion of the recorded seismograms with the weighted Green's function derivatives the moment rate tensor is found.

2.4 Nonlinear Inversion for the Average Aero'th Order Moment Tensor and Source-Time Function (Step 2)

In order to obtain the average zero'th order moment tensor and the source rate function the final step (step 2) in the method by Sileny et al. [59] (see also Campus et al. [8]) includes global minimization of the normalised residual-L_2-norm of the difference between the moment rate tensor obtained in step 1 and the factored moment tensor (i.e. the constant moment tensor, M_{ij}, multiplied by the source time function, $\dot{m}(t) \geq 0$) with respect to M_{ij} and $\dot{m}(t) \geq 0$. The normalized residual-L_2-norm is:

$$N_r = \frac{(\sum_{i=1}^{3} \sum_{j=i}^{3} \int_0^T [\dot{M}_{ij}(t) - M_{ij}\dot{m}(t)]^2 dt)^{1/2}}{(\sum_{i=1}^{3} \sum_{j=i}^{3} \int_0^T [\dot{M}_{ij}(t)]^2 dt)^{1/2}}, \tag{20}$$

where T is the duration of the rupture process.

2.5 Synthetic Seismograms and Kernel Functions.

The synthetic seismograms for a horizontally, layered earth can be obtained by modal summation (Panza [46], Panza and Suhadolc [47], Florsch et al. [20]). The method used in this work is described by Schwab and Knopoff [55], [56], [57], [58]), Schwab [53], Panza [46], Panza and Suhadolc [47] (Rayleigh modes), and Florsch et al. [20] (Love modes). The first-order approximation of anelasticity of the exact method by Schwab [54] and Schwab and Knopoff [56], [57], [58] is applied: i.e. Futterman's results based on causality analysis (Futterman [22]) and the variational methods to get the complex phase velocity for surface waves (Takeuchi and Saito [62], Schwab and Knopoff [57]). The approximation allows sedimentary Q values (Q about 20). The method used provide synthetic seismograms with accuracy of three significant figures for distances greater than the wavelength in the phase velocity range between zero and the S-wave velocity in the half space (Panza et al. [48], Panza and Suhadolc [47]). The seismic point source model used here is described by Ben-Menahem [4] and Ben-Menahem and Harkrider [5].

3 Data, Initial Source Parameters, and Structural Model

3.1 Data

The strong-motion data are obtained from The Friuli Accelerometric Network (Costa et al. [9], [10], [11], [12]).

3.2 Source Parameters

The earthquake locations and polarities for the five studied earthquakes in Friuli Venezia Giulia are obtained by Osservatorio Geophysico Sperimentale (OGS) [41], [42], [43], [44], [45], Renner et al. [51], and Renner [50]. Figure 1 shows the station locations of the Friuli Accelerometric Network and the source locations. Table 1 shows the hypocentral parameters of the five earthquakes and Table 2 provides the station co-ordinates.

Table 1. Source Parameters. All information is from OGS. The uncertainties dt, $dxdy$, and dz are the origin time error, and the horizontal and vertical location errors, respectively.

Locality/Label	Date (h:m:s)	dt (s)	Md	Latitude (deg)	Longitude (deg)	dxdy (km)	Depth (km)	dz (km)
TRASAGHIS/EQ1	19940221/07:31:16.4	.3s	3.1	46.322N	13.058E	1.3	7.	3.1
TRASAGHIS/EQ2	19940222/04:14:59.9	.2s	1.9	46.302N	13.056E	1.3	7.	2.3
GEMONA/EQ6	19950725/11:53:53.9	.2s	3.	46.257N	13.137E	.3	8.3	.5
CLAUT/EQ9	19960227/11:13:45.5	.1s	4.	46.312N	12.562E	.5	12.2	1.4
CLAUT/EQ10	19960413/13:00:22.4	.1s	4.3	46.315N	12.563E	.4	13.	.8

3.3 Structure

The structural model used (Table 3) is the structure FRIUL7W proposed by Mao and Suhadolc [33] (see Engell-Sørensen [19] for the relation between velocity and attenuation).

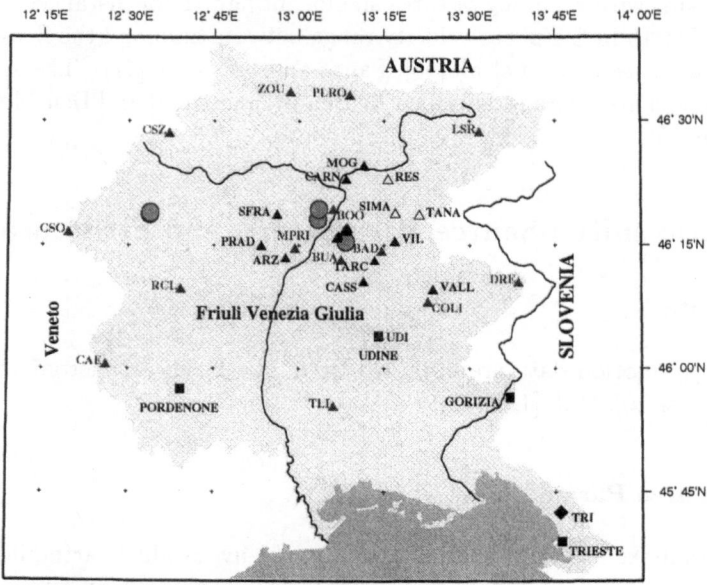

Fig. 1. Station and Earthquakes analysed in Friuli Venezia Giulia, Italy. Triangles: Friuli Accelerometric Network (see Table 1) and Friuli Seismometer Network. The stations GECE, GERO, GETM are not labelled. Circles: analysed earthquakes (EQ1,EQ2,EQ6,EQ9,EQ10) (see Table 2).

Table 2. Station Co-ordinates, Friuli Accelerometric Network.

Station	Latitude (deg)	Longitude (deg)	Height (km)	No. Comp.
GERC/PRAD	46.2481	12.8888	0.520	3
GECE	46.2816	13.1370	0.250	3
GERO	46.2764	13.1314	0.192	3
CARN	46.3798	13.1383	0.270	3
VIL_	46.2564	13.2814	0.608	3
GETM	46.2670	13.1157	0.188	3
MOG	46.4078	13.1894	0.387	3
VALL	46.1582	13.3930	0.666	3

Table 3. Structure FRIUL7W of the Friuli area with P- and S-velocities, and P- and S-wave attenuation.

Thickness (km)	Density (kg/m^2)	P-Wave Vel. (km/s)	P-Wave Att. (s/km)	S-wave Vel. (km/s)	S-Wave Att. (s/km)
.04	2.0	1.5	.007	0.6	.04
.06	2.3	3.5	.002	1.8	.009
.2	2.4	4.5	.0004	2.5	.002
.7	2.4	5.55	.0002	3.05	.0008
2.	2.6	5.88	.0002	3.2400	.0008
.1	2.6	5.7	.0004	3.14	.002
.2	2.6	5.65	.0004	3.1	.002
.2	2.6	5.6	.0004	3.06	.002
1.	2.6	5.57	.0004	3.03	.002
.5	2.6	5.55	.0004	3.02	.002
1.	2.6	5.57	.0004	3.03	.002
.2	2.6	5.6	.0004	3.06	.002
.2	2.6	5.65	.0004	3.1	.002
.1	2.6	5.7	.0004	3.14	.002
4.5	2.6	5.88	.0002	3.25	.0008
.1	2.6	6.1	.00008	3.4	.0004
.1	2.6	6.2	.00008	3.5	.0004
.1	2.6	6.3	.00008	3.6	.0003
.7	2.6	6.45	.00008	3.75	.0003
2.5	2.6	6.47	.00008	3.77	.0003
5.	2.6	6.5	.00008	3.8	.0003
6.	2.6	6.55	.00008	3.82	.0003
4.	2.75	7.	.00007	3.85	.0003
7.5	2.8	6.5	.0003	3.75	.001
4.	2.85	7.	.0001	3.85	.0006
3.	3.2	7.5	.00007	4.25	.0003
1.5	3.4	8.	.00006	4.5	.0003
9.	3.45	8.2	.00006	4.65	.0003

4 Moment Tensor Inversion and Source Depth Retrieval

The main purpose of the following analysis is to evaluate the resolution and the estimates of source depth and moment rate-tensor obtained from three- component strong motion data. The earthquake data are selected from the Friuli Accelerometric Network [10], [11], [12]. First the 10 earthquakes recorded on more than two three-component stations were selected. From these 10 earthquakes those were selected which fulfil the two demands:

1. We need at least two three-component stations in order to resolve five moment-rate tensor components as function of time (the isotropic component is assumed to be zero) and the source depth.
2. The epicentre distance has to be larger than the source depth (from OGS bulletin) (a demand for the straight forward application of the mode summation method we use for forward modelling).

We remain with five earthquakes in the period 1994-1996 having epicentre distances of more than 14 km from two or three stations. Table 1 and Figure 1 show the earthquakes: EQ1 and EQ2 have been recorded by two three-component stations, EQ6, EQ9, and EQ10 have been recorded by three three-component stations. For the source depth and moment tensor retrieval we minimise the misfit between the recorded and synthetic ground acceleration. The velocity structure is not changed during the inversion.

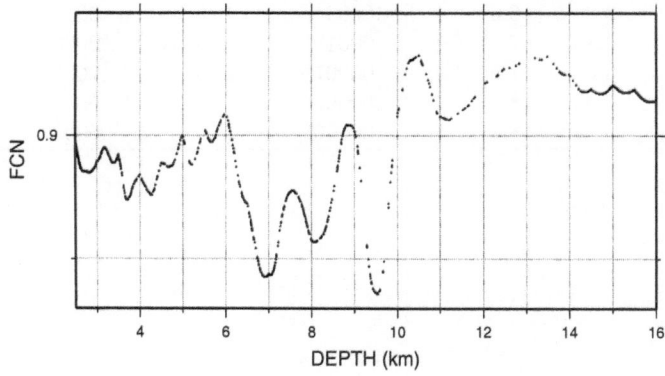

Fig. 2. Earthquake EQ6. Object Function (FCN) as Function of Source Depth.

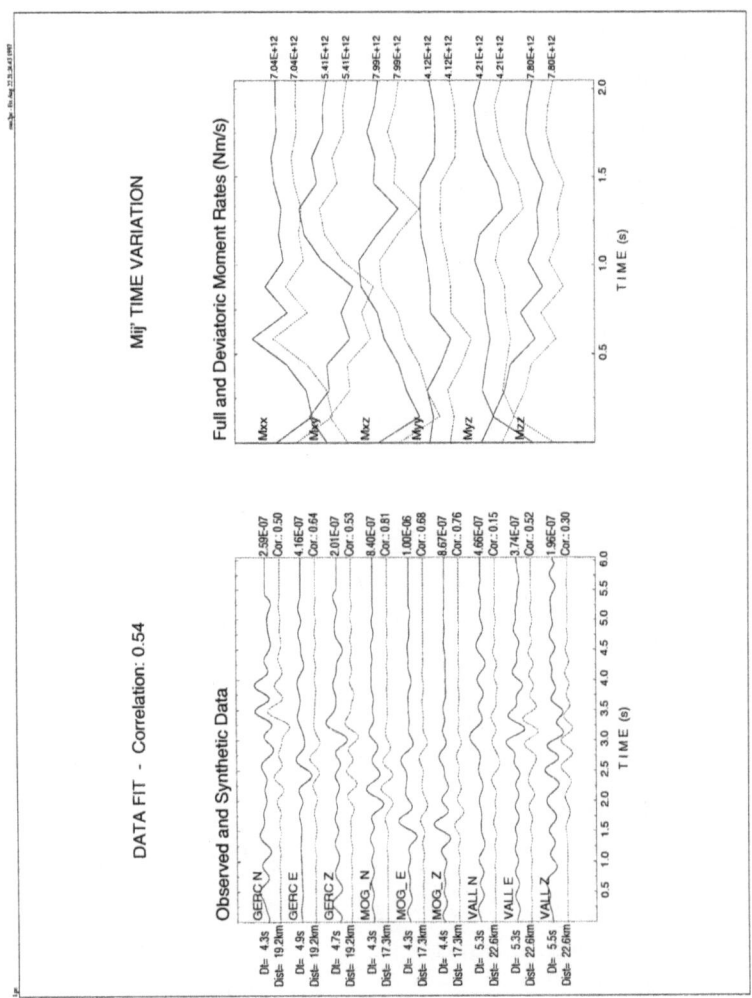

Fig. 3. Earthquake EQ6. Data Fit and Moment-Rate Tensor. The amplitudes of synthetics and data are in m/s^2. Left: Observed data (above) and synthetic seismograms obtained for the "best" moment-rate tensor-solution (below). Right: Estimated moment-rate tensor with no constraints (above) and with a deviatoric moment rate tensor constraint (below). Origin time is at 0.1 s to the right and shown by "Dt" to the left. Also displayed: Distance between source and station (left). Maximum moment-rate amplitudes in Nm/s (right). Maximum absolute value of the seismograms in m/s^2 and correlation coefficient between observed and synthetic seismograms (left).

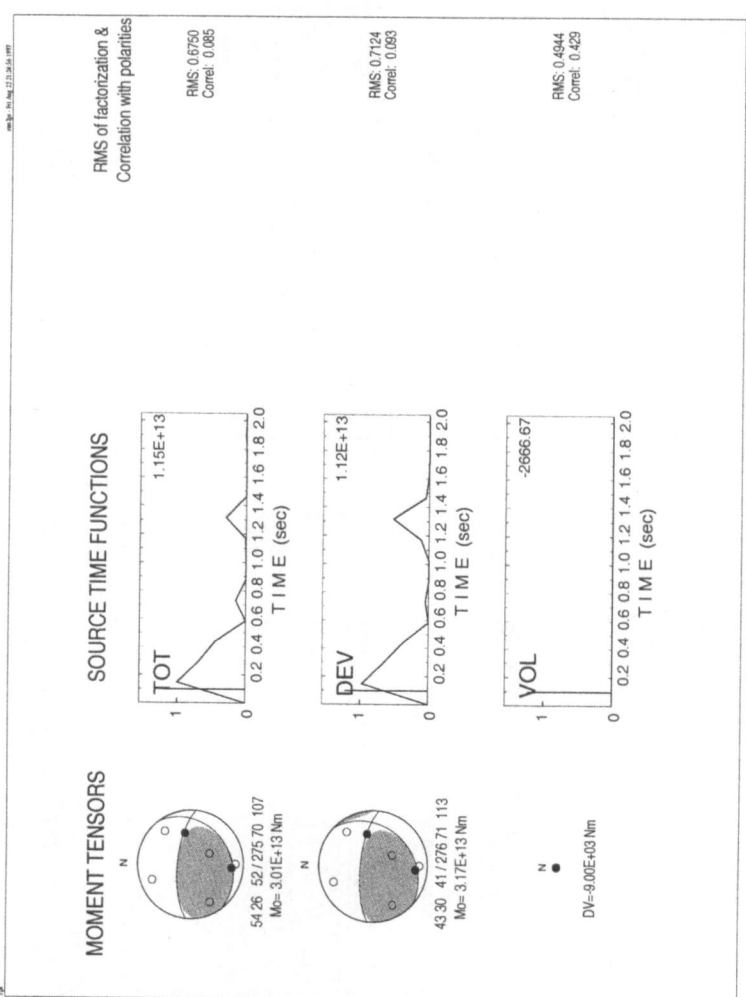

Fig. 4. Earthquake EQ6. Factored moment tensor and source time function. Non-constrained inversion result with damping=0.01. Equal area stereographic projection on lower hemisphere of fault-plane solution obtained using maximum (T-axis) and minimum (P-axis) eigenvalues of factored moment tensor and area of P- wave compression stress, obtained using T-axis of factored moment tensor and the eigen values. Also shown: Strike, dip, and rake, respectively, of the two double-couple solutions obtained assuming that the largest and smallest eigenvectors of the moment tensor are the T and P axes, respectively, global seismic moment, M_0 (Nm), observed polarities: black:compression, white:dilatation. The common source time function (right) and its maximum value (Nm/s).

Figure 2 shows the object function as function of source depth for earthquake EQ6. For the analysed earthquakes within the area covered by the stations the object functions show many local minima, whereas the two last earthquakes out of the station network are characterised by large global minima. We examine all local minima and find for the first three earthquakes, that one (and only one) local minima, which in addition is within the error estimate of the OGS bulletin (see Table 1) is giving large percentage of double-couple in the factored moment tensor. We therefore chose those minima for further analysis. For the two last earthquakes, the global minima are consistent with the OGS locations, so we chose those solutions, although the factored moment tensors have equal percentage of double-couple and CLVD contribution. The standard deviation of the source depth estimate given in Table 4 is calculated from the first order derivative in the minimum, and does therefore represent the linear approximation in the local minima only, and not the true standard deviations.

In Table 4 σ_f represents the data noise only and does not take into account the "noise" due to inaccurate structural model and epicentre. However even if we scale the moment-rate standard deviations with a factor 10 higher than the data noise we get standard deviations (see Table 4) less than 10 per cent of the largest components of the moment-rate tensor. For that reason we accept all the five moment-rate tensors and evaluate the factored fault mechanisms and source time functions.

The moment-rate tensor (in Nm/s) is given for one of the 13 triangles giving large amplitude in the factored source-time function, and the standard deviation of the moment rate components is calculated as an average for the largest of the 13 triangles. The numbers given first (and if only one value is given) are for moment-tensor inversion with large damping ($d=0.01$) as used in the search for source depth and the numbers given second are for small damping ($d=0.001$). The percentage of the moment tensor solution related to volumetric, double-couple, and CLVD sources are given [27]. The volumetric component was assumed to be zero. M_W is calculated using Kanamoris formula (Bullen and Bolt [7]: $M_W = 2 \cdot 2/3 \log_{10} M_0 - 10.7$. θ, δ, λ indicate strike, dip, and rake, respectively. No.E.V. shows the number of eigen values in the $\mathbf{A}$ matrix (equation 14) larger than the maximum eigen value multiplied by $\sqrt{d}$ (total number of eigen values is 65). σ_f is the average standard deviation of the observed data (in 10^{-9} m/s^2) and σ_z is the standard deviation of z (km) (DFP method).

Figure 3 (EQ6) and equivalent figures for the other analysed earthquakes show good correlation between synthetic acceleration and strong motion data for EQ1, EQ2, and EQ6, whereas mainly the station VALL for EQ9 and EQ10 has little positive or negative correlation with the synthetic signals, possibly because the waves passes not only the foothills of the Alps, but also the Friuli plain. Figure 4 shows the factored moment-tensor for EQ6 for the source depth solution obtained as described above (see also the corresponding values in Table 4). The choice between the best factored moment tensor and the next best in a least squares sense is based on (1) the root mean squares of the difference between the moment-rate tensor and the source time function multiplied by the factored

Table 4. Friuli Venezia Giulia Source Depth and Moment-Rate Tensor Results Using few Three-Component Stations.

Model Par.	EQ1		EQ2		EQ6	
(x)	x	σ_x	x	σ_x	x	σ_x
z	5.230	.188	6.198	0.669	8.066	.098
M_{11}	.492/.811	.005/.008	-.670/-1.69	.007/.015	-.704/-.767	.004/.008
M_{12}	.722/.846	.004/.005	.0366/-.229	.013/.009	.346/.49	.003/.006
M_{13}	.412/.344	.006/.010	.962/2.41	.009/.020	-.799/-1.33	.004/.009
M_{22}	-.184/-.135	.004/.006	.325/.993	.007/.014	-.076/-.137	.004/.007
M_{23}	.546/.787	.005/.007	.563/.767	.007/.012	-.234/-.349	.004/.007
M_{33}	-.307/-.676	-	.345/.698	-	.780/.903	-
M_0(tot.)	3.87		3.54		3.01	
M_W	3.025		3.00		2.95	
Vol. Cmp.	-		-		-	
D-C Cmp.	98.2		91.2		81.0	
CLVD Cmp.	1.80		8.76		19.	
θ,δ,λ	82,10,169		218,16,9		54,26,52	
θ,δ,λ	/183,88,80		/119,87,106		/275,70,107	
No.E.V.	46/60		46/58		49/59	
σ_f	3.03		5.25		3.50	

Model Par.	EQ9		EQ10	
(x)	x	σ_x	x	σ_x
z	10.359	.158	12.102	.219
M_{11}	.821/1.14	.051/.123	-.403/-.351	.048/.109
M_{12}	1.86/-1.	.037/.064	-2.75/-1.56	.034/.070
M_{13}	-3.51/-9.01	.063/.124	1.91/4.88	.049/.092
M_{22}	3.07/5.31	.048/.088	-6.17/-15.3	.037/.080
M_{23}	4.10/6.74	.041/.066	-3.15/-5.39	.031/.061
M_{33}	-3.89/-6.72	-	6.57/18.9	-
M_0(tot.)	20.7		19.5	
M_W	3.51		3.49	
Vol. Cmp.	.0		.0	
D-C Cmp.	37.5		43.6	
CLVD-Cmp.	62.5		56.4	
θ,δ,λ	138,37,-142		164,35,70	
θ,δ,λ	/16,68,-60		/8,58,103	
No.E.V.	44/57		34/51	
σ_f	18.8		14.9	

moment tensor, (2) the origin time agreement with the OGS bulletin, (3) the percentage of double-couple solution in the factored moment tensor, and (4) the agreement with polarities (as reproted by OGS bulletins). Two (EQ2, and EQ6) of the five double-couple fault-plane solutions mark the plate motion in the area giving North-South direction of compression stress and about East-West strike direction for one of the two planes. These two fault-plane solutions correspond well with the focal mechanisms obtained in the four years after the large Gemona earthquake of 1976 (Renner et al. [51]). The other three source mechanisms have North-South striking fault planes. The source-time functions show various shapes: The two largest earthquakes (EQ9 and EQ10) have duration's of the main signal of about 0.15 s, and show small initial signals followed by the large signal. EQ6 has a large initial signal of total duration 0.30 s which is followed by small signals, whereas EQ 1 and EQ2 have several individual signals, where the last is the largest in both cases.

We can conclude that the five earthquakes show mainly double-couple fault-plane solutions and source-time functions with several source signals. We did not consider the velocity model while obtaining these results, mainly due to poor knowledge of uncertainties of the applied velocity model. Hence one reason for the apparent secondary sources and large percentage of CLVD component in four of the five events might be a sligtly wrong one-dimentional velocity structure. Other reasons may be: (1) effects from details in the structural velocity-model, (2) reflections from nearby inhomogeneities, (3) secondary sources. Initial tests have shown that assuming shorter time signals give large false global minima in the object function, hence the later apparent source effects cannot be avoided. In order to improve the resolution of the secondary sources, which may occur few kilometres away from the estimated source depth a refinement of the method is needed which allows backslip and more than one source location. We did not here consider uncertainties of the direction of slip, tension axis, or pressure axis in the fault-plne solutions found here. The reason is that the main purpose has been to study the primary source parameters: moment-rate tensor and source depth. It is, however, straight forward to apply the Monte-Carlo method with the Metropolis criterion for nonlinear optimization for also this purpose (see next section).

5   Global Monte Carlo Inversion for Source Depth and Moment Tensor

In order to obtain true standard deviations of the source depth estimate we can formulate the probabilistic inverse problem by defining a probabilistic distribution in the model space. The probabilistic distribution combines *a priori* information with data information. In the three tests analysed here no *a priori* information is used, however we describe the theory for including such information. The theory linking data with model parameters is linear for some model parameters (moment rate tensors) and nonlinear for others (source depth). The marginal *a posteriori* probability in the model space is hence easy to describe in all directions except one (the source depth direction). The *a posteriori* probability distribution is obtained by randomly generating models in the moveclass and using the Metropolis acceptance criteria for each new model (see Mosegaard and Tarantola [38], Metropolis et al. [37]). If a new model is accepted, the centre of the new search area (within the moveclass) is moved to that model. The moveclass is an area within boundaries. In order to avoid boundary problems, the moveclass is made periodic by using a tangens function. We seek information on the resolution power of the data and do not consider the maximum likelihood model estimate as sufficient. Hence the *a posteriori* probability of each model parameter (source depth and moment rate tensor) is found by tabulating the estimated models and counting the number of times they occur within equal intervals (i.e. we make a histogram for the *a posteriori* distribution). Also the average value, standard deviation, median, and average deviation are calculated for model parameters in which the problem is nonlinear. For the mean and median estimates the model parameters in which the problem is linear and their standard deviations are also found. The posterior probability density is given by the data likelihood function and the *a priori* probability density by Bayes's rule (see Tarantola [63], Engell-Sørensen [18], Mosegaard and Tarantola [38]).

As test examples we have selected the earthquakes EQ1, EQ2, and EQ6. We have selected every 100'th point in a Monte Carlo search in source depth intervals (see Table 5). After a number of calculations of the object function (e.g. total of 34196 calculations for EQ6) we obtain the posterior values of mean value, standard deviation, median value, and average deviation seen in Table 5. Table 5 also shows the value of source depth at the global minimum, and the analytical standard deviation in the global minimum, which is based on the covariance matrix and linear approximation of the object function in the global minimum (DFP method). It is seen that both the mean values and the median values give realistic error estimates, whereas the linear approximation gives too small error estimates. For EQ6 both the mean and the median values are shifted from the global minimum towards source depths for EQ6, which are in agreement with the two large local minima seen for shallower source depths in Figure 2. However, the shifts are small compared to the standard deviations.

Table 5 includes the OGS locations and standard deviations. We see that the new Monte Carlo estimates agree well with those values. These good results give reasons to believe, that the introduction of *a priori* information in a Monte Carlo search as described above, would provide very reasonably source depth estimates and hence reasonably moment-rate tensors in the *a posteriori* mean or median source depth estimates.

6 Discussion

We have seen in the analyses with real data, that the main problem is the wave form fitting in order to obtain the source depth. The global minima do not necessarily represent the best (i.e. "most realistic") source mechanism, and even within a source-depth interval determined *a priori* by arrival time information

Table 5. Monte Carlo search for source depth for earthquakes EQ1, EQ2, and EQ6. All other values are as in Figure 2.

Source Depth (km)	EQ1	EQ2	EQ6
Search interval	3-11	4-14	5-13
Mean value	5.264	5.815	9.51
Standard deviation	.599	.595	1.09
Median value	5.222	5.803	9.51
Average derivation	.486	.488	.891
Global minimum value	5.233	5.846	9.530
Standard deviation in global minimum (DFP)	.055	.104	.069
OGS bulletin value	7.0	7.0	8.3
OGS bulletin standard deviation	3.1	2.3	.5

a secondary local minimum gives the most realistic fault mechanism. Another main problem is the determination of the number of triangles to be estimated. If we use 13 triangles, as preferred here in order to obtain a "long" wave form fit, we allow source duration's of two seconds for point sources. This means that also secondary or small primary sources, if they exist, are included in the moment-rate tensor. Since such sources might have other locations than the primary sources, we would obtain better average fault-mechanisms by allowing more than one source depth in the synthetic seismograms.

7 Conclusions

In synthetic tests not presented in this work, it has been seen, that three three-component stations are able to obtain better determined moment-rate tensor than five vertical component stations. By using two and three three-component strong motion data and well estimated hypocentres from bulletins (within 1-2 km) we have in this paper been able to obtain very well determined source depths and moment-rate tensors, which give fault-plane solutions in agreement with the regional plate motion. The synthetic tests showed that the source depth is determined with less wide object functions while using three-component stations, than while using vertical stations only. This might indicate, that also the determination of the source depth can be made as well or better by using few three component stations. However, the strong motion results showed, that good *a priori* locations are needed. Finally we have seen that Monte Carlo search based on the Metropolis algorithm provides realistic source depth estimates and standard deviations.

Acknowledgements

The Research was done under the EEC Program: HCM based on Contract No. ERBCHBGCT940645 under the heading: Tomography and 3-D Wave form Modelling of Complex Structures and Earthquake Sources Along the Alpine Belt. The

116

work documented was carried out at the Department of Earth Sciences, University of Trieste, Italy. The data used in this study was recorded by the Friuli Accelerometric Network. The work was guided by Professor G. F. Panza, University of Trieste to whom one author (LES) express her sincere thanks for good help and inspiring opinions and suggestions. We also thank Dr. G. Costa, Dr. S. Kravanja, Dr. K. Aoudia, Dr. H. Dufumier, A. Peresan, F. Marrara, Dr. F. Vaccari, Dr. A. Sarao, S. Ceschia, Dr. P. Suhadolc, G. Chimera, Dr. A. Michelini, Dr. Z. Du, Dr. F. Romanelli, and G. D'Adamo for making the used data available and for discussions during the period of the work. We thank Jacob Mørch Pedersen, Ødegaard A/S for helpful discussions on the Monte Carlo optimization.

References

1. Aki, K., Richards, P. G.: Quantitative Seismology. Freeman, San Francisco (1980)
2. Backus, Mulcahy. Moment tensors and other phenomenological descriptions of seismic sources-I. Continuous displacements. Geophys. J. R. Astr. Soc. **46** (1976) 341–361
3. Backus, Mulcahy. Moment tensors and other phenomenological descriptions of seismic sources-II. Discontinuous displacements. Geophys. J. R. Astr. Soc. **47** (1976) 301–329.
4. Ben-Menahem, A. Radiation of Seismic Surface Waves from Finite Moving Sources. Bull. Seismol. Soc. Am. **51** (1961) 401–435.
5. Ben-Menahem, A., Harkrider, D. G. Radiation patterns of seismic surface waves from buried dipolar point sources in a flat stratified Earth. J. Geophys. Res. **69** (1964) 2605–2620
6. Bukchin, B. G. Determination of stress glut moment of total degree 2 from teleseismic surface wave amplitude spectra. Tectonophysics **248** (1995) 185–191.
7. Bullen, K. E., Bolt, B. A. An Introduction to the Theory of Seismology. Cambridge University Press, Cambridge. (1985).
8. Campus, P., Suhadolc, P., Panza, G. F., Šílený, J. Complete moment tensor retrieval for weak events: application to orogenic and volcanic areas. Tectonophysics **261** (1996) 147–163.
9. Costa, G., Fäh, D., Suhadolc, P., Paparelli, F., Smit, P., Mayer-Rosa, D., Panza, G. F. The Friuli (NE Italy) Accelerometric Network: Recording and Estimation of Strong-Ground Motion. European Seismological Commission, Papers presented at the XXV General Assembly, September 9-14 (1996) Reykjavik, Iceland.
10. Costa, G., Panza, G. F., Suhadolc, P. The Friuli Strong Motion Network, 1993-94. Dipartimento di Scienze della Terra, Universita' degli Studi di Trieste. (1996)
11. Costa, G., Panza, G. F., Suhadolc, P. The Friuli Strong Motion Network, 1995. Dipartimento di Scienze della Terra, Universita' degli Studi di Trieste. (1996)
12. Costa, G., Panza, G. F., Suhadolc, P. The Friuli Strong Motion Network, 1996. Dipartimento di Scienze della Terra, Universita' degli Studi di Trieste. (1997)
13. Dziewonski, A. M., Gilbert, F. Temporal variation of the seismic moment tensor and the evidence of precursive compression for two deep earthquakes. Nature (London) **247** (1974) 185–188.
14. Dziewonski, A. M., Woodhouse, J. H. An experiment in systematic study of global seismicity: centroid-moment tensor solutions for 201 moderate and large earthquakes of 1981. J. Geophys. Res. **88** (1983) 3247–3271.

15. Dziewonski, A. M., Chou, T.-A., Woodhouse, J. H. Determination of earthquake source parameters from wave form data for studies of global and regional seismicity. J. Geophys. Res. **86** (1981) 2825–2852.
16. Doornbos, D. J. Seismic moment tensors and kinematic source parameters. Geophys. J. R. Astr. Soc. **69** (1982) 235–251.
17. Eadie, W. T., Drijard, D., James, F. E., Roos, M., Sadoulet, B. Statistical Methods in Experimental Physics, North-Holland Public. Co., Amsterdam. (1971)
18. Engell-Sørensen, L. Inversion of arrival times of microearthquake sources in the North Sea using a 3-D velocity structure and prior information. Part I. Method. Bull. Seism. Soc. Am. **81** (1991) 1183–1194.
19. Engell-Sørensen, L. North Sea Earthquake Source Parameters. Ph. D. Thesis, University of Bergen, Institute of Solid Earth Physics, Norway. (1993)
20. Florsch, N., Fäh, D., Suhadolc, P., and Panza, G. F. Complete Synthetic Seismograms for High-Frequency Multimode SH-waves. Pageph **136** 4 (1991) 529-560.
21. Frankel, Venneberg. Rupture process of the M_S=6.6 Superstition Hills, California, earthquake determined from strong-motion recordings: application of tomographic source inversion. Bull. seism. Soc. Am. **79** (1989) 515–541.
22. Futterman, W. I. Dispersive body waves. J. Geophys. Res. **67** (1962) 1587–1610.
23. Gilbert, F., Buland, R. An enhanced deconvolution procedure for retrieving the seismic moment tensor from a sparse network. Geophys. J. R. Astron. Soc. **47** (1976) 251–255.
24. Gilbert, F., Dziewonski, A. M. An application of normal mode theory to the retrieval of structural parameters and source mechanisms from seismic spectra. Philos. Trans. R. Soc. London, Ser. A **278** (1975) 187–269.
25. Hartzel, S. H., Heaton, T. H. Inversion of strong ground motion and teleseismic wave form data for the fault rupture history of the 1979 Imperial Valley, California earthquake. Bull. seism. Soc. Am. **73** (1983) 1553–1583.
26. James, F., Roos, M. Minuit, Cern Computer Centre Program Library, Updated version of CERN/DD internal report 75/20 (1975). in Computer Physics Computations **10** (1983) 343–367.
27. Jost, M. L., Herrmann, R. B. A student's guide and review of moment tensors. Seism. Res. Lett. **60** (1998) 36–57.
28. Kanamori, H., Given, J. W. Use of long-period surface waves for rapid determination of earthquake-source parameters. Phys. Earth Planet. Inter. **27** (1981) 8–31.
29. Kanamori, H., Stewart, G. S. Mode of the strain release along the Gibbs fracture zone, Mid-Atlantic Ridge. Phys. Earth Planet. Inter. **11** (1976) 312–332.
30. Keilis-Borok, V. I., Yanovskaya, T. B. Bull. Acad. Sci., U.S.S.R., Geophy. Ser. (English translation) **11** (1962) 1532–1539.
31. Kikuchi, Kanamori. Inversion of complex body waves. Bull. Seism. Soc. Am. **72** (1982) 491–506.
32. Kravanja, S., Panza, G. F., Šílený, J. Robust retrieval of time-dependent seismic moment tensor for point-sources. Geophysical J. Int. (1999). in Press.
33. Mao, W. J., Suhadolc, P. Simultaneous inversion of velocity structures and hypocentral locations: application to the Friuli seismic area NE Italy. Pageoph **138** (1992) 267–285.
34. Masters, G., Gilbert, F. Source retrieval from a sparse long-period network. EOS **60** (1979) 829 (abstract).
35. McCowan, D. W. Moment tensor representation of surface wave sources. Geophys. J. R. Astron. Soc. **44** (1976) 595–599.
36. Mendiguren, J. A. Inversion of surface wave data in source mechanism studies. J. Geophys. Res. **82** (1977) 889–894.

37. Metropolis, N., Rosenbluth, A. W., Rosenbluth, M. N., Teller, A. H., Teller, E. Equation of state calculations by fast computing machines. J. Chem. Phys. 1 (6) (1953) 1087–1092.
38. Mosegaard, K., Tarantola, A. Monte Carlo sampling of solutions to the inverse problems. J. Geophys. Res. 100 (1995) 12431–12447.
39. Nabelek, J. L. Determination of earthquake source parameters from inversion of body waves, PhD thesis, MIT. (1984).
40. Ohlson, A. H., Apsel, R. J. Finite faults and inverse theory with applications to the 1979 Imperial Valley earthquake. Bull. seism. Soc. Am. 72 (1982) 1969–2001.
41. Osservatorio Geofisico Sperimentale (OGS) Bollettino di Geofisica Teoretica ed Applicata, The NorthEastern Italy Seismometric Network Catalogue. (1995)
42. Osservatorio Geofisico Sperimentale (Trieste) and Centro Ricerche Sismologiche (Udine) (OGS). Bollettino della Rete Sismometrica del Friuli Venezia Giulia. (1988)
43. Osservatorio Geofisico Sperimentale (Trieste) and Centro Ricerche Sismologiche (Udine) (OGS). Bollettino della Rete Sismometrica del Friuli Venezia Giulia. 13 (1 and 2). (1994)
44. Osservatorio Geofisico Sperimentale (Trieste) and Centro Ricerche Sismologiche (Udine) (OGS). Bollettino della Rete Sismometrica del Friuli Venezia Giulia. 14 (1 and 2). (1995)
45. Osservatorio Geofisico Sperimentale (Trieste) and Centro Ricerche Sismologiche (Udine) (OGS). Bollettino della Rete Sismometrica del Friuli Venezia Giulia. 15 (1 and 2). (1996)
46. Panza, G. F. Synthetic seismograms: the Rayleigh waves modal summation. J. Geophys. 58 (1985) 125–145.
47. Panza, G. F., Suhadolc, P. Complete strong motion synthetics. Ed. Bolt, A. B., Academic Press, Orlando, FL. (1987) pp. 135–204.
48. Panza, G. F., Schwab, F., Knopoff, L. Channel and crustal Rayleigh waves. Geophys. J. R. Astron. Soc. 30 (1973) 273–280.
49. Press, W. H., Flannery, B. P., Teukolsky, S. A., Vetterling, W. T. Numerical Recipes, Cambridge University Press, Cambridge. (1986)
50. Renner, G. The Revision of the Northeastern Italy Seismometric Network Catalogue. In: Bollettino di Geofisica Teorica ed Applicata, Supplement to VOL. XXXVII, N. 148-December 1995. (1995)
51. Renner, G., Rebez, A., Slejko, D. Revisione dei dati Sismometrici della rete dell'Italia Nord-Orientale (1977-1987): Analisi Preliminare, in Atta del 13' Convegno Annuale del Gruppo Nazionale di Geofisica della Terra Solida, Roma 28-30 Novembre 1994. (1994)
52. Robinson, E. A. Multichannel Time Series Analysis with Digital Computer Programs. Goose Pond Press, Texas. (1983)
53. Schwab, F. Surface-wave Dispersion Computations: Knopoff's Method. Bull. Seismol. Soc. Am. 60 (1970) 1491–1520.
54. Schwab, F. Mechanism of Anelasticity. Geophys. J. 95 (1988) 261–284.
55. Schwab, F., Knopoff, L. Surface Wave Dispersion Computations. Bull. Seismol. Soc. Am. 60 (1970) 321–344.
56. Schwab, F., Knopoff, L. Surface waves on multilayered anelastic media. Bull. Seismol. Soc. Am. 61 (1971) 893–912.
57. Schwab, F., Knopoff, L. Fast surface wave and free mode computations. In: Methods in computational physics, 11, B. A. Bolt, ed.: (1972) pp. 87–180. New York: Academic Press.
58. Schwab, F., Knopoff, L. Love waves and the torsional free modes of a multilayered anelastic sphere. Bull. Seismol. Soc. Am. 63 (1973) 1107–1117.

59. Šílený, J., Panza, G. F. Inversion of seismograms to determine simultaneously the moment tensor components and source time function for a point source buried in a horizontally layered medium. Studia geoph. et geod. **35** (1991) 166–183.

60. Šílený, J., Campus, P., Panza, G. F. Seismic moment tensor resolution by waveform inversion of a few local noisy records-I. Synthetic tests. Geophys. J. Int. **126** (1996) 605–619.

61. Sipkin, S. A. Estimation of earthquake source parameters by inversion of waveform data: synthetic waveforms. Phys. Earth planet. Inter. **30** (1982) 242–259.

62. Takeuchi, H., Saito, M. Seismic surface wave. In: Methods in computational physics Vol. 11, B.A. Bolt (ed) (1972) pp. 217–295. New York. Academic Press.

63. Tarantola, A. Inverse Problem Theory, Methods for Data Fitting and Model Parameter Estimation, Elsevier, Amsterdam. (1987)

64. Trehu, A. M., Nabelek, J. L., Salomon, S. C. Source characterization of two Reykjanes Ridge earthquakes: surface waves and moment tensors; P wave forms and nonorthogonal nodal planes. J. geophys. Res. **86** (1981) 1701–1724.

65. Walter W. R. Source Parameters of the June 29, 1992 Little Skull Mountains earthquake from complete regional waveforms at a single station. Geophys. Res. Lett. **20** (1993) 403–406.

Resolution of Cole-Cole Parameters Based on Induced Polarization Data

Flemming Effersø

Department of Earth Sciences, University of Aarhus
Finlandsgade 8, DK-8200 Århus N, Denmark

Abstract. Induced polarization data are commonly interpreted using a Davidson-Cole dispersion model. The Davidson-Cole model is composed of five intrinsic parameters that characterize the spectral behaviour of the earth resistivity. Recently, several workers have reported the time-constant τ, one of the intrinsic parameters, to be strongly related to hydraulic conductivity.

In the present work one Davidson-Cole dispersion, which is considered to be representative for glacial soils, has been analysed with two different noise and assumptions of prior information on the model parameters. Models were analysed with regard to (1) noise in directions of smallest singular values, (2) uncertainty in model parameters, and (3) significance of prior information.

The study indicates that the time-constant τ can be determined fairly well under low noise levels such as in laboratory, whereas under noise conditions typical for field measurements the determination is poor. Furthermore, this study indicates that increasing the noise level by a factor of 10 for frequencies above 1 Hz, increases the uncertainty in τ markedly.

1. Introduction

The induced polarization (IP) method as used in geophysics is based on the fact that the resistivity of earth materials is generally found to be a function of frequency of applied current (Bodmer et al, 1968). The IP phenomena are of electrochemical origin and are caused either by metallic mineral particles in a rather poorly conducting rock matrix or by differences in the ion concentrations in the pore space or at the interface between the matrix and pore space (Sumner, 1976). When an electric dc-current is applied to an earth volume, local ion accumulations can emerge in the interstitial water (pore fluid) filling up the voids of the earth material. These variations in ion concentrations originate from minerals filling up partially or completely some of the micropores of the earth material. Upon termination of applied current, diffusion voltages that result from these impressed concentration gradients decay with time as ions redistribute themselves to equilibrium (Vinegar and Waxman, 1984). These voltage decays are referred to as "overvoltage" or the induced polarization effect. Similarly, impressing an alternating current to the earth volume will result in an IP effect being dependent on frequency of applied current.

A Cole-Cole dispersion model (Cole and Cole, 1941) is commonly used for parameterization of IP data collected in either time-domain or frequency-domain. The Cole-Cole model is composed of five intrinsic parameters that characterize the spectral behaviour of the earth resistivity.

Several workers (Pelton et al., 1978; Vanhala and Peltoneimi, 1992; Vanhala, 1997) have found the time constant (one of the Cole-Cole parameters) to be strongly dependent upon grain size for both glacial soils and ore bodies of disseminated minerals. Due to the fact that hydraulic conductivity is strongly dependent upon grain size, parameterization of IP data by Cole-Cole parameters may provide valuable information on variations in hydraulic conductivity of mixed layers of sand/clay sediments (e.g. glacial soils and alluvial deposits).

The present study addresses the parameter resolution of Cole-Cole parameters for glacial soils based on frequency-domain spectral IP (spectral IP) data in the presence of noise.

2. The Forward Solution

In the subsequent sections it is assumed that the IP response is measured by a four-electrode array composed of transmitter and receiver dipoles. Furthermore, a half-space of a homogeneous electrical resistivity is assumed.

In recent works (e.g. Vanhala, 1997), the Davidson-Cole model (Pelton et al., 1983) has been used in the interpretation of SIP data. In this model, the frequency dependent resistivity ρ is related to five intrinsic parameters through the relation

$$\rho(\omega) = \rho_0 \left\{ 1 - m \left[1 - \frac{1}{\left(1 + (i\omega\tau)^c \right)^a} \right] \right\} \qquad (1)$$

where ω is angular frequency, ρ_0 is dc-resistivity, m is chargeability, c is frequency dependence, τ is time constant , and a is a symmetry parameter. The model is referred to as a Cole-Cole model if a is fixed to one, that is the phase spectrum is symmetric around the phase peak.

3. Resolution Analysis

Inversion of phase spectrum data only is sufficient because in a minimum phase shift function, the phase spectrum determines the amplitude spectrum unambiguously by means of the Hilbert transform, once the DC value of ρ, i.e. ρ_0, is known (Vanhala, 1997). ρ_0 is determined from a low frequency (1 Hz) measurement. Thus, only phase data are considered in the subsequent.

The linear approximation to the covariance of the estimation error is given by (Jackson, 1979; Menke, 1989)

$$\mathbf{C}_{est} = (\mathbf{A}^T \mathbf{C}_{obs}^{-1} \mathbf{A} + \mathbf{C}_{prior}^{-1})^{-1} \qquad (2)$$

where $\mathbf{A}$ is a matrix containing the partial derivatives, $\mathbf{C}_{obs}$ is the covariance matrix for the observational errors, and $\mathbf{C}_{prior}$ is a model covariance matrix. In this study, both observation errors and model parameter are considered uncorrelated in which case $\mathbf{C}_{obs}$ and $\mathbf{C}_{prior}$ become diagonal matrices.

Further insight may be obtained by the means of singular value decomposition analysis. Jackson (1972) suggests transformation in model space based on error covariances, $\mathbf{C}_{obs}$, as will give likely variation intervals for individual model parameters, $\mathbf{C}_{prior}$.

Two matrices may be defined

$$\mathbf{C}_{obs} = \mathbf{E}\mathbf{E}^T$$
$$\mathbf{C}_{prior} = \mathbf{G}\mathbf{G}^T \tag{3}$$

and the transformation to primed and double primed coordinates can then be written as

$$\mathbf{A}' = \mathbf{E}^{-1}\mathbf{A}\mathbf{G} = \mathbf{U}\mathbf{S}\mathbf{V}^T \tag{4}$$
$$\mathbf{x}'' = \mathbf{V}^T\mathbf{x}' = \mathbf{V}^T\mathbf{G}^{-1}\mathbf{x}$$
$$\mathbf{y}'' = \mathbf{U}^T\mathbf{y}' = \mathbf{U}^T\mathbf{E}^{-1}\mathbf{y} \tag{5}$$

where $\mathbf{S}$ is a (pxp) diagonal matrix containing the $p=\text{rank}(\mathbf{A}')$ non-zero singular values, $\mathbf{V}$ is a (mxp) matrix containing singular vectors that span a subspace in the model space in primed coordinates, and $\mathbf{U}$ is a (nxp) matrix containing the corresponding singular vectors in data space. In this case of phase data, p will be four and not five, because the DC-resistivity, ρ_0, is contained in the null space, $\mathbf{V}_0$.

In double primed coordinates the forward solution becomes

$$\mathbf{y}''_{obs} = \mathbf{A}''\mathbf{x}''_{true} + \mathbf{e}''_{obs} = \mathbf{S}\mathbf{x}''_{true} + \mathbf{e}''_{obs}$$

where $\mathbf{x}''_{true}$ is the true model vector. The least squares solution may then be calculated to obtain

$$\mathbf{x}''_{est} = (\mathbf{S}^T\mathbf{S} + \mathbf{I})^{-1}(\mathbf{S}^T\mathbf{y}''_{obs} + \mathbf{x}''_{prior}) \tag{6}$$

which can be decoupled into scalar equations

$$x''_{est,i} = \frac{s_i}{s_i^2 + 1} y''_{obs,i} + \frac{1}{s_i^2 + 1} x''_{prior}$$

$$= \frac{s_i^2}{s_i^2 + 1} x''_{true,i} + \frac{1}{s_i^2 + 1} (x''_{prior} + e''_{obs}) \quad i = 1, ..., \text{rank}(\mathbf{A}') \tag{6a}$$

$$x''_{est,i} = x''_{prior} \quad i = \text{rank}(\mathbf{A}') + 1, ..., \dim(\mathbf{A}') \tag{6b}$$

From equation (6a and 6b) it is seen that the signal and noise break even for degrees of freedom where $s_i \approx 1$, whereas the signal dominates when $s_i > 1$ and the noise dominates when $s_i < 1$.

4. Noise Model

In field measurements, the reading uncertainty is varying with frequency. For frequencies lower than 1 Hz the noise is low, typically less than a few milliradians (mrad). For frequencies above 1 Hz the noise is increasing and may exceed 5-20 mrad at 64-124 Hz, depending on the conditions at the field site (Vanhala and Peltoniemi, 1992). Under favourable conditions in laboratory, the reading uncertainty can be as less as 1 mrad over the frequency range from some mHz to some kHz (Vanhala and Soininen, 1995).

In field measurement, frequencies higher than 10 Hz are usually omitted due to electromagnetic coupling between transmitter and receiver dipoles. The IP effect and the electromagnetic coupling are inherently intertwined, but the "pure" IP response dominates the "pure" electromagnetic response below 10 Hz (Wait and Gruszka, 1986). Inductive and capacitive coupling between transmitter and receiver cables are normally reduced by keeping a large separation between them and by placing them perpendicular to each other. Thus, any noise contributions due to coupling effects are considered negligible.

Geologic noise is defined as responses due to features of the geologic setting that are not included in the physical model used for parameterization of the data. In this context, where grounded potential electrodes are involved in the measurements, heterogeneities in resistivity near the electrodes may cause significant geologic noise. This error is usually assumed to be 5% in the apparent resistivity.

In summary, two noise models will be considered

Noise model A, low noise level (laboratory environment)

    reading uncertainty   : 1 mrad at all frequencies

    geologic noise   : 5% uncertainty in the dc resistivity estimatere ρ_0.

Noise model B, high noise level (field measurements)

    reading uncertainty   : 1 mrad below 1 Hz and 10 mrad above 1 Hz

    geologic noise   : 5% uncertainty in the dc resistivity estimatere ρ_0.

The resolution analysis based on these noise models will generally result in optimistic estimates of parameter resolution as some of the noise contributions have been disregarded.

5. Relevant Davidson-Cole Models

As previously mentioned, the relevant Davidson-Cole models in this study are those representative for glacial soils. Fig. 1 shows three phase spectra determined from measurements in laboratory on three soil samples representative of glacial soils. Application of small devices in laboratory allows a broad frequency range from 0.01 Hz to 10 kHz to be used. As seen, the signal is ranging from about 0.5 mrad to about 20 mrad, which is fairly modest compared to the ambient noise level in field measurements. For each soil sample, estimated Cole-Cole models are presented in the table on the plot. As the frequency range is broad, it is necessary to apply one Davidson-Cole and one Cole-Cole model to fit the low and the high frequency part of the responses. However, if only the frequency range from 0.01 Hz to 10 Hz is considered one Davidson-Cole model is sufficient to fit the data.

The resolution analysis in this paper will be based on one of the Davidson-Cole models, $[\rho_0, m, c, \tau, a]^T = [100\ \Omega\cdot\text{m}, 0.079, 0.720, 1.500\ \text{s}, 0.400]^T$, occurring in the plot. This model is hence assumed to be representative for glacial soils so that conclusions can be based there upon.

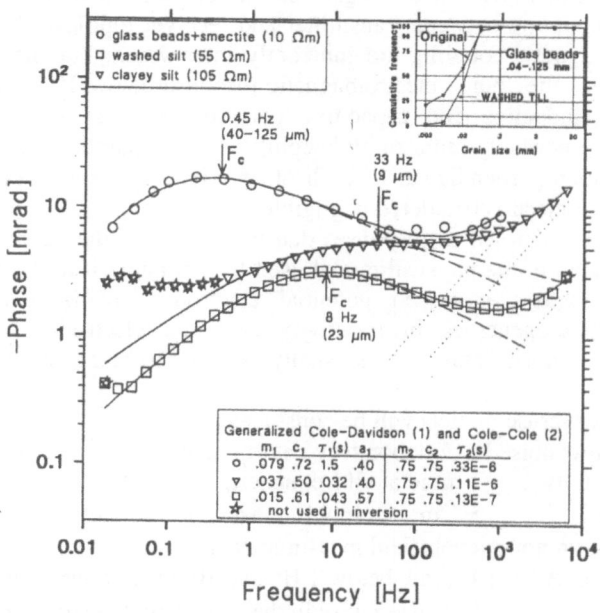

Fig.1   Phase spectra obtained from measurements of three soil samples in laboratory (from Vanhala, 1997).

As ρ_0 is assumed to be known with 5% uncertainty in the resistivity (100 $\Omega\cdot$m), m, c, and a range from 0 to 1, and τ rarely exceeds a few seconds for glacial soils, the following weak prior $\mathbf{C}_{obs} = [5^2, 1, 1, 5^2, 1]^T$ is applied for transformation to primed coordinates, see equations (3) to (5).

6. Results

In the resolution analysis two frequency ranges in binary steps [0.0125, Hz, 0.250 Hz,...,6.40 Hz] and [0.0156, 0.0312 Hz,...,8 Hz] are assembled so that the full frequency range is $F_{as} = [0.0125\ \text{Hz}, 0.0156\ \text{Hz}, ..., 6.40\ \text{Hz}, 8\ \text{Hz}]$. These frequencies are chosen because (1) they are applied in commercially available instruments, and (2) field measurements are generally not subject to electromagnetic coupling in this range. The resolution analysis is performed for varying subsets of F_{as} centred at 0.3 Hz. Thus, the narrowest frequency range contains 8 frequencies in the range from 0.1 Hz to 1.0 Hz.

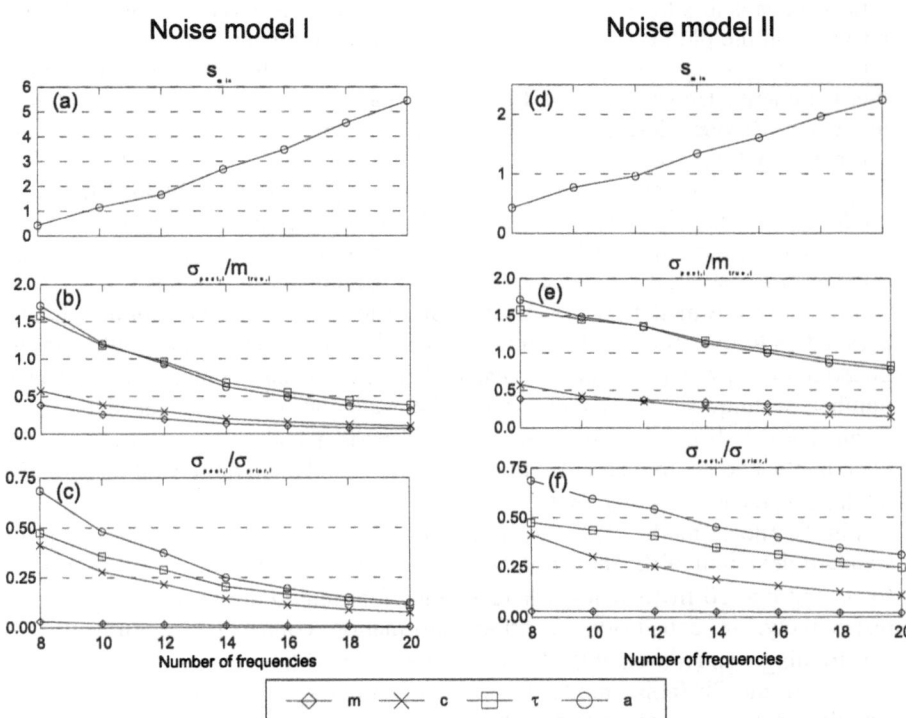

Fig. 2   (a) and (d) Smallest singular value obtained from SVD of **A'**. (b) and (e) Uncertainties in model parameters given as the ratio of posterior standard deviation to true model parameter. (c) and (f) Ratio of posterior standard deviation to a priori standard deviation for model parameters.

Fig.2a (noise model I) shows the smallest singular values, s_{min}, obtained from SVD of **A**. As seen, s_{min} attains a value of about 1.0 when 10 frequencies are used, indicating that the signal-to-noise (S/N) is very poor for degrees of freedom relating to s_{min} when the number of frequencies is less than 10. However, at 20 frequencies the smallest singular value is on the order of 5.5, suggesting a good S/N in this case.

Fig. 2b (noise model I) shows ratios of posterior standard deviation of the model parameter, σ_{post}, to the true model parameter, m_{true}, where σ_{post} is determined from equation (2). As seen, the uncertainties in m and c and the uncertainties in τ and a are close to each other over the entire range of pre-selected number of frequencies. The uncertainties in m and c are on the order of 50% at 8 frequencies, whereas τ and a are completely undetermined with uncertainties greater than 150%. For 20 frequencies, the highest number of frequencies applied, the uncertainties in all parameters have dropped below 40% with uncertainties in m and c on the order of 10%, and τ and a on the order of 40%. In conclusion, data constrain m and c quite well, but still τ and a are not completely constrained by data.

Fig. 2c (noise model I) shows the ratio of posterior standard deviation to a priori standard deviation, σ_{prior} of the model parameter. σ_{prior} is based on $\mathbf{C}_{prior}$. This ratio is always less than or equal to $2^{-1/2}$ when s_{min} attains a value of one, which is also seen in the plot. It is noticed that a maintains the highest ratio over the entire interval, indicating that a priori information contributes the most to estimation of this parameter. However, ratios for all four parameters are 25 % at the highest number of frequencies, showing that data do strongly contribute to parameter estimation in this case.

Fig. 2d (noise model II) shows the smallest singular values obtained for noise model II. As seen s_{min} attains a value of 1.0 when 12 frequencies are used. At 20 frequencies s_{min} attains a value close to 2.0, indicating that degrees of freedom relating to s_{min} have a poor S/N at any pre-selected number of frequencies.

Fig. 2e (noise model II) shows ratios of posterior standard deviation of model parameters to true model parameter. As seen, the uncertainties in m and c and the uncertainties in τ and a are close to each other over the entire range of pre-selected number of frequencies. m and c are quite well constrained by the data at 20 frequencies with uncertainties near 25%, whereas τ and a are constrained with uncertainties no better than 80%. In conclusion, m and c are quite well constrained by data, whereas τ and a are not.

Fig. 2f (noise model II) shows ratios of posterior standard deviation to prior standard deviation of the model parameter. As seen, the ratios are lower than 15% for m and c at 20 frequencies, whereas ratios for τ and a are near 25% and 30%, respectively. In conclusion, a priori information contributes significantly to constraining τ and a even at the higher number of used frequencies.

Fig. 3a and 3b (noise model I) show the data singular vector and the model singular vector that correspond to the singular value at 20 frequencies in Fig. 2a. As seen, the model vector consists largely of τ and a, indicating that these parameters are the least well-determined parameters. This is even more pronounced for noise model II (Fig. 3c and 3d) where the τ and a components are larger.

7. Discussion

One Davidson-Cole model, which was assumed to be representative for glacial soils, has been analysed for two different noise model I and II, respectively, and assumptions of prior information on the model parameters. Models were analysed with regard to (1) noise in directions of smallest singular values, (2) uncertainty in model parameters, and (3) significance of prior information. Different number of frequencies, corresponding to different frequency ranges centred at 0.3 Hz, were considered, i.e. 8, 10, ..., 20 frequencies where 8 frequencies corresponds to the frequency range from 0.1 Hz to 1.0 Hz and 20 frequencies corresponds the range from 0.01 Hz to 8 Hz. These frequencies are typically available in commercial instruments.

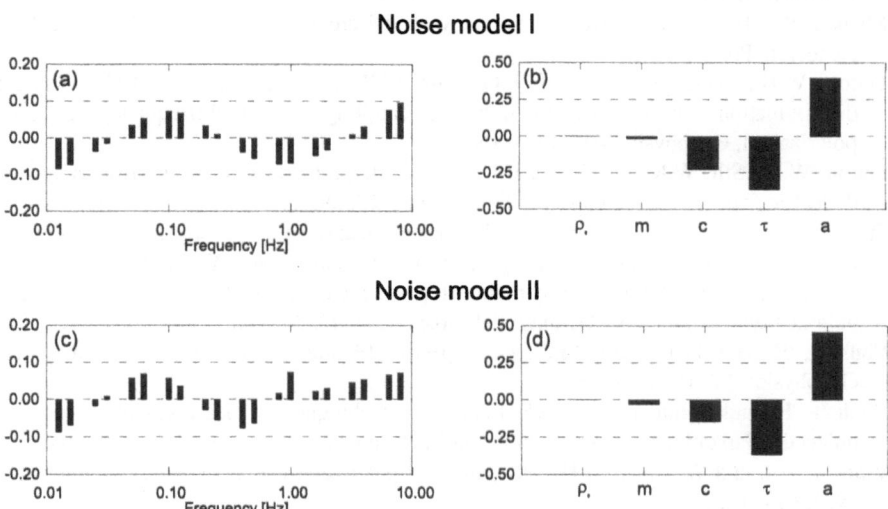

Fig. 3   (a) and (c) data singular vectors, and (b) and (d) model singular vectors corresponding to the singular values at 20 frequencies in Fig. 2c (noise model I) and Fig. 2f (noise model II).

For both noise models I and II, a large number of frequencies were required to reduce the uncertainties in all parameters below 50%. For both noise models, m and c were the best determined parameter, whereas the uncertainties in τ and a were substantially larger. For noise model I, τ and a were fairly well determined at 20 frequencies, whereas for noise model II they were poorly determined at 20 frequencies.

For both noise models, the model singular vector corresponding to the smallest singular value at 20 frequencies was largely composed of τ and a. For noise model I, the smallest singular was on the order of 5.5, whereas for noise model II it was on the order of 2.0. In result, τ and a were subject to a good S/N in the case of noise model I, but quite poor in the case of noise model II.

In conclusion, the results of this study indicate that the time-constant τ, which is of particular interest, can be determined fairly well under low noise levels such as in laboratory, whereas under noise conditions typical for field measurements the determination is poor. Furthermore, this study indicates that increasing the noise level by a factor of 10 for frequencies above 1 Hz, increases the uncertainties in τ and a significantly.

128

References

Bodmer, R., Ward, S.H. and Morrison, H.F., 1968, On induced polarization and groundwater, Geophysics, **33**, 805 – 821.

Cole, K.S. and Cole, R.H., 1941, Dispersion and absorption in dielectrics, I. Alternating current characteristics, Journal of Chemical Physics, **9**, 341-351.

Jackson, D.D. 1972, Interpretation of Inaccurate, Insuffient and Inconsistent Data, Geophys. J. Roy. Astron. Soc., **28**, 97 – 110.

Jackson, D.D. 1979, The use of a priori data to resolve non-uniqueness in linear inversion, Geophys. J. Roy. Astron. Soc., **57**, 137 – 157.

Menke, W. 1989. Geophysical data analysis: Discrete inverse theory. Revised ed., Academic Press Limited.

Pelton, W.H., Ward, S.H., Hallof, P.G., Sill, W.R. and Nelson, P.H., 1978, Mineral discrimination and removal of inductive coupling with multifrequency induced-polarization, Geophysics, **43**, 588 – 609.

Pelton, W.H., Sill, W.R. and Smith, B.D., 1983, Interpretation of complex resistivity and dielectric data, part 1. Geophysical Transactions, **29**, 297-330.

Sumner, J.S., 1976, Principles of induced polarization for geophysical exploration. Developments in Economic Geology, 5, Elsevier Scientific, New York. 255 pp.

Vanhala, H., 1997, Mapping oil-contaminated sand and till with the spectral induced polarization (SIP) method, Geophysical Prospecting, 45, 303-326.

Vanhala, H. and Peltoniemi, M., 1992, Spectral IP studies of Finnish ore prospects, Geophysics, **57**, 1545 – 1555.

Vanhala, H. and Soininen, H., 1995, Laboratory technique for measurement of spectral induced polarization response of soil samples, Geophysical Prospecting, 43, 655-687.

Vinegar, H.J. and Waxman, M.H. 1984, Induced-polarization of shaly sands, Geophysics, **49**, 1267 – 1287.

Wait, J.R. and Gruszka, T.P., 1986, On electromagnetic coupling "removal" from induced polarization surveys, Geoexploration, **24**, 21 – 27.

The Resolution Function in Linearized Born and Kirchhoff Inversion

Leiv-J. Gelius* and Isabelle Lecomte**

* Department of Geophysics, University of Oslo
P.O.Box 1022 Blindern, N-0315 Oslo, Norway
** NORSAR, P.O. Box 51, N-2027 Kjeller, Norway

Abstract. We consider the problem of constructing the depth image of a heterogeneous model (f.ex. the subsurface) based on surface measurements. Due to a limited-aperture and band limited signals, the output from a general inversion scheme will be a distorted or blurred image. If the transmitter/receiver layout, the signal bandwidth and the (initial) background model are all known, this distortion can be computed employing the concept of resolution functions. For a general heterogeneous model this function will be space-variant. Hence, knowing the resolution function such image distortions can in principle be corrected for.

In this paper we discuss two different classes of inversion methods: the Kirchhoff technique which is in favour of reflecting boundaries, and the Born-diffraction technique which relates equally well to a discontinous model (weak contrast). We show that for both classes of methods the resolution function can be expressed by the same integral over the model's Fourier space. The key parameter is the socalled resolution vector (or scattering wavenumber), which at a particular image point is defined by the incident and scattered ray directions. For a model in favour of reflecting boundaries only scattering ray directions corresponding to Snell's law contribute asymptotically. For such a model we introduce a modified resolution function which we denote the reflector spread function. This function gives information about the resolving power of the inversion algorithm with respect to two nearby reflectors.

1. Introduction

Integral equation approaches to acoustic imaging within the Born or Kirchhoff approximation give compact inversion algorithms, using pre-calculated Green's functions. They moreover give direct access to a fundamental parameter: the scattering wavenumber (Beylkin et al., 1985). This parameter controls the resolution of the inversion result. Though the formalism of scattering wavenumber exists for heterogeneous backgrounds, most of the work carried out on resolution are using homogeneous backgrounds, especially for 3-D applications in survey planning (von Seggern, 1994; Wapenaar, 1997; Lavely, 1997; Vermeer, 1997). To obtain the scattering wavenumbers needed for the resolution function in an arbitrary model, asymptotic solutions of the wave equation are employed, i.e. obtained by using efficient methods for Green's function computations like Wavefront Construction (WFC) (Vinje et al., 1993). Within the Born approximation, Lecomte and Gelius (1998) have already demonstrated how the resolution functions can be computed for general 2-D and 3-D models and acosutics waves based on WFC. Here we give a uniform approach showing that the resolution function within both the Born and the Kirchhoff approximations can be expressed by the same integral over the model's Fourier space. Moreover, we demonstrate that knowing the resolution function image distortions caused by limited view and bandwidth can in principle be corrected for.

2. Born Scattering Model

For a more complete discussion of the Born scattering model we refer to Miller et al. (1987), Gelius et al. (1991) and Hamran and Lecomte (1993).

We assume monochromatic acoustic waves in a 3-D model propagating according to the Helmholtz equation

$$\left[\nabla^2 + k^2(\vec{r})\right]p(\vec{r},\omega) = 0 \quad , \quad k(\vec{r}) = \omega / c(\vec{r}) \tag{1}$$

where $\vec{r}$ is the position vector of a typical point in space and ω is the angular frequency. Next, we introduce a *background* velocity field $c_0(\vec{r})$ which is assumed to be smooth. The deviations between $c(\vec{r})$ and $c_0(\vec{r})$ then describe the discontinuities (scattering features) of the medium. Eq.(1) can now be rewritten as

$$\left[\nabla^2 + k_0^2(\vec{r})\right]p(\vec{r},\omega) = \gamma(\vec{r})p(\vec{r},\omega) \quad , \quad k_0(\vec{r}) = \omega / c_0(\vec{r}) \tag{2}$$

where γ is the *scattering potential* defined by

$$\gamma(\vec{r}) = k_0^2(\vec{r}) - k^2(\vec{r}) \tag{3}$$

The exact solution of (2) is

$$p_s(\vec{r}_g,\omega) = \int p(\vec{r}',\omega)\gamma(\vec{r}')G(\vec{r}_g,\vec{r}',\omega)d^3\vec{r}' \tag{4}$$

where p_s is the scattered wavefield measured by a receiver at $\vec{r}_g$ and given as

$$p_s(\vec{r}_g,\omega) = p(\vec{r}_g,\omega) - p_0(\vec{r}_g,\omega) \tag{5}$$

with p_0 being the background wavefield which is a solution of (2) when $\gamma = 0$. Moreover, in (4) G is the Green's function which represents a solution of (2) when the right-hand side of the equation is replaced by a point source at $\vec{r}'$. Assume now weak scattering, i.e.

$$\left|c(\vec{r}) - c_0(\vec{r})\right| < (<)c_0(\vec{r}) \tag{6}$$

This implies that we can make use of the *Born approximation* where the total field p is replaced by the background field p_0 in Eq.(4):

$$p_s(\vec{r}_g,\omega) \cong \int p_0(\vec{r}',\omega)\gamma(\vec{r}')G(\vec{r}_g,\vec{r}',\omega)d^3\vec{r}' \tag{7}$$

We now apply dynamic ray theory to represent the background wavefield and the Green's function in Eq.(7), .i.e.

$$G(\vec{r}_g,\vec{r}',\omega) \cong A_g(\vec{r}_g,\vec{r}') \cdot e^{i\omega\tau_g(\vec{r}_g,\vec{r}')} \tag{8}$$

where A_g represents the amplitude and τ_g the traveltime (parabolic approximation), and correspondingly

$$p_0(\vec{r}',\vec{r}_s,\omega) = s(\omega)G(\vec{r}',\vec{r}_s,\omega) \cong s(\omega)A_s(\vec{r}',\vec{r}_s)e^{i\omega\tau_s(\vec{r}',\vec{r}_s)} \tag{9}$$

In (9) we have assumed a point source at $\vec{r}_s$ and a general source signature $s(\omega)$. By combining Eqs. (7)-(9) we finally obtain

$$p_s(\vec{r}_g,\vec{r}_s,\omega) \cong s(\omega)\int a(\vec{r}_g,\vec{r}_s,\vec{r}')\gamma(\vec{r}')\exp[i\omega\tau(\vec{r}_g,\vec{r}_s,\vec{r}')]d^3\vec{r}' \tag{10}$$

where a is the combined amplitude and τ is the total traveltime from $\vec{r}_s$ via image point $\vec{r}'$ to $\vec{r}_g$, i.e.

$$\begin{aligned}a(\vec{r}_g,\vec{r}_s,\vec{r}') &= A_s(\vec{r}',\vec{r}_s) \cdot A_g(\vec{r}_g,\vec{r}')\\\tau(\vec{r}_g,\vec{r}_s,\vec{r}') &= \tau_s(\vec{r}',\vec{r}_s) + \tau_g(\vec{r}_g,\vec{r}')\end{aligned} \tag{11}$$

3. Kirchhoff Scattering Model

For a more complete discussion of the Kirchhoff scattering model the reader is refered to Bleistein (1987), Pedersen et al. (1988) and Tygel et al. (1993).

We consider a 3-D model. Suppose a time-harmonic scalar field is known across a surface S. According to the Kirchhoff theory, the singly scattered field away from the surface is given by

$$p_s(\vec{r}_g,\vec{r}_s,\omega) = \iint_S \hat{n}(\vec{r}')[p_s(\vec{r}',\vec{r}_s,\omega)\nabla'G(\vec{r}_g,\vec{r}')-G(\vec{r}_g,\vec{r}')\nabla'p_s(\vec{r}',\vec{r}_s,\omega)]dS \tag{12}$$

where $\hat{n}$ is the outer normal to S and G is the Green's function associated with the propagation from the surface point $\vec{r}'$ to the observation point $\vec{r}_g$. Moreover, ∇' denotes a gradient with respect to the $\vec{r}'$ variables. Introduce now the *generalized Kirchhoff theory* where the integrand in Eq. (12) is obtained from dynamic raytracing, in which we trace the source field from $\vec{r}_s$ to $\vec{r}'$ and trace the Green's function from $\vec{r}'$ to $\vec{r}_g$. This procedure is also valid if the media between the various layers are inhomogeneous. The *incident* wavefield p_i can be written as

$$p_i(\vec{r}',\vec{r}_s,\omega) = s(\omega)G(\vec{r}',\vec{r}_s) \cong s(\omega)A_s(\vec{r}',\vec{r}_s)e^{i\omega\tau_s(\vec{r}',\vec{r}_s)} \qquad (13a)$$

Hence, the boundary values of p_s becomes:

$$p_s(\vec{r}',\vec{r}_s,\omega) = R(\vec{r}',\theta)p_i(\vec{r}',\vec{r}_s,\omega) \quad , \quad \nabla'p_s(\vec{r}',\vec{r}_s,\omega) = -R(\vec{r}',\theta)\nabla'p_i(\vec{r}',\vec{r}_s,\omega) \qquad (13b)$$

In Eq.(13b) R is the angular dependent reflection coefficient and θ is the angle between the normal and the ray connecting the source with the point $\vec{r}'$.

Substituting Eqs. (8) and (13) into Eq. (12), and neglecting the gradients of the amplitudes A_S and A_g, we find that

$$p_s(\vec{r}_g,\vec{r}_s,\omega) \cong i\omega \cdot s(\omega)\iint\limits_S R(\vec{r}',\alpha) \cdot A_s(\vec{r}'\vec{r}_s) \cdot A_g(\vec{r}_g,\vec{r}') \cdot \hat{n}(\vec{r}') \cdot \qquad (14)$$
$$\nabla'\tau(\vec{r}_g,\vec{r}_s,\vec{r}') \cdot \exp\!\left[i\omega\tau(\vec{r}_g,\vec{r}_s,\vec{r}')\right]\!dS$$

where τ denotes the traveltime from $\vec{r}_s$ to $\vec{r}'$ on the surface S and then back to $\vec{r}_g$. Next, we assume that the main contribution of the backscattered field is due to specular reflection, i.e.

$$\hat{n}(\vec{r}') \cdot \nabla'\tau(\vec{r}_g,\vec{r}_s,\vec{r}') \cong -\left|\nabla'\tau(\vec{r}_g,\vec{r}_s,\vec{r}')\right| \qquad (15)$$

according to Snell's law. Moreover, we introduce the *singular function* $\xi(\vec{r}')$ of a smooth surface S (Bleistein, 1984). For each point on S, we assign a distribution $\delta(d)$, where d measures the normal distance from S. If $f(\vec{r})$ is a differentiable function, we require that

$$\xi(\vec{r}) = \delta(d) \Rightarrow \iiint\limits_V f(\vec{r}) \cdot \xi(\vec{r})dV = \iint\limits_S f(\vec{r})dS \qquad (16)$$

That is, the effect of the singular function on a volume integral is to reduce it to a surface integral over the support surface of the singular function.

We now rewrite Eq. (14) as a volume integral using the singular function of the surface and the condition in Eq. (15), i.e.

$$p_s(\vec{r}_g,\vec{r}_s,\omega) \cong -i\omega \cdot s(\omega)\int\left|\nabla'\tau(\vec{r}_g,\vec{r}_s,\vec{r}')\right|a(\vec{r}_g,\vec{r}_s,\vec{r}') \cdot R(\vec{r}',\theta)\xi(\vec{r}') \cdot \qquad (17)$$
$$\exp\!\left[i\omega\tau(\vec{r}_g,\vec{r}_s,\vec{r}')\right]\!d^3\vec{r}'$$

where we also have introduced the combined amplitude a as defined in Eq. (11).

4. Scattering Wavenumber and Resolution Function

By comparing Eqs. (10) and (17) we realize that both scattering models can be described using a single (combined) equation:

$$p_s(\vec{r}_g,\vec{r}_s,\omega) \cong s(\omega)\int w(\vec{r}_g,\vec{r}_s,\vec{r}',\omega)\cdot a(\vec{r}_g,\vec{r}_s,\vec{r}')\cdot\Gamma(\vec{r}')\cdot$$
$$\exp\left[i\omega\tau(\vec{r}_g,\vec{r}_s,\vec{r}')\right]d^3\vec{r}' \tag{18}$$

where w is a weight factor.

We have explicitly:

$$\Gamma(\vec{r}') \equiv \gamma(\vec{r}') \quad , \qquad w(\vec{r}_g,\vec{r}_s,\vec{r}',\omega) \equiv 1 \qquad\qquad Born$$
$$\Gamma(\vec{r}') \equiv R(\vec{r}',\theta)\xi(\vec{r}') , \; w(\vec{r}_g,\vec{r}_s,\vec{r}',\omega) = -i\omega\left|\nabla'\tau(\vec{r}_g,\vec{r}_s,\vec{r}')\right| \quad Kirchhoff \tag{19}$$

Assume now:

- the scattering (reflection) process is of a local reaction type
- local plane-wave contribution (far-field assumption)
- bacground velocity model $c_0(\vec{r})$ is known.

These assumptions imply that we can consider the contribution from a specific part of the model without taking into account interactions with other parts of the model except very locally. Hence, consider now the response from such a localized part of the model at $\vec{r}$, and introduce the Taylor expansions

$$a(\vec{r}_g,\vec{r}_s,\vec{r}')w(\vec{r}_g,\vec{r}_s,\vec{r}',\omega) \cong a(\vec{r}_g,\vec{r}_s,\vec{r})w(\vec{r}_g,\vec{r}_s,\vec{r},\omega)$$
$$\tau(\vec{r}_g,\vec{r}_s,\vec{r}') \cong \tau(\vec{r}_g,\vec{r}_s,\vec{r}) + \nabla\tau(\vec{r}_g,\vec{r}_s,\vec{r})\cdot(\vec{r}'-\vec{r}) \tag{20}$$

The first approximation in (20) simulates a local plane wave, whereas the second approximation supports local reaction. Inserting (20) in (18) gives

$$p_s(\vec{r}_g,\vec{r}_s,\omega) \cong s(\omega)w(\vec{r}_g,\vec{r}_s,\vec{r},\omega)a(\vec{r}_g,\vec{r}_s,\vec{r})e^{i\omega\tau(\vec{r}_g,\vec{r}_s,\vec{r})}\cdot$$
$$\int\Gamma(\vec{r}')e^{i\omega\nabla\tau\cdot(\vec{r}'-\vec{r})}d^3\vec{r}' \tag{21}$$

Introduce the *scattering wavenumber vector* $\vec{K}$:

$$\vec{K} = -\omega\nabla\tau(\vec{r}_g,\vec{r}_s,\vec{r}) = -\omega\left[\nabla\tau_s(\vec{r},\vec{r}_s) + \nabla\tau_g(\vec{r}_g,\vec{r})\right] = -\vec{k}_s + \vec{k}_g \tag{22}$$

In order to arrive at the results in Eq. (22) we have made use of the Eikonal equation. Figur 1 shows the ray interpretation of $\vec{K}$, in terms of Green's function ray paths and directions. Note that the Green's functions are computed in the background model which is assumed to be known.

Combining Eqs. (21) and (22) give

$$p_s(\vec{r}_g,\vec{r}_s,\omega) \cong s(\omega)w(\vec{r}_g,\vec{r}_s,\vec{r},\omega)a(\vec{r}_g,\vec{r}_s,\vec{r})e^{i\omega\tau(\vec{r}_g,\vec{r}_s,\vec{r})}.$$

$$e^{i\vec{K}\cdot\vec{r}}\int\Gamma(\vec{r}')e^{-i\vec{K}\cdot\vec{r}'}d^3\vec{r}' = \tag{23}$$

$$s(\omega)w(\vec{r}_g,\vec{r}_s,\vec{r},\omega)a(\vec{r}_g,\vec{r}_s,\vec{r})e^{i\omega\tau(\vec{r}_g,\vec{r}_s,\vec{r})}\cdot e^{i\vec{K}\cdot\vec{r}}\cdot\tilde{\Gamma}(\vec{K})$$

where $\tilde{\Gamma}(\vec{K})$ is the three-dimensional Fourier transform of the scattering potential (Born) or reflectivity function (Kirchhoff).

By inverse Fourier transform we have

$$\Gamma(\vec{r}) = \int\tilde{\Gamma}(\vec{K})e^{i\vec{K}\cdot\vec{r}}d^3\vec{K} \tag{24}$$

Eq.(23) inserted in Eq.(24) gives:

$$\Gamma(\vec{r}) = \int\frac{D(\vec{r}_g,\vec{r}_s,\omega)}{s(\omega)w(\vec{r}_g,\vec{r}_s,\vec{r},\omega)a(\vec{r}_g,\vec{r}_s,\vec{r})}\cdot e^{-i\omega\tau(\vec{r}_g,\vec{r}_s,\vec{r})}d^3\vec{K} \tag{25}$$

which is the *linearized inversion* equation. In Eq.(25) we have replaced the local scattering contribution p_s by the actual measurement data D (response from total model), which is a valid approximation within the local reaction assumption. To obtain a more practical inversion algorithm we introduce a parametrization of the acquisition surface, i.e.

$$\vec{r}_s = \vec{r}_s(\vec{\xi}) \quad, \quad \vec{r}_g = \vec{r}_g(\vec{\xi}) \tag{26}$$

where the parameter vector $\vec{\xi}$ is given by

$$\vec{\xi} = \{\xi_1,\xi_2\} \tag{27}$$

This parametrization can be used to write Eq.(25) on its final form

$$\Gamma(\vec{r}) = \int\frac{D(\vec{r}_g,\vec{r}_s,\omega)}{s(\omega)w(\vec{r}_g,\vec{r}_s,\vec{r},\omega)a(\vec{r}_g,\vec{r}_s,\vec{r})}\cdot e^{-i\omega\tau(\vec{r}_g,\vec{r}_s,\vec{r})}\cdot J(\vec{K}|\vec{\xi},\omega)d^2\xi d\omega \tag{28}$$

where J is the Jacobian for the coordinate transformation between the model space $(\vec{K})$ and the acquisition space $(\vec{\xi},\omega)$, i.e.

$$J(\vec{K}|\vec{\xi},\omega) = \omega^2\cdot\det\begin{vmatrix}\nabla\tau(\vec{r}_g,\vec{r}_s,\vec{r})\\ \dfrac{\partial}{\partial\xi_1}\nabla\tau(\vec{r}_g,\vec{r}_s,\vec{r})\\ \dfrac{\partial}{\partial\xi_2}\nabla\tau(\vec{r}_g,\vec{r}_s,\vec{r})\end{vmatrix} \tag{29}$$

All the quantities in (29) can be determined from dynamic raytracing (Gelius, 1995). However, since we will only employ

- a finite number of source and receiver positions, and
- a finite frequency band

in a seismic experiment, Eq. (28) will only recover an estimate $< \Gamma >$ of the inversion parameters. From Eq.(22) we see that by varying the incident and scattered field directions $\vec{K}$ spans out the Fourier space of Γ. In addition, by varying the frequency ω the scattering wavenumber vector also changes its length. However, we only recover $\vec{K}$ within a finite band $[\vec{K}]$. Thus, Eq. (24) should be replaced by

$$< \Gamma(\vec{r}) > = \int_{[\vec{K}]} \varepsilon(\vec{K}) \tilde{\Gamma}(\vec{K}) e^{i\vec{K} \cdot \vec{r}} d^3 \vec{K} \tag{30}$$

where $\varepsilon(\vec{K})$ is a *directivity* parameter. If we are in a Born scattering model no prefered scattering directions exist. However, in a Kirchhoff scattering model only backscattered energy which fall along directions defined by Snell's law will dominate. This implies that (at least asymptotically):

$$\textit{Born}: \quad \varepsilon(\vec{K}) = 1 \quad \textit{for all possible directions } \vec{K} \tag{31}$$

$$\textit{Kirchhoff}: \varepsilon(\vec{K}) = 0 \quad \textit{for } \vec{K} \textit{ directions not described by Snell's law} \tag{32}$$

We also have the Fourier relationship

$$\tilde{\Gamma}(\vec{K}) = \int \Gamma(\vec{r}') e^{-i\vec{K} \cdot \vec{r}'} d^3 \vec{r}' \tag{33}$$

Combining Eqs. (30) and (33):

$$< \Gamma(\vec{r}) > = \int \Gamma(\vec{r}') \left[\int_{[\vec{K}]} \varepsilon(\vec{K}) e^{i\vec{K} \cdot (\vec{r} - \vec{r}')} d^3 \vec{K} \right] d^3 \vec{r}' = \int \Gamma(\vec{r}') RF(\vec{r}, \vec{r}') d^3 \vec{r}' \tag{34}$$

where *RF* is the *resolution function* which can be written explicitly as

$$RF(\vec{r}, \vec{r}') = \int_{[\vec{K}]} \varepsilon(\vec{K}) e^{i\vec{K} \cdot (\vec{r} - \vec{r}')} d^3 \vec{K} = \int_{[\vec{\xi}, \omega]} \varepsilon(\vec{K}) e^{i\vec{K} \cdot (\vec{r} - \vec{r}')} J(\vec{K}|\vec{\xi}, \omega) d^2 \vec{\xi} d\omega \tag{35}$$

For a Born scattering model described by Eqs. (31) and (35) the resolution function is a standard *point spread function*. For a Kirchhoff scattering model described by Eqs. (32) and (35) the resolution function can be interpreted as a *reflector spread function* (cfr. Eq. (51)).

5. A 2.5-D Assumption

Up til now we have assumed 3-D wave propagation in a 3-D model. In this section we will consider the 2.5-D forward problem, i.e. 3-D wave propagation in a 2-D model. We assume that the vector $\vec{r}_j$ represents a general 3-D position vector. Since the model is two dimensional, we have

$$\Gamma(x, y, z) \equiv \Gamma(x, z) \tag{36}$$

so that Eq. (18) becomes (omitting the prime marks without any loss of generality)

$$p_s(\vec{r}_g, \vec{r}_s, \omega) = s(\omega) \iint \Gamma(x, z) \left[\int w(x, y, z; \vec{r}_g, \vec{r}_s, \omega) a(x, y, z; \vec{r}_g, \vec{r}_s) \cdot \right. \\ \left. e^{i\omega\tau(x, y, z; \vec{r}_g, \vec{r}_s)} dy \right] dxdz \tag{37}$$

This result can be simplified by evaluating the y-integral using the method of stationary phase of single integrals. The following asymptotic formula then applies:

$$\int g(\eta) e^{ikf(\eta)} d\eta \sim \sqrt{\frac{2}{k|f''(\eta_{st})|}} g(\eta_{st}) e^{ikf(\eta_{st}) + i \operatorname{sgn} f''(\eta_{st})\pi/4} \tag{38}$$

where the stationary point η_{st} is defined by

$$f'(\eta_{st}) = 0 \tag{39}$$

In (37) we do not have an explicit expression for the traveltime function τ. We then apply results from ray theory to evaluate the integral asymptotically. We know from (11) that τ is a combined traveltime, i.e. a sum of the two contributions τ_s and τ_g. τ_s is the traveltime associated with a ray from a source point $\vec{r}_s$ to an image point $\vec{r}$, and τ_g represents the traveltime along a ray from the image point $\vec{r}$ to a receiver point $\vec{r}_g$. Both rays are given by the Eikonal equation (cfr. Figur 1), i.e.

$$\nabla \tau_s(\vec{r}, \vec{r}_s) = \frac{\vec{k}_s(\vec{r}, \vec{r}_s)}{\omega} \quad , \quad \nabla \tau_g(\vec{r}_g, \vec{r}) = -\frac{\vec{k}_g(\vec{r}_g, \vec{r})}{\omega} \tag{40}$$

where $\vec{k}_s$ and $\vec{k}_g$ are the wavenumber vector associated with the two rays. If we now apply the result in (39) to (40), we obtain the following criterion for the stationary point y_{st}:

$$\left. \frac{\partial \tau}{\partial y} \right|_{y=y_{st}} = \frac{\partial \tau_s}{\partial y} + \frac{\partial \tau_g}{\partial y} = \frac{k_{sy}}{\omega} - \frac{k_{gy}}{\omega} \equiv 0 \tag{41}$$

where k_{sy} and k_{gy} are the y-components of the two wavenumber vectors $\vec{k}_s$ and $\vec{k}_g$. We see from Eq. (41) that the following solution is possible:

$$k_{sy} = k_{gy} \tag{42}$$

Since the solution should reflect the fact that the two rays are independent of each other, this implies that the components in (42) should be *equal to zero*. Hence, we have:

$$y_{st} = y_s = y_g \tag{43}$$

With this result in mind, we can combine (37) and (38) to obtain finally the 2.5-D forward model

$$p_s(\vec{x}_g, \vec{x}_s, \omega) = s(\omega) \int w(\vec{x}_g, \vec{x}_s, \vec{x}, \omega) a(\vec{x}_g, \vec{x}_s, \vec{x}) \mu(\vec{x}_g, \vec{x}_s, \vec{x}, \omega) \Gamma(\vec{x}) \cdot$$
$$\exp\!\left[i\omega\tau(\vec{x}_g, \vec{x}_s, \vec{x})\right] d^2\vec{x} \tag{44}$$

where $\vec{x}$ is now a 2-D position vector. In Eq. (44) μ is a geometry factor defined as

$$\mu(\vec{x}_g, \vec{x}_s, \vec{x}, \omega) = \sqrt{\frac{2\pi i}{\omega \partial^2 \tau(\vec{x}_g, \vec{x}_s, \vec{x}) / \partial y^2}} \tag{45}$$

where $\partial^2 \tau / \partial y^2$ is the out-of-plane wavefront curvature in the 2-D model.

Based on the result in Eq. (44) we can establish an inverse equation for a 2.5-D model (compare with (25)):

$$< \Gamma(\vec{x}) > = \int \frac{D(\vec{x}_g, \vec{x}_s, \omega)}{s(\omega) w(\vec{x}_g, \vec{x}_s, \vec{x}, \omega) a(\vec{x}_g, \vec{x}_s, \vec{x}) \mu(\vec{x}_g, \vec{x}_s, \vec{x}, \omega)} \cdot e^{-i\omega\tau(\vec{x}_g, \vec{x}_s, \vec{x})} d^2\vec{K} \tag{46}$$

and the corresponding resolution function (compare with (35)):

$$RF(\vec{x}, \vec{x}') = \int_{[\vec{K}]} e^{i\vec{K} \cdot (\vec{x} - \vec{x}')} d^2\vec{K} = \int_{[\vec{K}]} e^{i\vec{K} \cdot (\vec{x} - \vec{x}')} J(\vec{K}|\xi, \omega) d\xi d\omega \tag{47}$$

6. The Point Spread Function (Born Scattering Model)

Figur 2a shows a plot of the available set of scattering wavenumbers for a typical image point assuming a Born scattering model (i.e. $\varepsilon = 1$ for all possible $\vec{K}'s$), taken from Lecomte and Gelius (1998). Horizontal and vertical resolution can be roughly estimated by calculating $2\pi / \Delta K_x$, respectively $2\pi / \Delta K_z$ from Fig. 2a.

A larger frequency band gives larger ΔK_x and ΔK_z, and hence a better resolution. A more direct interpretable quantity is the resolution function RF given by Eq. (47). A complete coverage in the (K_x, K_z)-domain would give RF close to a delta pulse indicating that a point scatterer (or more precisely: a vertical slice through a line scatterer since we are in 2-D) is recovered as a ‚point'. The more limited the coverage and the coarser the sampling, the larger is the deviation of the RF from this ideal point model. Employing the scattering wavenumber model in Fig. 2a we obtain the point-spread function shown in Fig. 2b. It shows the following characteristics found in general: there is a clear low-resolution (LR) axis perpendicular to a high-resolution (HR) axis. Note that these axis are not necessarily coincident with the horizontal and vertical axis. The direction of the high-resolution axis can be found to a good approximation by adding all the different scattering wavenumber vectors together forming a *combined* vector which will fall along the HR-direction.

7. The Reflector Spread Function (Kirchhoff Scattering Model)

In cases where the model is in favour of reflecting boundaries, which is the underlying assumption of a Kirchhoff scattering model, the backscattered wave energy will mainly fall along certain directions in space which fullfill Snell's law (at least correct in an asymptotic sence). Hence, for *common-source* data there is only one source-receiver combination which really contributes at a given image point (if it belongs to a reflecting boundary in the model), which implies that only one wave scattering number direction $\vec{K}_{as}$ contributes to the resolution function. This wavenumber vector is defined as

$$\vec{K}_{as} = \omega \left| \nabla \tau \right| \hat{n} = \omega \frac{2 \cos \theta(\vec{x})}{c_0(\vec{x})} \hat{n} \tag{48}$$

where $\hat{n}$ is the outward normal to the interface, $c_0(\vec{x})$ is the background velocity at the image point on the reflector and $\theta(\vec{x})$ is the Snell's reflection angle at the same point. We see from (48) that for a reflecting interface the important resolution correction should be a compensation for distortions in a direction *normal* to the reflector. Normally, we plot the inversion result as *vertical* depth traces so its even more appropriate to consider the vertical distance

$$d = (\vec{x}' - \vec{x}) \cdot \hat{j} \tag{49}$$

where the dip angle α of the reflector is given by

$$\cos \alpha = \hat{n} \cdot \hat{j} \tag{50}$$

Combining Eqs. (47)-(49) give the 1-D *vertical resolution* function

$$RF_n(d) = |\nabla\tau| \int_{[\omega]} e^{i\omega|\nabla\tau|d\hat{n}\cdot\hat{j}} \, d\omega = \frac{2\cos\theta}{c_0} \int \exp\left[i\omega\frac{2\cos\theta\cos\alpha}{c_0}d\right]d\omega \qquad (51)$$

We interpret (51) as a *reflector spread function*, which gives information about the resolving power of the inversion algorithm with respect to two nearby reflectors.

In cases with small and moderate dips we can approximate θ with the angle which the source ray makes with the vertical at the image point considered and set $\cos\alpha$ equal to one. Eq. (51) is a 1-D integral (expressed by the vertical coordinate d), which can be applied trace by trace. Hence, the resolution effects can in principle be corrected for employing a 1-D space-variant deconvolution for each output (image) trace.

8. Reflector Imaging and a Born Scattering Model

It is reasonable to assume that Eq. (51) will give better predictions for the distortions which will appear in a reflector model after Kirchhoff inversion than in the case of Born inversion. The reason is that Born inversion only will give reliable results for a reflector in cases with weak contrasts and small reflection angles. This fact can be illustrated by a simple example. Let us consider a two-layer model with a horizontal reflector at a depth z_R. A constant velocity c_0 is assumed in the upper layer and a constant velocity c_1 in the lower one. For a Born scattering model, we have from Eq. (10) (omitting the primes for simplicity)

$$p_s(\vec{r}_s, \vec{r}_g, \omega) = s(\omega)\omega^2 \int \frac{\vartheta \cdot H(z - z_R)}{c_0^2} a(\vec{r}_s, \vec{r}_g, \vec{r}) \exp\left[i\omega\tau(\vec{r}_s, \vec{r}_g, \vec{r})\right] d^3\vec{r} \cong$$

$$- s(\omega)i\omega \int \frac{\vartheta \cdot H(z - z_R)}{c_0^2 \nabla\tau \cdot \hat{n}} \nabla_{\hat{n}}\left\{a(\vec{r}_s, \vec{r}_g, \vec{r}) \exp\left[i\omega\tau(\vec{r}_s, \vec{r}_g, \vec{r})\right]\right\} d^3\vec{r}$$

$$(52)$$

where

$$\vartheta = \left(\frac{c_0}{c_1}\right)^2 - 1 \qquad (53)$$

H is the unit step function and $\hat{n}$ is the outward normal to the reflector (z-direction). Integration by parts of Eq. (52) and assuming specular reflection gives the same expression as in Eq. (17), the only difference is that R is replaced by R_B where:

$$R_B = -\frac{\vartheta}{4\cos^2\theta} \qquad (54)$$

with θ being the Snell's reflection angle. It is now straightforward to show that R_B is a good approximation to the angle-dependent acoustic reflection coefficient for weak contasts (i.e. $|\vartheta| < (<)1$) and small angles ($\cos\theta \cong 1$).

9. An Example: Resolution Compensation in a Reflector Model

The phase distortions predicted by Eq. (51) (i.e. stretching of pulses) have been investigated by Tygel et al. (1994) within a Kirchhoff formulation. Moreover, the (peak) amplitude part of the reflector spread function in (51) has been discussed earlier by Bleistein (1987). But neither of these two works give the complete resolution function shown here, or try to correct for such resolution effects.

We demonstrate in this section a simple example of resolution compensation employing the formula in Eq. (51). The geometry is taken from Tygel et al. (1994) and it involves a horizontal reflector at 600m depth, with a p-wave velocity of 4000m/s at the upper layer and 4100m/s in the lower one, and common-source data. Synthetic data were generated using dynamic ray theory. Figur 3b shows the image obtained of the reflector after Kirchhoff inversion (employing Eq. (46)). We can easily see the pulse stretch discussed earlier by Tygel et al. Fig. 3a shows the corresponding resolution function (or reflector spread function) computed using Eq. (51). Based on knowledge of this function we can correct for resolution effects and obtain the improved image in Fig. 3c. Finally, Fig. 3d shows the (peak) amplitude variation along the reflector (lower curve corresponding to the image in (b) and the middle curve to the corrected image in (c)) together with the actual reflectivity curve (upper). We can easily see the improvement in parameter inversion after resolution compensation has been applied.

10. Concluding Remarks

A uniform approach to the concept of resolution function valid for both Born and Kirchhoff linearized inversion has been presented. Moreover, for a model in favour of reflecting interfaces which is often the case when considering the subsurface on a larger scale, we have also introduced the concept of reflector spread function which better describes the distortions in such models. Finally, employing a simple synthetic test model we have shown that parameter inversion (i.e. estimation of angle-dependent reflectivity) was improved if resolution compensation was applied.

References

Beylkin, G., Oristaglio, M., and Miller, D., 1985. Spatial resolution of migration algorithms, in Berkhout, A.J., Ridder, J., and van der Waals, L.F., eds., *Acoust. Imag.* **14**, 155-167.

Bleistein, N., 1984. Mathematical methods for wave phenomena, Academic Press Inc.

Bleistein, N., 1987. On the imaging of reflectors in the earth, *Geophysics* **52**, 931-942.

Gelius, L.-J., 1995. Generalized acoustic diffraction tomography, *Geophys. Prosp.* **43**, 3-29.

Gelius, L.-J., Johansen, I., Sponheim, N., and Stamnes, J.J., 1991. A generalized diffraction tomography algorithm, *J. Acoust. Soc. Am.* **89**, 523-528.

Hamran, S.-E., and Lecomte, I., 1993. Local plane-wavenumber diffraction tomography in heterogeneous backgrounds. Part I: Theory, *J. Seism. Expl.* **2**, 133-146.

Lavely, E., Gibson, R.L., and Tzimeas, C., 1997. 3-D Seismic Survey Design for Optimal Resolution, *Extended Abstract 67th SEG*, 31-34.

Lecomte, I., and Gelius, L.-J., 1998. Have a look at the resolution of prestack depth migration for any model, survey and wavefields, *SEG Extended Abstract* **SP2.3**.

Miller, D., Oristaglio, M., and Beylkin, G., 1987. A new slant on seismic imaging: migration and integral geometry, *Geophysics* **52**, 943-964.

Pedersen, H.M., Gelius, L.-J., and Stamnes, J.J., 1989. 3D seismic modelling of edge diffractions, *Geophys. Prosp.* **37**, 639-646.

Tygel, M., Schleicher, J., and Hubral, P., 1994. Pulse distortion in depth migration, *Geophysics* **59**, 1561-1569.

Vermeer, G.J.O., 1997. Factors affecting spatial resolution, *Extended Abstract 67th SEG*, 27-29.

Vinje, V., Iversen, E., and Gjøysdal, H., 1993. Traveltime and amplitude estimation using wavefront construction, *Geophysics* **58**, 564-575.

Von Seggern, D., 1994. Depth-imaging resolution of 3-D seismic recording patterns, *Geophysics* **59**, 564-576.

Wapenaar, C.P.A., 1997. 3-D migration of cross-spread data: Resolution and amplitude aspects, *Geophysics* **62**, 1220-1225.

SVD Analysis of a 3D Inverse Thermal Model

Lykke Gemmer and Søren B. Nielsen

Department of Earth Sciences, University of Aarhus
Finlandsgade 8, 8200 Århus N, Denmark

Abstract. The steady-state heat equation is used to calculate the thermal structure of the lithosphere in the Danish and surrounding areas. The variable parameters in the inversion are the background heat flow and the upper crustal heat production rate. The data are surface heat-flow values. The optimal model is obtained by iterative least squares inversion. An infinite number of combinations of crustal heat production rate and background heat flow yield the observed surface heat-flow field within the uncertainties of the data. This non-uniqueness of the thermal model is analysed by singular value decomposition of the matrix of partial derivatives. It is concluded that the surface heat flow cannot resolve the long wavelength components of crustal heat production and background heat flow without addition of prior information. However, local anomalies in shallow heat sources can be detected by the method.
In the Danish and surrounding areas, the surface heat flow decreases towards the north-east. This variation is modelled by a combination of a lower background heat flow and a lower upper crustal heat production rate, whereas more local heat-flow anomalies are modelled by local variations in crustal heat production rate.

1. Introduction

The steady-state thermal structure of the lithosphere depends on the thermal parameters and on the thermal boundary conditions. In this paper, the steady heat equation is used in a least squares inverse procedure to estimate the optimal lithospheric thermal structure of the Danish and surrounding areas. The available data is the surface heat flow. A priori parameter information comprises the thermal parameter values of different rock types of the lithosphere. The variable parameters are the background heat flow and the upper crustal heat production rates.

Non-uniqueness is a general problem in thermal reconstruction. An infinite number of combinations of crustal heat production rate and background heat flow yield the same surface heat-flow pattern within the uncertainties of the data. For this reason, inverse estimation of the thermal structure requires a priori information. Singular value decomposition (SVD) (Golub and Reinsch, 1970; Jackson, 1972) reveals which parameter combinations are resolved by data and which depend on prior information. The inverse solution, and its analysis by SVD, is studied in a synthetic example and in a thermal model of the Danish and surrounding areas.

2. Method

2.1 Forward Procedure

The forward procedure is based on the three-dimensional steady-state heat equation (Carslaw and Jaeger, 1959)

$$\nabla \cdot (\Lambda \nabla T) = -A \tag{1}$$

Λ is the thermal conductivity tensor, T is temperature, and A is the heat production rate. The boundary conditions are the surface temperature, T_0, and the background heat flow at a depth of 100 km. The thermal parameters of the model are given in Table 1. They are all constant, except for the background heat flow and the upper crustal heat productions, which vary laterally. The surface temperature is 8°C.

Equation 1 is solved using the finite element method (Akin, 1989) in three dimensions. The model consists of 36*36*10 elements, each 20 km long and wide and of varying height. Altogether the model is 100 km thick, thus covering the main part of the lithosphere. The elements are assigned material properties so that the upper two element layers represent the sediments. The two layers below comprise the upper crust. Beneath, the lower crust is contained in two element layers, and at the bottom of the model the upper mantle is represented by four element layers. The heights of the elements are adjusted so that the geological layer boundaries correspond to the respective element boundaries (Figure 1).

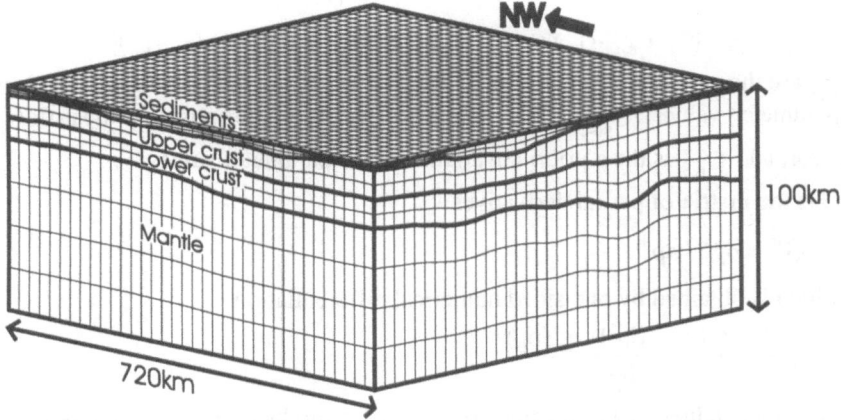

Fig. 1. 3-D finite element grid. The heights of the elements are adjusted so that the element boundaries coincide with the geological layer boundaries.

Table 1. Thermal parameters used in the forward model.

Parameter	Sediments	Upper crust	Lower crust	Mantle
Conductivity (W/m/°C)	2.5	3.0	2.5	4.0
Heat production rate ($\mu W/m^3$)	1.1	Varied laterally	0.2	0.01

2.2 Inverse Procedures

The optimal values of the background heat flow and the upper crustal heat production rates are obtained by joint least squares inversion of data and prior information (Tarantola and Valette, 1982). The heat production rate is related to the rock type, and it may therefore fluctuate more locally than e.g. the background heat flow. For this reason, a shorter correlation length is used for the heat production rate, and a larger for the background heat flow. The applied correlation lengths and standard deviations are given in Table 2.

On the basis of the correlation lengths, the variable parameters are discretised as the background heat flow in the four corners of the model and the natural logarithm of the upper crustal heat production rate in 25 locations, as indicated in Figures 2a and b. The background heat flow is interpolated from the four corner points to the bottom nodal points using kriging, and the heat production rates are interpolated to the integration points in the upper crust also by kriging. Hereby they model continuous background and heat production fields. Data in the algorithm are the surface heat-flow values from the forward model at the 81 locations indicated in Figure 2c. The data errors are assumed to be uncorrelated, whereas the parameters are correlated through the spatial Gaussian correlation function (Tarantola and Valette, 1982)

$$C(r^i, r^j) = \sigma^2 \exp\left(-\frac{1}{2}\frac{(r^i - r^j)^2}{\Delta^2}\right) \tag{2}$$

σ is the standard deviation, Δ is the correlation length, and $(r^i - r^j)^2$ is the squared distance between the correlated points.

The optimal model parameters are obtained by iteration using the least squares equation (Tarantola and Valette, 1982)

$$p_{k+1} = p_0 + \left(G_k^T \cdot C_{d_0 d_0}^{-1} \cdot G_k + C_{p_0 p_0}^{-1}\right)^{-1} \cdot G_k^T \cdot C_{d_0 d_0}^{-1} \cdot \left[d_0 - g(p_k) + G_k \cdot (p_k - p_0)\right] \tag{3}$$

p_{k+1} are the parameter values after $k+1$ iterations, p_0 are the initially assumed (prior) parameter values, $C_{d_0 d_0}$ is the covariance matrix of the observed heat-flow data error, and $C_{p_0 p_0}$ is the a priori covariance of the parameter values.

G is the matrix of partial derivatives

$$G^{ij} = \partial g^i / \partial p^j \tag{4}$$

which is obtained by one-sided numerical differentiation

$$\frac{\partial g^i}{\partial p^j} = \frac{g^i(p_0 + \delta p^j) - g^i(p_0)}{\delta p^j} \tag{5}$$

The prior values of the parameters are a heat production rate of 1 μW/m^3 and a background heat flow of 45 mW/m^2.

Table 2. Standard deviations and correlation lengths used in inversion

Parameter	Standard deviation, σ	Correlation length, Δ (km)
Heat production rate (ln A)	0.5	200
Background heat flow	10 mW/m^2	1000
Surface heat flow	5 mW/m^2	0

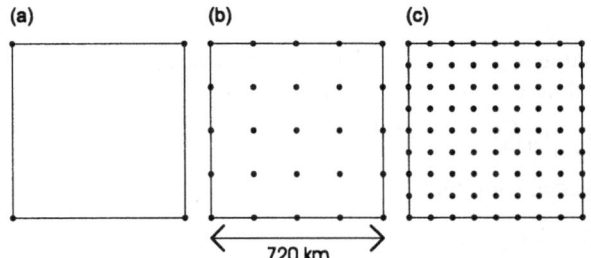

Fig. 2. Locations of the parameter and data points. (a) Background heat-flow parameters. (b) Upper crustal heat production parameters. (c) Surface heat-flow data.

2.3 Singular Value Decomposition (SVD)

The inverse thermal problem is non-unique because an infinite number of combinations of crustal heat production rate and background heat flow will model the same surface heat-flow field within the given uncertainties.

To study which parameter combinations may be resolved, and how the necessary amount of information is added to the system without over-stabilising it, SVD of the matrix of partial derivatives is performed. The SVD method assumes a linear problem. The constraint that the heat production rate should be larger than zero introduces a non-linearity to the problem. The non-linearity is neglected in this study.

In order to take the data and parameter covariances properly into account in the SVD, it is necessary to decompose each covariance matrix (e.g. Jackson, 1972)

$$C_{d_0 d_0} = D_{d_0 d_0} D_{d_0 d_0}{}^{T} \tag{6}$$

$$C_{p_0 p_0} = D_{p_0 p_0} D_{p_0 p_0}{}^{T} \tag{7}$$

where $D_{d_0 d_0}$ and $D_{p_0 p_0}$ may be taken as simple Cholesky factors.
The SVD of the matrix

$$G' = D_{d_0 d_0}{}^{-1} G D_{p_0 p_0} \tag{8}$$

is (e.g. Hansen, 1994)

$$G' = USV^{T} = \sum_{i=1}^{n} u_i \sigma_i v_i^{T} \tag{9}$$

where n is the number of singular values, S is the diagonal matrix of singular values, σ_i. U and V consist of the singular vectors, u_i and v_i. U contains the vectors in data space, u_i. These are related to the model space vectors in V, v_i, via the singular values (e.g. Hansen, 1994)

$$G' v_i = \sigma_i u_i \tag{10}$$

The transformations

$$d'' = U^{T} D_{d_0 d_0}{}^{-1} d$$

$$p''_{prior} = V^T D_{p_0 p_0}{}^{-1} p$$

define the "double primed" coordinates in data space and model space.
The corresponding version of the estimation Equation (3) is

$$p_{optimal}'' = p_{prior}'' + \left(S^T S + I\right)^{-1} S^T \left[\left(d'' - g(p_k'')\right) + S\left(p_k'' - p_{prior}''\right)\right] \qquad (11)$$

Hereby the optimal solution is

$$p_{optimal} = D_{p_0 p_0} V p_{optimal}'' \qquad (12)$$

The diagonal elements of $(S^T S + I)^{-1} S^T$ are $\sigma_i(\sigma_i^2 + 1)^{-1}$. These factors express to which degree the data signal influences the parameter estimations in the final solution and which singular vector components are influenced by prior information alone. For singular values much larger than one, the observed data influence the solution strongly. For singular values smaller than one, the a priori model becomes more important in the solution.

3. Modelling Results and Discussion

The temperature structure is studied for a synthetic model and for a model of Denmark and surrounding areas.

3.1 Synthetic Model

The synthetic model consists of horizontal layer boundaries. In the model, the sediments have a thickness of 10 km, and the crystalline crust is 25 km thick with the upper/lower crust transition in the middle of the crust. The upper mantle part of the model is 65 km thick. In the forward model of the synthetic example, the upper crustal heat production rate is 1 $\mu W/m^3$ close to the edges of the model and increases up to 1.8 $\mu W/m^3$ in the centre of the model (Figure 3a). The background heat flow has a constant value of 45 mW/m². The surface heat-flow field corresponding to the model is shown in Figure 3b.

(a)   (b)

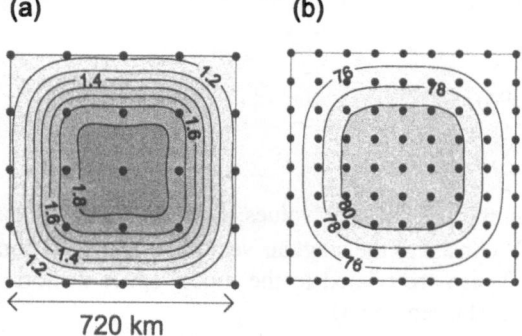

720 km

Fig. 3. Synthetic model. Dots indicate the locations of data and parameter points. (a) Upper crustal heat production rate ($\mu W/m^3$). (b) Surface heat flow (mW/m²) of the forward model.

SVD is performed on the transformed G matrix (Equation 8) obtained after five iterations over Equation 3. SVD is applied twice to the synthetic model. In the first case, the correlation lengths of both parameters are set to 1 m, whereby the parameters are in effect uncorrelated. In the second case, the correlation lengths of Table 2 are included.

The singular values of the two synthetic models are plotted in Figure 4. The singular values of the case of uncorrelated parameters vary from 13 to 0.03. It is notable that the last four singular values are significantly smaller than the first 25 values. For this reason, particularly 4 singular vector components are difficult to recover.

For the model with correlated parameters, the singular values decay from 29 to 0.005. The number of singular values below 1 is 10, which may be interpreted as the number of degrees of freedom now mainly constrained by the correlation prior. Furthermore, it is notable that for correlated parameters the 'jump' in singular values observed for the uncorrelated model has almost vanished. Thus, the distinction between well resolved parameter vectors and poorly resolved parameter vectors is not so significant when the parameters are correlated.

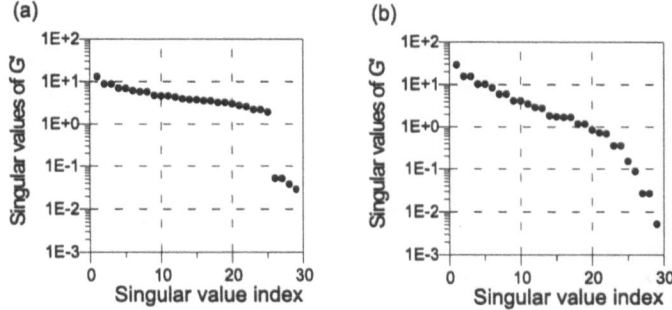

Fig. 4. Singular values of the synthetic example. (a) Uncorrelated parameters. (b) Correlated parameters.

Figure 5 shows the data and model vectors corresponding to the 6 largest singular values of the model of correlated parameters. The first vector set is characterised by an almost constant data vector modelled by almost constant heat production and background heat-flow values. This shows that the „sum" of constant background heat flow and constant crustal heat production is particularly well resolved by data, and that this degree of model freedom couples to a constant heat-flow level.

The second and the third singular triplets are characterised by a high surface heat flow in one half of the model and a low surface heat flow in the other half. The parameter vectors model this long wavelength signal in surface heat flow as a compromise between long wavelength variations in the background heat flow and in crustal heat production rate of the same polarity. Thus, anomalies in the two parameter types of the same polarity will produce a significant signal in the data space.

The 4th, 5th and 6th singular vectors are characterised by surface heat-flow anomalies of a shorter wavelength. The anomalies are caused by local anomalies in heat production rate and an almost constant background heat-flow field. This shows that the short wavelength variations in surface heat flow caused by shallow sources are well resolved by the applied parameterisation.

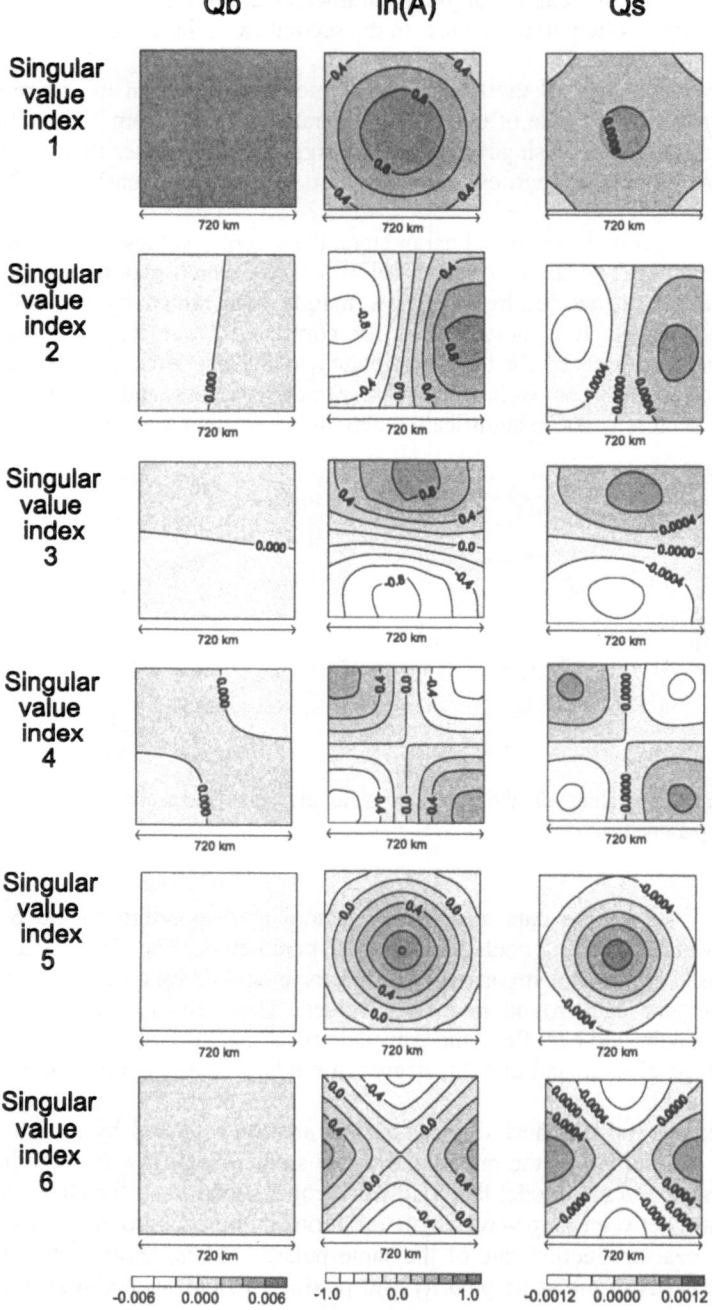

Fig. 5. Data and model vectors corresponding to the 6 largest singular values of the synthetic example with correlated parameters. Qb: Background heat-flow parameter vectors. ln(A): Parameter vectors of the natural logarithm of the upper crustal heat production rate. Qs: Surface heat-flow data vectors.

Figure 6 shows the data and model vectors corresponding to the four smallest singular values. The singular triplets 27 and 28 are similar to triplets 2 and 3 in Figure 5, but now the interference between the background trend and the crustal heat production trend is destructive. Likewise, singular triplet 29 represents destructive interference of the quadrupole patterns. The associated small singular values reflect the inability of data to resolve these degrees of freedom. The associated complicated surface heat-flow vectors represent very low amplitude patterns. Some details in these patterns may actually be artefacts due to the approximations in the numerical computation of partial derivatives.

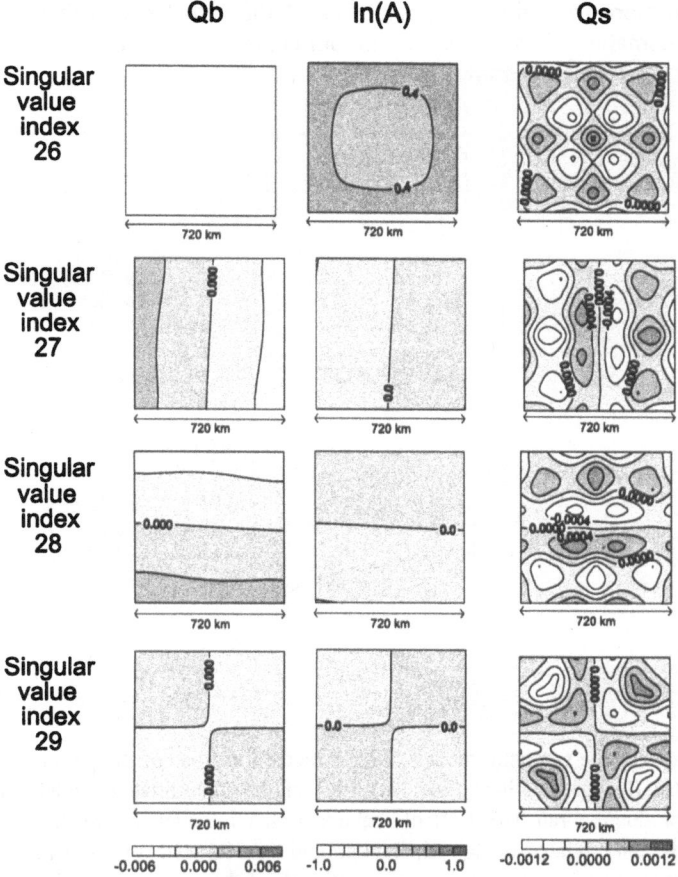

Fig. 6. Data and model vectors corresponding to the 4 smallest singular values of the synthetic example with correlated parameters. Qb: Background heat-flow parameter vectors. ln(A): Parameter vectors of the natural logarithm of the upper crustal heat production rate. Qs: Surface heat-flow data vectors.

The optimal parameters of the two models are shown in Figure 7. In both cases, the surface heat flow is reproduced within the standard deviations. In the model with uncorrelated parameters, the estimated background heat flow has a constant value of 48.6 mW/m². The upper crustal heat production rates vary from 0.9 to 1.4

$\mu W/m^3$. This produces a surface heat-flow field that is within the uncertainties of the data.

In the model with correlated parameters, the estimated background heat flow is close to constant at a value of 46.8 mW/m^2. The estimated upper crustal heat production rates vary from 1 $\mu W/m^3$ to 1.7 $\mu W/m^3$ and are thus higher than the prior values and closer to the values of the forward model than in the uncorrelated case. However, the tabular shape of the heat production function in the forward model is best recovered in the uncorrelated case. It can therefore be concluded that the effects of the introduction of correlated parameters are parameter values that may deviate more from the priors and thus fit the actual data values. In return parameter anomalies are modelled as smooth curves because they must obey the covariances between the parameters.

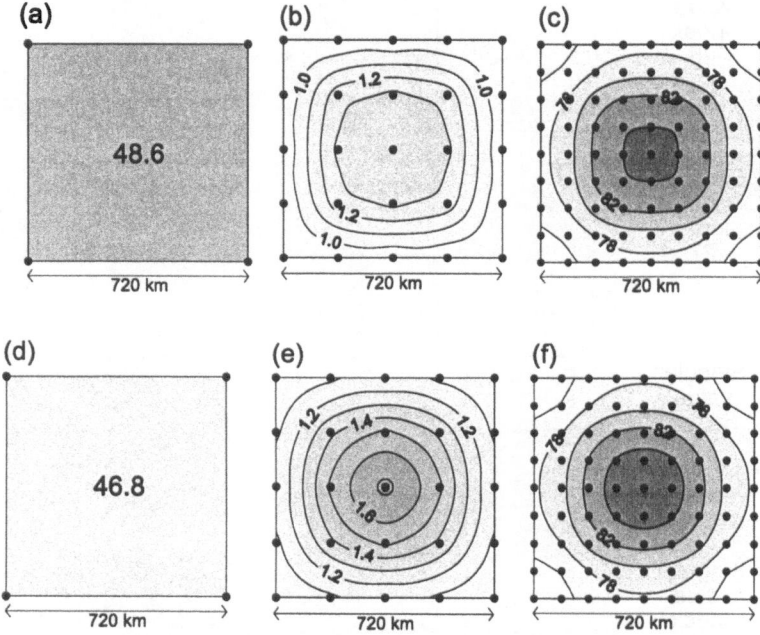

Fig. 7. Solutions of the synthetic models. Dots indicate locations of the parameter and the data points. (a) Background heat flow (mW/m²) of the uncorrelated model. (b) Upper crustal heat production rate (μW/m³) of the uncorrelated model. (c) Surface heat flow (mW/m²) of the uncorrelated model. (d) Background heat flow (mW/m²) of the correlated model. (e) Upper crustal heat production rate (μW/m³) of the correlated model. (f) Surface heat flow (mW/m²) of the correlated model.

3.2 Model of the Danish and Surrounding Areas

The location of the model of the Danish and surrounding areas is indicated in Figure 8a. Besides Denmark, the area covers the southernmost part of Norway, south-western Sweden, North Germany and a part of the North Sea. The sediment thicknesses and the depths to the crust-mantle boundary are contoured in Figures

151

8c and d. These are characterised by lateral variations, and it is particularly significant that, in the areas where the crust-mantle transition is relatively shallow (around the Norwegian-Danish Basin, the North German Basin and the Central Graben), the largest sediment thicknesses are observed. In the Baltic Shield and at the Ringkøbing-Fyn High, the crust-mantle transition is deeper, and the sediment thickness is small.

The observed surface heat flow of the region (Hurtig et al., 1992; Balling, 1995) is shown in Figure 8b. We sample the data values from this map as a regular pattern even though the map is based on boreholes with an irregular distribution. In the study area the density of drillings is quite large, so this procedure is warranted. However, the procedure presented would readily work also for irregularly distributed measurements of surface heat flow.

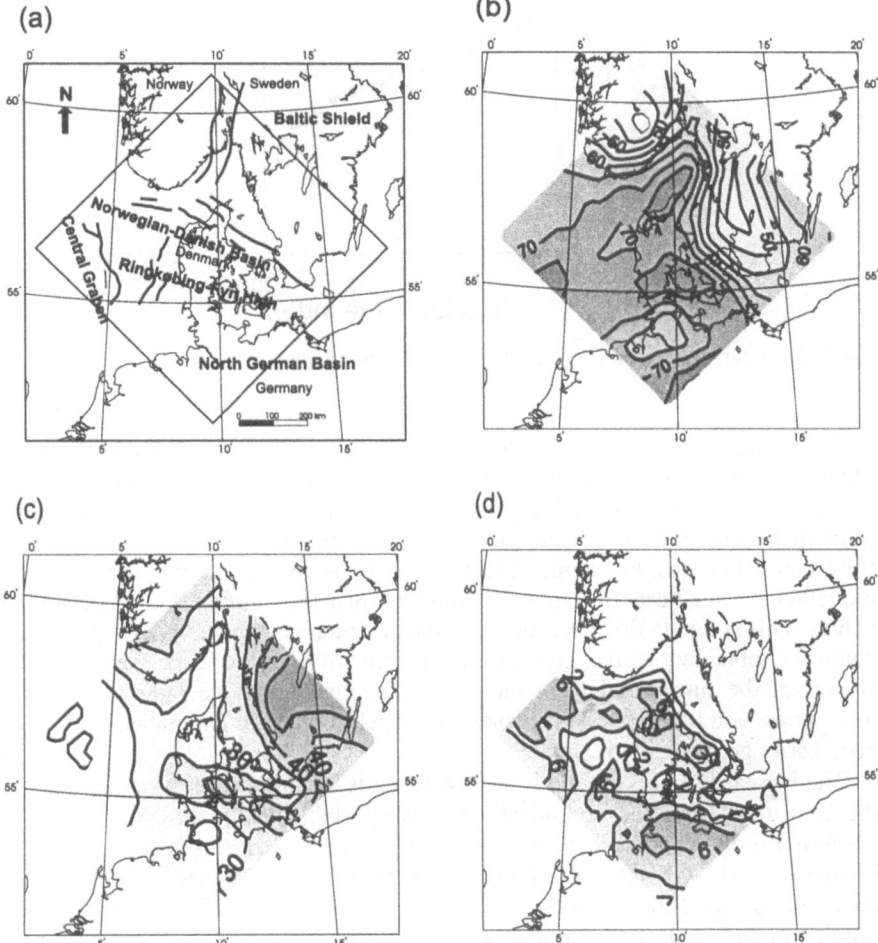

Fig. 8. Model of the Danish and surrounding areas. (a) Studied area and main geological regions of the area. (b) Surface heat flow (mW/m^2). Compiled from Hurtig et al. (1992) and Balling (1995). (c) Moho depths (km). Compiled from Thybo (1997), Kinck et al. (1993), Hurtig et al. (1992), and Pedersen (1990). (d) Basement depths (km). Compiled from Vejbæk and Britze (1994), Vejbæk (1997) and Kockel (1995).

152

The heat-flow values vary from about 50 mW/m² in the Baltic Shield to 70-80 mW/m² in the south-western part of the area. The parameters are assigned the standard deviations and correlation lengths given in Table 2. The a priori background heat-flow values are 45 mW/m², and the prior upper crustal heat production rates are 1 μW/m³.

SVD is performed on the transformed matrix of partial derivatives corresponding to the model obtained after five iterations of Equation 3. The singular values are shown in Figure 9. They vary between 22 and 0.04 and, like the correlated case of the synthetic model, the values decrease without significant jumps. 12 singular values are below 1.

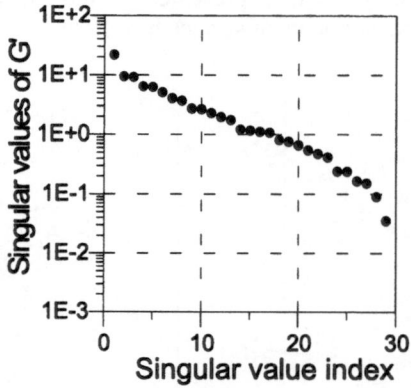

Fig. 9. Singular values of the model of the Danish and surrounding areas.

The solution (Figure 10) shows that the lower surface heat flow in the Baltic Shield has been modelled by a combination of a lower background heat flow and a lower upper crustal heat production rate. Superimposed on the regional trend, the estimated upper crustal heat production rates fluctuate to make up for local heat-flow anomalies and to compensate for variations in the crustal thickness. The modelled surface heat flow shows a smoother distribution than the observed data values. This is due to the correlations of the parameters. A better data reproduction could be obtained if the standard deviations of the the data were reduced. However, the uncertainties on the measurements of surface heat flow do not justify this, and it would just provoke the model to produce anomalies that are not necessarily geologically realistic.

The increase in background heat flow towards south-west may be explained by the transition from the tectonically relatively stable Baltic Shield to areas in western Europe that have been affected by several tectonic events during the Phanerozoic (EUGENO-S Working Group, 1988). In particular, part of the extra background heat flow could be a remnant of a thermal anomaly associated with doming in the North Sea in Jurassic (Ziegler, 1988).

The modelled upper crustal heat production rate is characterised by values up to 3 μW/m³ in the Central Graben where the crust is thin and the layer of sediments thick. Geologically this value is relatively high (Balling, 1995) which could indicate that the estimated heat production rate in the sediments of 1.1 μW/m³ is too low. However, a more likely explanation is that the background heat flow in this area should be higher, which could have been obtained by the model if the standard deviation on the background heat flow had been increased. This fact only serves to illustrate the fundamental ambiguity involved in extrapolating the near surface thermal structure to greater depths.

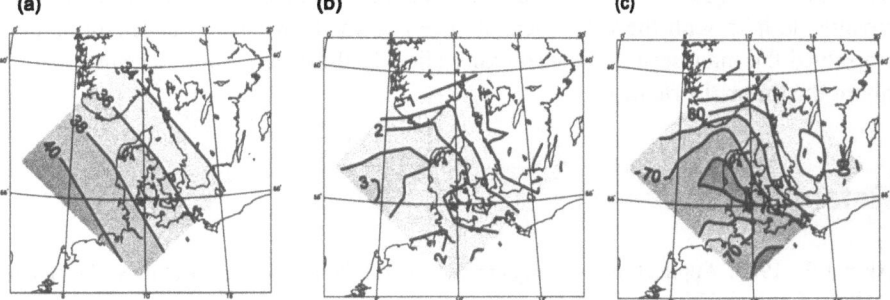

Fig. 10. Solution of the model of the Danish and surrounding areas. (a) Background heat flow (mW/m^2). (b) Upper crustal heat production rate ($\mu W/m^3$). (c) Surface heat flow (mW/m^2).

4. Conclusion

SVD has been used to quantify the non-uniqueness in lithospheric thermal modelling. In the three study cases, some of the singular values are below 1.0 indicating that the corresponding singular vector components are constrained by prior information and not by observations. When the parameters are uncorrelated, in particular the four vector components that correspond to the estimates of the absolute values of the parameters cannot be resolved. Correspondingly, the solution takes values that are close to the a priori given values. The correlation of the parameters smoothens the decrease pattern of the singular values. At the same time it decreases the number of singular vectors to be resolved from the data. The solution of the correlated model deviates more from the prior model and shows a smoother pattern than does the solution of the uncorrelated model. This is due to the additional constraint of the parameters being correlated. However, both solutions are within the standard deviations of the parameters.

In the model of the Danish and surrounding areas, the lower surface heat flow in the Baltic Shield is modelled as a combination of a lower upper crustal heat production rate and a lower background heat flow. Local heat-flow anomalies are modelled by variations in upper crustal heat production rate. These results are in general agreement with the geological setting of the area. The fundamental ambiguity of the results emphasises the necessity of adding prior knowlegde about the distribution of heat sources when extrapolating the near surface thermal structure to greater depths.

154

Acknowledgements

The authors would like to thank Bo Holm Jacobsen for constructive and invaluable help with the theoretical studies and for many useful suggestions that improved the manuscript. We also thank Niels Balling for stimulating discussions about the thermal structure.

References

Akin J.E., 1989. Application and Implementation of Finite Element Methods. Academic Press Limited. 372 pp.

Balling N., 1995. Heat flow and thermal structure of the lithosphere across the Baltic Shield and northern Tornquist Zone. Tectonophysics 244, 13-50.

Carslaw H.S. and Jaeger J.C., 1959. Conduction of heat in solids. Oxford University Press. Second edition. 510 pp.

EUGENO-S Working Group, 1988. Crustal structure and tectonic evolution of the transition between the Baltic Shield and the North German Caledonides (The EUGENO-S Project). Tectonophysics 150, 253-348.

Golub G.H. and Reinsch C., 1970. Singular Value Decomposition and Least Squares Solutions. Numer. Math. 14, 403-420.

Hansen, P.C., 1994. Regularization tools: A Matlab package for analysis and solution of discrete ill-posed problems. Numerical Algorithms 6, 1-35.

Hurtig E., Cermák V., Haenel R. and Zui V., 1992. Geothermal Atlas of Europe. Hermann Haack Verlagsgesellschaft mbH.

Jackson D.D., 1972. Interpretation of Inaccurate, Insufficient and Inconsistent data. Geophys. J. R. astr. Soc. 28, 97-109.

Kinck J.J., Husebye E.S., Larsson F.R., 1993. The Moho depth distribution in Fennoscandia and the regional tectonic evolution from Archean to Permian times. Precambrian Research 64, 23-51.

Kockel F., 1995. Structural and palaeogeographical development of the German North Sea Sector. Beiträge zur regionalen Geologie der Erde 26. Gebrüder Borntraeger, Berlin. 96 pp.

Pedersen T., 1990. Crustal thicknesses, crustal reflectivity and tectonic subsidence in Skagerak. In: Balling N., Nielsen O.B., Korstgård J.A., Nielsen S.B. (Eds.) Proceedings of Basin Workshop Aarhus 1989. GeoSkrifter No 35 1990, 65-74.

Tarantola A. and Valette B., 1982. Generalized nonlinear inverse problems solved using the least squares criterion. Reviews of Geophysics and Space Physics 20, 219-232.

Thybo H., 1997. Geophysical characteristics of the Tornquist Fan area northwest TESZ: Indication of late Carboniferous to early Permian dextral transtension. Geol. Mag. 134, 597-606.

Vejbæk O.V., 1997. Dybe strukturer i danske sedimentære bassiner. Dansk Geologisk Forenings Nyheds- og Informationsskrift 4, 1-31.

Vejbæk O.V. and Britze P., 1994. Geologisk kort over Danmark. Top præ-Zechstein. Danmarks Geologiske Undersøgelse. Kortserie nr 45.

Ziegler P.A., 1988. Evolution of the Arctic-North Atlantic and the Western Tethys. AAPG Memoir 43. The American Association of Petroleum Geologists, Oklahoma. 198 pp.

Optimization Tools for Tikhonov Regularization of Nonlinear Equations Using the L-curve and its Dual

Mårten Gulliksson and Per-Åke Wedin

Department of Computing Science, Umeå University, S-901 87 Umeå, Sweden
Email: marten@cs.umu.se
WWW home page: http://www.cs.umu.se/~marten

Abstract. We consider the regularization of the finite dimensional nonlinear system of equations $f(x) = 0$. The regularization is performed by formulating a Tikhonov problem with an unknown regularization parameter. The nonlinear L-curve is the size of the solution considered as a function of the size of the residual (when the regularization parameter is changed). The dual curve connected to the nonlinear L-curve is defined as the value of the minimization function as a function of the regularization parameter. We show that the L-curve is a strictly decreasing convex function and its dual is strictly increasing and concave. The connection between these two curves and other natural regularization formulations is presented as well as a thorough analysis concerning logarithmic scales and corners of the L-curve.
Given approximations of the solution of the Tikhonov problem we define upper and lower piecewise linear approximations of the L- and a-curve called shadow curves. These shadow curves are thoroughly investigated. Finally, we present ways to update the shadow curves and their use to identify good regularized solutions.

1 Introduction

We will consider the regularization of the nonlinear system of equations

$$f(x) = 0, \quad f : \mathbb{R}^n \to \mathbb{R}^m. \tag{1}$$

Let $\nu(f) \geq 0$ and $y(x) \geq 0$ be given convex functions that attain their minima for $f = 0$ and $x = x_c$, respectively. Our intention is to find a parameter vector x such that the residual

$$t(x) = \nu(f(x)) \tag{2}$$

is small, while the size of $x - x_c$ measured by $y(x)$ is not large. To achieve that consider the optimization problem

$$\min_x \mathcal{T}(x, \lambda), \quad \mathcal{T}(x, \lambda) = t(x) + \lambda y(x), \quad \lambda \geq 0. \tag{3}$$

Problem (3) is often called the Tikhonov problem, see Engl et al. [4]. The main difficulty is to choose the regularization parameter λ in order to find a good (hopefully optimal in some sense) solution to the Tikhonov problem (3). One possible method to choose the regularization parameter is the L-curve that has been used for linear problems see Hansen [8]. We make the following definition of the L-curve that is a generalization of the linear L-curve, Hansen [8], to the nonlinear problems treated here.

Definition 1. *Let $x(\lambda)$ solve problem (3), i.e.,*

$$x(\lambda) = \arg_{\lambda \geq 0} \left\{ \min_x \mathcal{T}(x, \lambda) \right\}.$$

The L-curve is defined as the curve $(t(x(\lambda)), y(x(\lambda)))$.

A typical L-curve is shown in Figure 1. The L-curve has been studied extensively

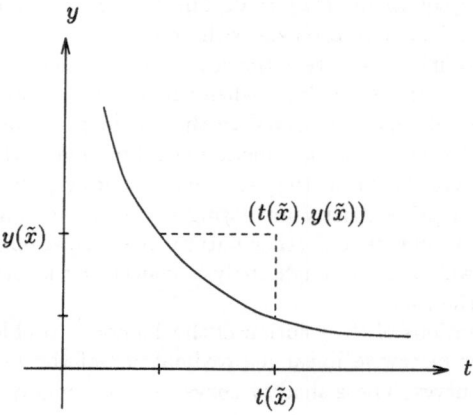

Fig. 1. L-curve.

for a linear function $f(x) = Ax - b$, see [9, 6, 12, 8] and in connection with nonlinear least squares it was used in [1]. As far as we know this is the first time the L-curve is studied in a more general context.

In the nonlinear case the usefulness of the L-curve for approximating a sequence of points $(t(x_i), y(x_i))$, $i = 1, \ldots$ is enhanced by the introduction of a curve $a(\lambda)$ related to the unconstrained problem (3).

Definition 2. *The a-curve is defined as the curve $(\lambda, a(\lambda))$ where*

$$a(\lambda) = \min_x \left\{ t(x) + \lambda y(x) \right\}, \quad \lambda \geq 0. \tag{4}$$

In Section 2 it is proved that the function $a(\lambda)$ is a strictly increasing strongly concave function.

Consider again an arbitrary point $\tilde{x} \in \mathbb{R}^n$. Then $t(\tilde{x}) + \lambda y(\tilde{x}) \geq \min t(x) + \lambda y(x) = a(\lambda)$ for any $\lambda \geq 0$. Hence, the straight line $(\lambda, t(\tilde{x}) + \lambda y(\tilde{x}))$ always lies

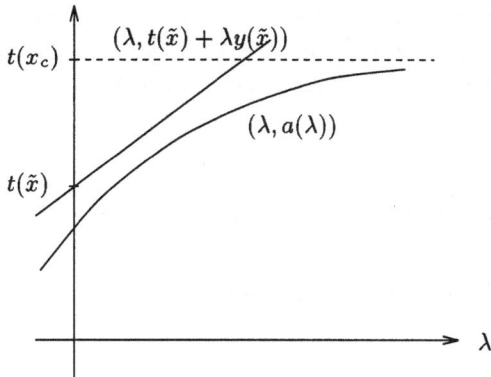

Fig. 2. The a-curve.

above the curve $(\lambda, a(\lambda))$ as shown in Figure 2. It will be shown that the point $(\tilde{\lambda}, a(\tilde{\lambda}))$ closest to the straight line has a tangent parallel to the straight line defined by $\tilde{x}$.

It is our opinion that the L-curve together with the complementary a-curve provide ideal tools for the organization of information about approximate solutions of the regularized problem (3). In this paper we will give the facts about the L-curve on which this opinion is based. We will do so without giving any detailed algorithms for solving (3). Such an algorithm depends heavily on the structure and size of the functions $f(x), t(x)$ and $y(x)$.

For ill-posed linear equations the L-curve in the log-log scale $(\tau(\lambda), \eta(\lambda)) = (\log(t(x(\lambda)), \log(y(x(\lambda))))$ has usually a distinct corner. Hansen and O'Leary [7] have developed an algorithm for identifying that corner, defined as the point where the curve $(\tau(\lambda), \eta(\lambda))$ has its greatest curvature. The corresponding choice of parameter λ has been criticized by several researchers, see e.g. [4,5,13]. Our intention in this paper is limited to extending algorithms for computing the corner defined by Hansen or Reginska for nonlinear equations. This is a nontrivial task. For a nonlinear function $f(x)$ the second derivatives in the logarithmic L-curve used by Hansen for a linear function f are complicated and depend on the third derivative of $f(x)$. Fortunately, we can do without these second derivatives when computing the curvature of the L-curve. Moreover, in Section 4 we prove that the corner, as defined by Reginska [12], minimizes $t(x(\lambda))y(x(\lambda))$ just as in the linear case.

Given any set of points $\{\tilde{x}_i\}$ they all will lie on or above the L-curve. Thus, by choosing the right subsequence of these we will construct an upper approximating L-curve. Constructing the upper approximating L-curve with care will make it inherit the convexity. Moreover, by taking a proper subset of the corresponding straight lines defined by $\{\tilde{x}_i\}$ we construct a piecewise linear continuous approximating curve above the a-curve. These upper approximating curves will be used to find the best possible approximate solution to (3) for a given set $\{x_i\}$.

Ideas similar to those presented here on the L- and a-curve in the linear case can be found in Kaufmann and Neumaier [10] together with an interesting application. For an excellent review of regularization in the linear case see Neumaier [11].

2 Local Properties of the L-Curve

We summarize the local results in a theorem. The proof can be found in Gulliksson and Wedin [3].

Theorem 1. *Take $t(x) \geq 0, y(x) \geq 0, \lambda \geq 0$ and let $x(\lambda)$ solve $\min_x t(x) + \lambda y(x)$. Define*

$$a(\lambda) = t(x(\lambda)) + \lambda y(x(\lambda)).$$

If $t(x)$ and $y(x)$ are differentiable functions then

$$\frac{da}{d\lambda} = y.$$

Further, $(t(x(\lambda)), y(x(\lambda)))$ defines y as a function of $t, y(t)$ with

$$\frac{dy}{dt} = -\frac{1}{\lambda}.$$

If $t(x)$ and $y(x)$ are twice differentiable and the Hessian $H = \nabla_x^2 t(x) + \lambda \nabla_x^2 y(x)$ is positive definite then $x(\lambda)$ is a differentiable function of λ with

$$\frac{dx}{d\lambda} = -H^{-1}\nabla_x y.$$

Take $\beta = (\nabla_x y)^T H^{-1} \nabla_x y > 0$ then $a(\lambda), y(x)$ are twice continuously differentiable with

$$\frac{d^2 a}{d\lambda^2} = -\beta < 0, \quad \frac{d^2 y}{dt^2} = (\lambda^3 \beta)^{-1} > 0.$$

Moreover, y and t are also differentiable functions of lambda with

$$\frac{dt}{d\lambda} = \lambda\beta > 0, \quad \frac{dy}{d\lambda} = -\beta < 0.$$

In summary, the geometrical properties of the functions are that

− y is a strictly convex, strictly decreasing function of t
− y is a strictly decreasing function of λ
− t is a strictly increasing function of λ
− a is a strictly concave, strictly increasing function of λ

3   Two Important Minimization Problems

The importance of the L-curve comes from the fact that there are two constrained optimization problems nicely related to problem (3), see Figure 3.

Firstly, $\mathcal{T}(x, \lambda) - \lambda\Delta$ is the Lagrange function for the constrained optimization problem

$$\min_x \ t(x)$$
$$s.t. \quad y(x) \le \Delta. \tag{5}$$

Problem (5) is often a natural regularization problem for $f(x) = 0$ when Δ is known or easy to estimate.

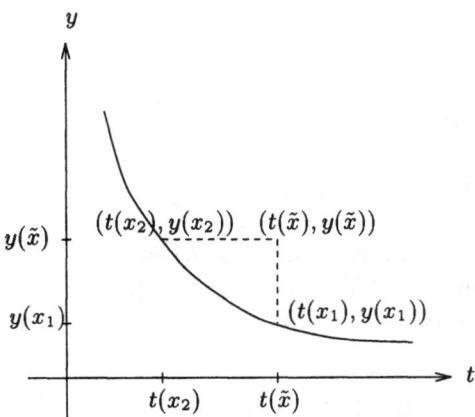

Fig. 3. Minimization properties and the L-curve.

Secondly, $y(x) + \gamma(t(x) - \delta)$ is the Lagrange function of the constrained optimization problem

$$\min_x \ y(x)$$
$$s.t. \quad t(x) \le \delta. \tag{6}$$

Just as for (5) this problem may be the natural choice when regularizing $f(x) = 0$ where the upper bound δ on the residual is known or easy to estimate.

However, (3) nicely takes into account both the restrictions in (5) and (6).

Theorem 2. *Any solution $\tilde{x}$ of (3) also solves (5) with $\Delta = y(\tilde{x})$ and (6) with $\delta = t(\tilde{x})$.*

Proof. The theorem follows directly by using the KKT-conditions on (3) and comparing them with the KKT-conditions for (6) and (5).

A difficulty that may occur for nonlinear problem is that problem (3) need not have a unique solution. We shall ignore these difficulties in this paper.

The following theorem is central for developing tools using the L-curve and the a-curve.

Theorem 3. *Take an arbitrary point $\tilde{x} \in \mathbb{R}^n$. Then the point $(t(\tilde{x}), y(\tilde{x}))$ lies on or above the L-curve and the line $t(\tilde{x}) + \lambda y(\tilde{x})$ is above the a-curve possibly with one tangent point.*

Proof. Let x_1 solve the constrained problem

$$\min_x y(x)$$
$$s.t. \quad t(x) \leq t(\tilde{x}).$$

Then $(t(x_1), y(x_1)) = (t(\tilde{x}), y(x_1))$ is the point on the L-curve vertically below $(t(\tilde{x}), y(\tilde{x}))$ as shown in figure 3. If x_2 solves the problem with constraints on $y(x)$,

$$\min_x t(x)$$
$$s.t. \quad y(x) \leq y(\tilde{x}).$$

then $(t(x_2), y(x_2)) = (t(x_2), y(\tilde{x}))$ is on the point on the L-curve that lies vertically to the left of $(t(\tilde{x}), y(\tilde{x}))$.

In Section 2 the L-curve was proven to be a strictly decreasing strongly convex function which gives the theorem.

The properties in Theorem 3 is most useful when constructing the L-curve corresponding to the truncated SVD, see Hansen [8]. We will use these properties extensively in constructing algorithms.

4 The Corner of the L-Curve

4.1 Using the Minimization Formulation of Reginska

In this section the local minimum introduced by Reginska [12] for a linear function $f(x) = Ax - b$ will be generalized to the nonlinear case. Take $\tau = \log(t(x(\lambda)))$, $\eta = \log(y(x(\lambda)))$. The basic idea is to rotate the log-log coordinate system $\pi/4$ rad and then compute the minima of the curve in the new coordinate system. Intuitively, minima in the rotated coordinate system corresponds to corners of the L-curve in the original coordinate system shown in figure 4. The rotation is performed as

$$\begin{bmatrix} \cos(\pi/4) & -\sin(\pi/4) \\ \sin(\pi/4) & \cos(\pi/4) \end{bmatrix} \begin{bmatrix} \tau \\ \eta \end{bmatrix} = \frac{1}{\sqrt{2}} \begin{bmatrix} \tau - \eta \\ \tau + \eta \end{bmatrix}$$

and we define the new function as

$$G(\theta) = \tau + \eta, \theta = \tau - \eta. \tag{7}$$

Theorem 4. *Define G as in (7) then*

$$\frac{dG}{d\theta} = \frac{-t + \lambda y}{t + \lambda y}$$

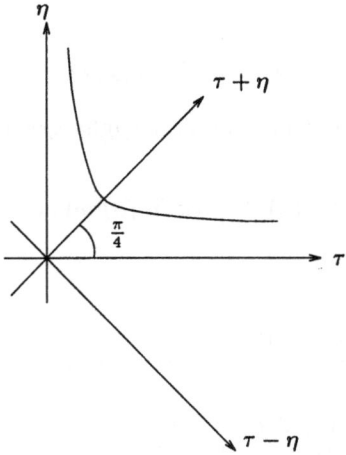

Fig. 4. Rotation of the logarithmic L-curve.

and $t = \lambda y$ when $dG/d\theta = 0$. Moreover,

$$\frac{d^2G}{d\theta^2} = \frac{t\lambda y}{(t + \lambda y)^2} \frac{d}{dt}(-t + \lambda y)$$

and this second derivative exists if $\nabla_x^2(t + \lambda y)$ is invertible.

Proof. From $d\tau = dt/t, d\eta = dy/y$ we have

$$\frac{dG}{d\theta} = \frac{ydt + tdy}{ydt - tdy} = \frac{y + t\frac{dy}{dt}}{y - t\frac{dy}{dt}} = \frac{y - t\lambda}{y + t\lambda}$$

since $dy/dt = -\lambda$.

Further, we have

$$\frac{dG'}{d\theta} = \frac{dG'}{dt} \frac{dt}{d\theta}$$

and using that

$$\frac{d\theta}{dt} = \left(\frac{dt}{t} - \frac{dy}{y}\right)\frac{1}{dt} = \frac{1}{t} - \frac{dy}{dt}\frac{1}{y} = \frac{1}{t} + \frac{1}{\lambda y}$$

together with

$$\frac{dG'}{dt} = \left(\frac{d}{dt}(-t + \lambda y)\right)(t + \lambda y)^{-1} + (-t + \lambda y)\frac{d}{dt}(t + \lambda y)^{-1}$$

$$= \left(\frac{d}{dt}(-t + \lambda y)\right)(t + \lambda y)^{-1}$$

gives the result stated in the theorem.

The last statement in the theorem follows from $d(-t + \lambda y)/dt = -2 - 1/(\lambda\beta)$, $\beta = (\nabla_x y)^T H^{-1} \nabla_x y$.

Now consider the function

$$F(\lambda) = t(x(\lambda))y(x(\lambda)). \tag{8}$$

We have the following theorem that is a slight generalization of a result by Reginska [12].

Theorem 5. *The function $F(\lambda)$ defined in (8) and $G(\theta)$ defined in (7) have the same local minima.*

Proof. We have

$$\frac{dF}{d\lambda} = t\frac{dy}{d\lambda} + \frac{dt}{d\lambda}y = \beta(-t + \lambda y), \quad \beta = \frac{1}{\lambda}\frac{dt}{d\lambda}.$$

giving that $G' = 0$ if and only if $F' = 0$.

Assume that $G'(\hat\theta) = 0$ then $t - \lambda y = 0$. At $\hat\theta$ we get

$$\frac{d^2F}{d\lambda^2} = \frac{d\beta}{d\lambda}(-t + \lambda y) + +\beta\frac{d}{d\lambda}(-t + \lambda y) = \beta\frac{d}{d\lambda}(-t + \lambda y)$$

if $\beta(\lambda)$ is differentiable. From Theorem 4 we have that $d^2G/d\lambda^2 > 0$ if and only if $d(-t+\lambda y)/dt > 0$ which is the same condition we have to put on $d^2F/d\lambda^2 > 0$ for a local minima.

To complete the proof we have to show that $\beta(\lambda)$ is differentiable. This is easily done by considering the differentiability of $\beta(\lambda)(-t + \lambda y)$ at $\hat\theta$.

4.2 The Corner Defined by the Durvature of the L-Curve

As noted in Hansen, O'Leary [7] the curvature of the logarithmic L-curve is defined as

$$\kappa = \frac{\frac{d\tau}{d\lambda}\frac{d^2\eta}{d\lambda^2} - \frac{d^2\tau}{d\lambda^2}\frac{d\eta}{d\lambda}}{\left[\left(\frac{d\tau}{d\lambda}\right)^2 + \left(\frac{d\eta}{d\lambda}\right)^2\right]^{3/2}} \tag{9}$$

For a linear function $f(x) = Ax - b$ the expression in (9) is also used for computing $\kappa(\lambda)$ and its maximum.

Theorem 1 implies that

$$\frac{d\tau}{d\lambda} = t^{-1}\frac{dt}{d\lambda} = t^{-1}\lambda\beta$$

and

$$\frac{d^2\tau}{d\lambda^2} = \beta t^{-1} + \lambda\left(\frac{d\beta}{d\lambda}t^{-1} - t^{-2}\lambda\beta^2\right)$$

where $\beta = \nabla_x y^T H^{-1}\nabla_x y$ and $H = \nabla_x^2 t + \lambda\nabla_x^2 y$. To calculate $\frac{d\beta}{d\lambda}$ we need to know $\frac{d}{d\lambda}\nabla_x^2 t$ and $\frac{d}{d\lambda}\nabla_x^2 y$. We have

$$\frac{d}{d\lambda}\nabla_x^2 t = \sum_{i=1}^n\left(\frac{\partial}{\partial x_i}\nabla_x^2 t\right)\frac{dx_i}{d\lambda}, \quad \frac{dx}{d\lambda} = -H^{-1}\nabla_x y.$$

In the linear least squares case when $t = \|Ax - b\|_2^2/2$ the term $d(\nabla_x^2 t)/d\lambda = 0$ and these third derivatives disappear. However, for f nonlinear the derivatives $\partial(\nabla_x^2 t)/\partial x_i$ do not disappear. Another approach to κ has to be chosen. The remedy is to prefer the L-curve described by the function $y(t)$ to the original definition of the curve $(t(x(\lambda)), y(x(\lambda)))$ defined with respect to the λ-parameter. The logarithmic L-curve can also be written as $(\tau, \eta(\tau))$ which gives the following expression for the curvature κ

$$\kappa = \frac{\frac{d^2\eta}{d\tau^2}}{\left(1 + \frac{d\eta}{d\tau}\right)^{3/2}} \tag{10}$$

which is equivalent with (9) when both expressions are defined. From $d\tau = dt/t$ and $d\eta = dy/y$ we get

$$\frac{d\eta}{d\tau} = \frac{dy}{dt}\frac{t}{y} = -\frac{t}{\lambda y} = -q, \quad q = \frac{t}{\lambda y}$$

and

$$\frac{d^2\eta}{d\tau^2} = -\frac{dq}{d\tau} = -t\frac{dq}{dt}$$

where

$$\frac{dq}{dt} = (\lambda y)^{-1} + t\left[y^{-1}\frac{d}{dt}\lambda^{-1} + \lambda^{-1}\frac{d}{dt}y^{-1}\right] = (\lambda y)^{-1} - \lambda^{-2}\beta^{-1}q + q(\lambda y)^{-1}$$

and hence

$$\frac{d^2\eta}{d\tau^2} = q(qy(\lambda\beta)^{-1} - (q+1))$$

giving

$$\kappa = \frac{q}{(1+q)^{-3/2}}(q(\lambda\beta)^{-1}y - (q+1)), \quad q = \frac{t}{\lambda y}. \tag{11}$$

5 Shadow Curves

5.1 Basic Ideas

The convexity of the L-curve and the concavity of the a-curve are direct consequences of the fact that the curves describe the solution of a sequence of optimization problems. It is natural to try to keep these properties when information at a finite point set $\mathcal{M}$ in $\mathbb{R}^n$ is used to approximate the functions $y(t)$ and $a(\lambda)$. We define the function $y_{sp}(t)$ as a shadow function of $y(t)$ if $y_{sp}(t)$ is a strictly decreasing convex function. To every shadow curve $y_{sp}(t)$ there should also exist a concave, strictly increasing shadow curve $a_{sp}(\lambda)$ of $a(\lambda)$.

5.2   The Polygon Shadow Curves

The first step towards smooth shadow curves is to find a subset $\{x_i\}_{i=1}^{p}$ in $\mathcal{M}$ such that $0 \leq t_1 < t_2 < \ldots < t_p$ and the function

$$y_{sp}(t) = y_i \frac{t_{i+1} - t}{t_{i+1} - t_i} + y_{i+1} \frac{t - t_i}{t_{i+1} - t_i}, t_i \leq t \leq t_{i+1}, i = 1, \ldots, p-1 \qquad (12)$$

is a strictly decreasing convex function for $t_1 \leq t \leq t_p$. If we add the points $(t_1, y), y \geq y_1$ to the points defined by $(t, y_{sp}(t))$ the set $\mathcal{M}$ defines points $((t(x_i), y(x_i))$ that are inside the convex set defined by (12) as these points are shown in Figure 5. For a given finite set $\mathcal{M}$ the polygon shadow curve con-

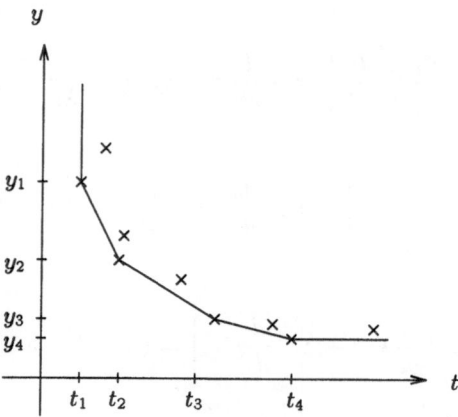

Fig. 5. The shape of y_{sp}-curve.

structed in this way is unique. But the point set $\{x_i\}_{i=1}^{p}$ need not be unique, since there may exist a point $\tilde{x} \in \mathcal{M}, \tilde{x} \neq x_i$ such that $t(\tilde{x}) = t(x_i)$ and $y(\tilde{x}) = y(x_i)$. Then x_i can be changed to $\tilde{x}$ without altering the shadow curve.

Now to the polygon shadow curve $a_{sp}(\lambda)$ of $a(\lambda)$ that corresponds to the polygon shadow curve $y_{sp}(t)$ constructed in (12). Define $\lambda_{i,i+1}$ as the point where the two straight lines $t_i + \lambda y_i$ and $t_{i+1} + \lambda y_{i+1}$ intersect. Hence, $t_i + \lambda_{i,i+1} y_i = t_{i+1} + \lambda_{i,i+1} y_{i+1}$ and

$$\lambda_{i,i+1} = -\frac{t_{i+1} - t_i}{y_{i+1} - y_i}. \qquad (13)$$

In other words $-1/\lambda_{i,i+1}$ is the slope of y_{sp} between t_i and t_{i+1}. Also define $\lambda_{p,\infty}$ as the point where the straight line $t_p + \lambda y_p$ cuts the asymptote $a = t(x_c)$, i.e.,

$$t_p + \lambda y_p = t(x_c), \quad \lambda_{p,\infty} = \frac{t(x_c) - t_p}{y_p}.$$

The definition of a_{sp} is now

$$a_{sp}(\lambda) = \begin{cases} t_1 + \lambda y_1, \ 0 \le \lambda \le \lambda_{1,2} \\ t_i + \lambda y_i, \ \lambda_{i-1,i} \le \lambda \le \lambda_{i,i+1} \\ t_p + \lambda y_p, \ \lambda_{p-1,p} \le \lambda \le \lambda_{p,\infty} \end{cases}$$

The function $a_{sp}(\lambda)$ is the unique strictly increasing concave function such that

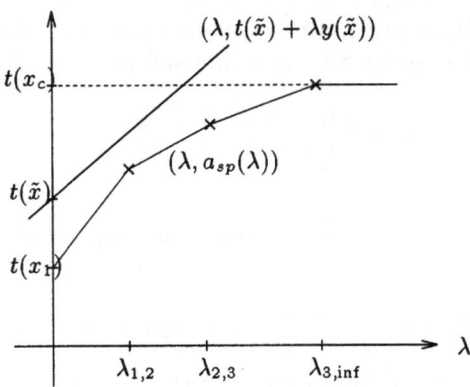

Fig. 6. The shape of the a_{sp}-curve.

for all points $\tilde{x} \in \mathcal{M}$ the straight lines $(\lambda, t(\tilde{x}) + \lambda y(\tilde{x}))$ lie above the curve $(\lambda, a_{sp}(\lambda))$.

5.3 Further Properties of the Shadow Curves

The following theorem summarizes the important properties and relations of the shadow functions y_{sp} and a_{sp}.

Theorem 6. *Given the definition of y_{sp} and a_{sp} we have the following four statements.*

1. *The points $\lambda_{i,j}$ satisfies $\lambda_{1,2} < \lambda_{2,3} < \ldots < \lambda_{p,\infty}$.*
2. *The shadow function $a_{sp}(\lambda)$ is an increasing function for $\lambda \le \lambda_{p,\infty}$.*
3. *The shadow function $a_{sp}(\lambda)$ is concave function.*

Proof. 1. Since y_{sp} by construction is convex and decreasing the slope of the straight line between t_{i-1} and t_i is greater than the slope of the line between t_i and t_{i+1}, i.e.,

$$\frac{y_{i-1} - y_i}{t_i - t_{i-1}} > \frac{y_i - y_{i+1}}{t_i - t_{i+1}}.$$

From the definition (12) we immediately get $1/\lambda_{i-1,i} > 1/\lambda_{i,i+1}$.

2. The shadow curve $a_{sp}(\lambda)$ is an increasing function since the slope of the straight line between $\lambda_{i-1,i}$ and $\lambda_{i,i+1}$ is $y_i > 0$.
3. The increasing function $a_{sp}(\lambda)$ is concave if the positive slopes are decreasing, i.e., $y_{i-1} > y_i$, a simple consequence of the fact that y_{sp} is decreasing.

The following simple lemma is the key to the connection between line segments in the L-curve and the a-curve.

Lemma 1. *Assume that $(t_i, y_i), (t_{i+1}, y_{i+1})$ and $(\tilde{t}, \tilde{y})$ are known and satisfy $t_i < t_{i+1}$, $t_i \leq \tilde{t} \leq t_{i+1}$ and $y_i > y_{i+1}$ with $\tilde{y} > 0$. Let $(t, l_i(t))$ be the straight line between (t_i, y_i) and (t_{i+1}, y_{i+1}). Correspondingly assume that the straight lines $(\lambda, t_i + \lambda y_i)$ and $(\lambda, t_{i+1} + \lambda y_{i+1})$ has a common point for $\lambda = \lambda_{i,i+1}$. Take*

$$a_i(\lambda) = \begin{cases} t_i + \lambda y_i, & \lambda \leq \lambda_{i+1} \\ t_{i+1} + \lambda y_{i+1}, & \lambda > \lambda_{i+1} \end{cases} .$$

Then

$$\tilde{y} - l_i(\tilde{t}) = \frac{1}{\lambda_{i,i+1}}(\tilde{t} + \lambda_{i,i+1}\tilde{y} - a_i(\lambda_{i,i+1})) \tag{14}$$

or

$$\tilde{y} - l_i(\tilde{t}) = \frac{y_i - y_{i+1}}{t_{i+1} - t_i}(\tilde{t} + \lambda_{i,i+1}\tilde{y} - a_i(\lambda_{i,i+1})).$$

Proof. We know that the slope of $l_i(t)$ is $-1/\lambda_{i,i+1}$. Therefore, we have

$$l_i(\tilde{t}) - y_i = -\frac{1}{\lambda_{i,i+1}}(\tilde{t} - t_i)$$

or by adding $(\tilde{y} - \tilde{y})/\lambda_{i,i+1}$

$$\tilde{y} - l_i(\tilde{t}) = \frac{1}{\lambda_{i,i+1}}(-t_i - \lambda_{i,i+1}y_i + \tilde{t} + \lambda_{i,i+1}\tilde{y}).$$

Using the fact that the slope of $l_i(t)$ also can be written as $(y_{i+1} - y_i)/(t_{i+1} - t_i)$ proves the theorem.

We are now able to state a central theorem.

Theorem 7. *A point $(\tilde{t}, \tilde{y})$, $\tilde{t} \geq t_1$, lies on the same side of the polygon shadow curve $y_{sp}(t)$ as the straight line $(\lambda, \tilde{t} + \lambda\tilde{y})$ lies of $a_{sp}(\lambda)$. In other words the following statements are true:*

$$\tilde{y} > y_{sp}(\tilde{t}) \Leftrightarrow \tilde{t} + \lambda\tilde{y} > a_{sp}(\lambda)$$

$$\tilde{y} = y_{sp}(\tilde{t}) \Leftrightarrow (\lambda, \tilde{t} + \lambda\tilde{y}) \text{ passes through a corner of } (\lambda, a_{sp}(\lambda))$$

$$\tilde{y} < y_{sp}(\tilde{t}) \Leftrightarrow (\lambda, \tilde{t} + \lambda\tilde{y}) \text{ intersects the polygon curve } a_{sp}(\lambda)$$

Proof. The second and third statement in the theorem follows directly from (14).

For the first statement consider the two points $(0, \tilde{t})$ and $(\lambda_{i,i+1}, \tilde{t} + \lambda_{i,i+1}\tilde{y})$ that define the straight line $(\lambda, \tilde{t} + \lambda\tilde{y})$. The point $(0, \tilde{t})$ lies above or on the shadow curve $(\lambda, a_{sp}(\lambda))$. From Lemma 1 it is known that the point $(\lambda_{i,i+1}, \tilde{t} + \lambda_{i,i+1}\tilde{y})$ lies on the same side of the curve $(\lambda, a_{sp}(\lambda))$ as the point $(\tilde{t}, \tilde{y})$ lies of the line segment between (t_i, y_i) and (t_{i+1}, y_{i+1}) which forms part of the curve $(t, y_{sp}(t))$.

5.4   Smooth Shadow Curves

Let a polygon shadow curve $(t, y_{sp}(t))$ be known and let $(t, y_{sm}(t))$ be a convex decreasing spline function that interpolates the polygon shadow curve at $(t_i, y_i), i = 1, \ldots, p$. The function y_{sm} is our smooth shadow function, and by definition it is twice differentiable. Define λ at a given point $(t, y_{sm}(t))$ from the derivative

$$\frac{dy_{sm}}{dt} = -\frac{1}{\lambda} \tag{15}$$

and set $a_{sm} = t + \lambda y_{sm}$ as our smooth shadow function corresponding to the a-curve. As before the differential $da_{sm} = dt + \lambda dy_{sm} + d\lambda \, y_{sm} = d\lambda \, y_{sm}$ giving

$$\frac{da_{sm}}{d\lambda} = y_{sm}.$$

Further,

$$\frac{d^2 a_{sm}}{d\lambda^2} = \frac{dy_{sm}}{d\lambda} = \frac{dy_{sm}}{dt}\frac{dt}{d\lambda} = -\lambda^{-1}\frac{dt}{d\lambda}$$

and by using 15 we have

$$\frac{d\lambda}{dt} = \lambda^2 \frac{d^2 y_{sm}}{dt^2}$$

and thus

$$\frac{d^2 a_{sm}}{d\lambda^2} = -\left(\lambda^3 \frac{d^2 y_{sm}}{dt^2}\right)^{-1} < 0.$$

6   Algorithmic Aspects

6.1   On the Computation of a Shadow Curve

When an optimization method is used to compute an approximate solution of the equation $f(x) = 0$ then usually a descent method is applied to the function $t(x)$. A starting point x_0 is given and iterates $x_1, x_2, \ldots$ are computed such that $t(x_{i+1})$ is sufficiently much smaller than $t(x_i)$ to guarantee convergence to a point $\hat{x}$ where $t(x)$ has a local minimum (or just $\nabla_x t(\hat{x}) = 0$).

We shall now outline a model algorithm that can be used to find a stable solution of $f(x) = 0$. In each step a new point x_{new} is computed such that a better shadow curve can be computed, i.e., $a_{new}(\lambda) \le a_{old}(\lambda)$ and $y_{new}(t) \le y(t)$ for the λ-value that is of interest.

Assume that a shadow curve $a_{sp}(\lambda)$ is known. One step of the algorithm can be described as follows.

1. Choose a regularization parameter λ_{new}. Assume that $\lambda_{i,i-1} \le \lambda_{new} \le \lambda_{i,i+1}$. If we want to choose λ_{new} smaller than $\lambda_{1,2}$ then choose the point where $t_1 + \lambda y_1$ is tangent to $a_{sp}(\lambda)$.
2. Set $x_{old} = x_i$ for which $y_i = y(x_i)$ and $a_{sp}(\lambda_{new}) = t_i + \lambda_{new} y_i$. If $\lambda_{new} = \lambda_{i,i+1}$ it may be favorable to start optimizing with $x_{old} = (x_i + x_{i+1})/2$, see Figure 7.

3. Compute x_{new} with $t(x_{new}) + \lambda_{new} y(x_{new})$ sufficiently much smaller than $a_{sp}(\lambda(x_{old}))$. A natural condition when to accept x_{new} is that an optimization algorithm designed to minimize $t(x) + \lambda_{new} y(x)$ should converge.
4. Determine new polygon shadow curves y_{sp} and a_{sp} that update the old shadow curves with the information given at $(t(x_{new}), y(x_{new}))$.

6.2 Cutting Corners

Let the shadow L-curve be computed from fairly few points $x_1, \ldots, x_p$ with $(t(x_1), y(x_1)), \ldots, (t(x_p), y(x_p))$ fairly close to the L-curve. Then there is a simple way to compute $p - 1$ additional points $z_1, \ldots, z_{p-1}$ that add extra information about the L-curve. The idea is as simple as it is useful.

Consider the a_{sp}-curve in Figures 5 and 6. By definition we have that $t_i + \lambda_{i,i+1} y_i = t_{i+1} + \lambda_{i,i+1} y_{i+1}$. Therefore, both x_i and x_{i+1} lie on the level surface $\{x : t(x) + \lambda_{i,i+1} y(x) = t(x_i) + \lambda_{i,i+1} y(x_i)\}$. Then, if the level curve is convex between x_i and x_{i+1} the point $\tilde{x}_i = (x_i + x_{i+1})/2$ will give $t(\tilde{x}_i) + \lambda_{i,i+1} y(\tilde{x}_i) \leq a_{sp}(\lambda_{i,i+1})$ and hence get a better shadow curve. However, when the level curve is concave we will not find a better point.

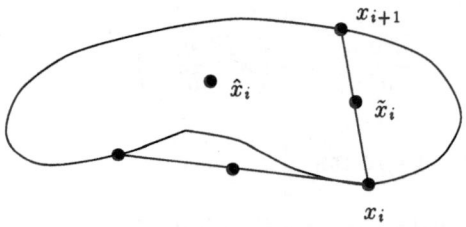

Fig. 7. Choice of new starting point.

Assume that $\hat{x}_i$ in Figure 7 minimizes $t(x) + \lambda_{i,i+1} y(x)$ and hence $a(\lambda_{i,i+1}) = t(\hat{x}_i) + \lambda_{i,i+1} y(\hat{x}_i)$. In the vicinity of $\hat{x}_i$ the level surface is convex. Hence, if the level surface is concave between x_i and x_{i+1} these points do not give very useful information about $a(\lambda_{i,i+1})$.

6.3 Determining Approximate Corners

The corner of the L-curve is of some interest when determining a good regularized solution see Hansen and O'Leary [7] in the linear case. In this section we will describe a simple way of determining approximations to such corners given the shadow L-curve. In Figure 4 the rotations of the L-curve is illustrated.

Even if the two functions give the same local minima it is seen from computational experiments that the minima to $G(\theta) = \tau + \eta$ from the shadow L-curve is easier to determine.

We now describe a simple corner detection based on the minimization of $G(\theta)$. The first step is to take the logarithm of the points (t_i, y_i) of the shadow L-curve and then rotate these points attaining new points (θ_i, G_i). Any three points

$$(\theta_j, G_j), \quad j = i - 1, i, i + 1, \quad i = 2, \ldots, p - 1$$

will define a parabola. If $G_{i-1} > G_i, G_i < G_{i+1}$ then G_i will be a candidate to a minimum to G, see Figure 8.

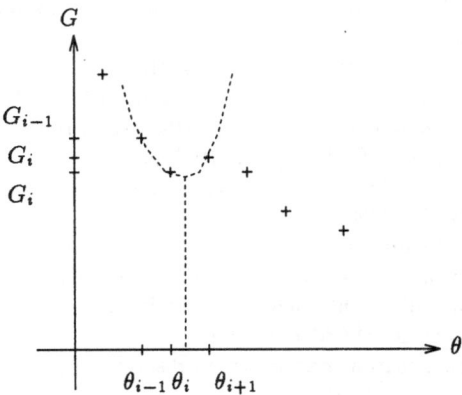

Fig. 8. Simple minimization.

Using $i = 2, \ldots, p - 1$ gives all possible parabolas and the minima of these may be used to either choose a suitable solution or to refine the shadow L-curve.

7 Conclusions and Further Work

We have introduced the generalization of the L-curve to nonlinear problems giving a tool to find a good solution to ill-posed nonlinear problems. Given any set of approximations to the ill-posed problem a shadow L- and a-curve may be defined that approximates the exact L- and a-curve.

However, much work remains. It is still not clear how to combine the shadow L- and a-curve in an algorithm using some special purpose optimization routine. We believe that such an algorithm will be problem dependent.

References

1. J. Eriksson. Optimization and regularization of nonlinear least squares problems. Technical Report UMINF 96.09 (Ph.D. Thesis), Dept. of Comp. Science, Umeå University, Umeå, Sweden, 1996.

2. M. E. Gulliksson and P.-Å Wedin. Algorithms for using the nonlinear L-curve. Technical Report Sumbitted to SIAM J. Optim., Dept. of Comp. Science, Umeå University, Umeå, Sweden, 1999.

3. M. E. Gulliksson and P.-Å Wedin. The nonlinear L-curve. Technical Report Sumbitted to SIAM J. Optim., Dept. of Comp. Science, Umeå University, Umeå, Sweden, 1999.

4. A. Neubauer H. Engl, M. Hanke. *Regularization of Inverse Problems.* Kluwer Academic Publishers, 1996.

5. M. Hanke. Limitations of the L-curve method in ill-posed problems. *BIT*, 36:2:287–301, 1996.

6. P. C. Hansen. Analysis of discrete ill-posed problems by means of the L-curve. *SIAM Review*, 34(4):561–580, 1992.

7. P. C. Hansen and D. P. O'Leary. The use of the L-curve in the regularization of discrete ill-posed problems. SIAM J. Sci. Comput. **14** (1993), 1487–1503.

8. P.C. Hansen. *Rank-Deficient and Discrete Ill-Posed Problems. Numerical aspects of linear inversion.* SIAM, Philadelphia, 1997.

9. R. J. Hanson and C. L. Lawson. *Solving least squares problems.* Prentice Hall, Englewood Cliffs, N. J., 1974.

10. L. Kaufman and A. Neumaier. PET regularization by envelope guided conjugated gradients. *IEEE Trans. Medical Imag.*, 15:385–389, 1996.

11. A. Neumaier. Solving ill-conditioned and singular systems: A tutorial on regularization. *SIAM Review*, 40(3):636–666, 1998.

12. T. Reginska. A regularization parameter in discrete ill-posed problems. *SIAM J. Sci. Comput.*, 17(3):223–228, 1996.

13. C. R. Vogel. Non-convergence of the L-curve regularization parameter selection method. *Inverse Problems*, 12:535–547, 1996.

The PP-TSVD Algorithm for Image Restoration Problems

Per Chr. Hansen, Michael Jacobsen, Jan M. Rasmussen, and Heino Sørensen

Department of Mathematical Modelling, Technical University of Denmark
Building 321, DK-2800 Lyngby, Denmark
Email: pch@imm.dtu.dk
Home page: http://www.imm.dtu.dk/~pch

Abstract. The PP-TSVD algorithm is a regularization algorithm based on the truncated singular value decomposition (TSVD) that computes piecewise polynomial (PP) solutions without any a priori information about the locations of the break points. Here we describe an extension of this algorithm designed for two-dimensional inverse problems based on a Kronecker-product formulation. We illustrate its use in connection with deblurring of digital images with sharp edges, and we discuss its relations to total variation regularization.

1 Introduction

In this work we focus on discretizations of linear inverse problems in the form of square systems $A\,z = b$ or overdetermined least squares systems $\min \|A\,z - b\|_2$. These systems, which we denote *discrete ill-posed problems*, represent a wealth of applications of inverse problems; see, e.g., [6, §1.2].

Our main "tool" from linear algebra for analysis as well as computations is the singular value decomposition (SVD) of the coefficient matrix A. If we assume that A is $m \times n$ with $m \geq n$, then the SVD takes the form

$$A = \sum_{i=1}^{n} u_i\,\sigma_i\,v_i^T, \tag{1}$$

where u_i and v_i are the left and right singular vectors which are orthonormal, and σ_i are the singular values which are nonnegative and appearing in non-increasing order. In terms of the SVD, discrete ill-posed problems are characterized by having a coefficient matrix A whose singular values decay gradually to zero (in practice: until they hit a level determined by the machine precision).

Standard methods for regularization of discrete ill-posed problems are *Tikhonov regularization*

$$\min \left\{ \|A\,z - b\|_2^2 + \lambda^2 \|z\|_2 \right\} \tag{2}$$

and *truncated SVD* (TSVD)

$$\min \|z\|_2 \quad \text{subject to} \quad \|A_k\,z - b\|_2 = \min, \tag{3}$$

where A_k is the TSVD matrix of rank k given by

$$A_k = \sum_{i=1}^{k} u_i \, \sigma_i \, v_i^T. \tag{4}$$

The Tikhonov and TSVD solutions are usually smooth, in the sense that they are continuous and tend to represent minimum energy. By replacing the norm $\|z\|_2$ in the above equations with the seminorm $\|L_p \, z\|_2$, where the matrix L_p is a discrete approximation to the pth derivative operator, we can produce solutions with, say, maximum flatness (for $p = 1$) or minimum roughness (for $p = 2$) — but the solutions remain continuous and smooth.

While smooth solutions are desirable in many applications, they are undesirable in other applications where the solutions are known to have discontinuities or steep gradients. Hence, other regularization algorithms must be used to compute such regularized solutions. One approach is to replace the 2-norm $\|z\|_2$ in Tikhonov's method with the norm $\|L_1 \, z\|_1$, i.e., the 1-norm of the first spatial derivative of the solution. This is called *total variation* (TV) regularization, and it is able to produce solutions with very steep gradients; see [15] for details. Due to the non-differentiability of the 1-norm, specialized optimization algorithms [16] must be used to compute the TV solutions.

A different approach, which is a modification of the TSVD method and which is based solely on tools from linear algebra, was proposed in [8] and [9]. The key idea is to replace the 2-norm $\|z\|_2$ in the TSVD method with the seminorm $\|L_p \, z\|_1$, where L_p is again an approximation to the pth derivative operator:

$$\min \|L_p \, z\|_1 \quad \text{subject to} \quad \|A_k \, z - b\|_2 = \min. \tag{5}$$

The change to the 1-norm has a dramatic effect on the computed solutions. As proved in [9], the solutions to (5) consist of *polynomial pieces*, and the degree of the polynomials is $p - 1$. Moreover, the number of break points is at most $k - p$, where k is the TSVD truncation factor in (4). The method is called the *PP-TSVD method*, and it has been used, e.g., in helioseismology [3] and inversion of gravity and magnetic data [12].

At this point we emphasize that we do not see our PP-TSVD algorithm as an "competitor" to existing algorithms for TV regularization. The PP-TSVD solutions have their own unique features, namely, being samples of piecewise polynomials, which may be useful in certain applications.

The original PP-TSVD algorithm was specifically developed for 1-D problems, and the extension to 2-D problems may not be obvious. The purpose of this work is to describe how the PP-TSVD is extended to treat 2-D inverse problems. Along this line, we also discuss important implementation details necessary to make the 2-D version useful for problems of realistic size. First we discuss some general issues in the treatment of discretizations of 2-D inverse problems in §2. Then we describe various discrete techniques for computing the 1-norm of the derivatives of the solution in §3. Next, in §§4–5 we describe the PP-TSVD algorithm and its implementation for two-dimensional problems. Various properties of 1-D and 2-D PP-TSVD solutions are discussed in §6, and finally we present some small image reconstructions in §7.

2 Discretization of 2-D Problems

In order to simplify our presentation, we limit our discussion to *2-D deconvolution problems whose variables separate*, i.e., we work with a 2-D first-kind Fredholm integral equation of the generic form

$$\int_0^1 \int_0^1 \kappa(x - x')\,\omega(y - y')\,f(x', y')\,dx'\,dy' = g(x, y), \qquad (6)$$

where κ and ω are functions. An example of such a problem is image deblurring with a Gaussian point spread function, for which

$$\kappa(t) = \omega(t) = \frac{1}{\sqrt{2\pi}\sigma}\,\exp\left(-\frac{1}{2}\left(\frac{t}{\sigma}\right)^2\right),$$

and which is used as a model for out-of-focus blur as well as atmospheric turbulence blur [1].

Assume now, also for ease of presentation, that we use the midpoint quadrature rule to discretize the integral equation. Thus, we approximate the integral over y' with a sum of n terms,

$$\int_0^1 \omega(y - y')\,f(x', y')\,dy' \approx n^{-1}\sum_{i=1}^n \omega(y - y'_\ell)\,\tilde{f}(x', y'_\ell) = \phi(x', y),$$

where y'_ℓ are the quadrature points and $\tilde{f}$ represents the approximate solution that we compute. Next, we approximate the integral over x' with another sum,

$$\int_0^1 \kappa(x - x')\,\phi(x', y)\,dx' \approx n^{-1}\sum_{i=1}^n \kappa(x - x'_k)\,\phi(x'_k, y) = \psi(x, y),$$

where x'_k are also quadrature points. Finally, we use collocation in the n^2 points (x_i, y_j),

$$\psi(x_i, y_j) = g(x_i, y_j), \qquad i, j = 1, \ldots, n,$$

which eventually will lead to a system of n^2 linear equations in the n^2 unknowns $\tilde{f}(x_i, y_j)$.

To derive this system we introduce the following four $n \times n$ matrices A, $\bar{A}$, F and G with elements

$$A_{ij} = n^{-1}\kappa(x_i - x'_k), \qquad \bar{A}_{j\ell} = n^{-1}\omega(y_j - y'_\ell)$$

$$F_{k\ell} = \tilde{f}(x'_k, y'_\ell), \qquad G_{ij} = g(x_i, y_j),$$

where all indices are in the range $1, \ldots, n$. Note that A and $\bar{A}$ consist of samples of the functions κ and ω, respectively, while F and G consist of samples of the approximate solution $\tilde{f}$ and the right-hand side g, respectively.

We now define an $n \times n$ matrix Φ that corresponds to the y'-integration, with elements

$$\Phi_{kj} = \phi(x'_k, y_j) = n^{-1} \sum_{\ell=1}^{n} \omega(y_j - y'_\ell)\, \tilde{f}(x'_k, y'_\ell), \qquad j, k = 1, \ldots, n,$$

and by carefully studying the indices of the above expression it follows that Φ can be written as

$$\Phi = F \bar{A}^T.$$

Similarly, we define an $n \times n$ matrix Ψ that corresponds to the x'-integration, with elements

$$\Psi_{ij} = \psi(x_i, y_j) = n^{-1} \sum_{k=1}^{n} \kappa(x_i - x'_k)\, \phi(x'_k, y_j), \qquad i, k = 1, \ldots, n,$$

and this matrix can be written as

$$\Psi = A \Phi = A F \bar{A}^T.$$

Collocation now corresponds to the requirement $\Psi = G$. We have thus shown that the discretization of the 2-D deconvolution problem leads to the linear relation $A F \bar{A}^T = G$ between the discrete solution F and the discrete data G.

Finally, to arrive at a standard system of linear equations, we introduce the Kronecker product and the "vec" notation. The Kronecker product $\bar{A} \otimes A$ of two $n \times n$ matrices A and $\bar{A}$ is defined as the $n^2 \times n^2$ matrix

$$\bar{A} \otimes A = \begin{pmatrix} \bar{a}_{11}A & \bar{a}_{12}A & \cdots & \bar{a}_{1n}A \\ \bar{a}_{21}A & \bar{a}_{22}A & \cdots & \bar{a}_{2n}A \\ \vdots & \vdots & & \vdots \\ \bar{a}_{n1}A & \bar{a}_{n2}A & \cdots & \bar{a}_{nn}A \end{pmatrix}$$

If X is an $n \times n$ matrix with column partitioning $X = (x_1, \ldots, x_n)$, then we define the vector $\mathrm{vec}(X)$ of length n^2 as

$$\mathrm{vec}(X) = \begin{pmatrix} x_1 \\ \vdots \\ x_n \end{pmatrix}.$$

The Kronecker product and the "vec" notation are connected via the following important relation

$$(\bar{A} \otimes A)\, \mathrm{vec}(X) = \mathrm{vec}\left(A X \bar{A}^T\right). \tag{7}$$

Thus we see that the discretized linear system that we derived above can be written in the following two alternative forms

$$A F \bar{A}^T = G \qquad \Longleftrightarrow \qquad (\bar{A} \otimes A)\, \text{vec}(F) = \text{vec}(G). \tag{8}$$

The rightmost form is a conventional system of linear algebraic equations with an $n^2 \times n^2$ structured coefficient matrix. It is by applying the original PP-TSVD algorithm to this problem that we shall derive the 2-D version of the algorithm.

3 Derivative Operators

In this section we derive some useful expressions for discretizations of derivative operators. We start with 1-D problems and a function $f = f(x)$ given in the interval $[0,1]$. Moreover, we assume that the n-vector z consists of equidistant samples of the function f, e.g., $z_i = f(ih)$, $i = 1, \ldots, n$, where $h = 1/n$ is the grid spacing. Then approximations to discretizations of the first and second derivative f' and f'' are given by $h^{-1} L_1 z$ and $h^{-2} L_2 z$, respectively, where the two matrices L_1 and L_2 are given by

$$L_1 = \begin{pmatrix} -1 & 1 & & \\ & \ddots & \ddots & \\ & & -1 & 1 \end{pmatrix}, \qquad L_2 = \begin{pmatrix} 1 & -2 & 1 & & \\ & \ddots & \ddots & \ddots & \\ & & 1 & -2 & 1 \end{pmatrix}, \tag{9}$$

and where L_1 is $(n-1) \times n$ and L_2 is $(n-2) \times n$. These definitions of L_1 and L_2 ensure that the two matrices have nontrivial null spaces whose basis vectors are exact discretizations of the functions that span the null spaces of the derivative operators:

$$L_1 \begin{pmatrix} 1 \\ \vdots \\ 1 \end{pmatrix} = 0, \qquad L_2 \begin{pmatrix} 1 \\ \vdots \\ 1 \end{pmatrix} = 0, \qquad L_2 \begin{pmatrix} 1 \\ \vdots \\ n \end{pmatrix} = 0.$$

Then it follows that a numerical approximation to the 1-norm of f can be computed as

$$\|f\|_1 = \int_0^1 |f(x)|\, dx \simeq h \|z\|_1$$

while numerical approximations to the 1-norm of the derivatives f' and f'' are given by

$$\|f'\|_1 \simeq \|L_1 z\|_1, \qquad \|f''\|_1 \simeq h^{-1} \|L_2 z\|_1.$$

For 2-D problems, we assume that the matrix F consists of samples of an underlying function $f = f(x, y)$ in the domain $[0,1] \times [0,1]$, sampled on a regular $n \times n$ mesh with grid spacing $h = 1/n$ in both directions. Using the matrices L_1 and L_2 defined in (9) we then obtain the following numerical approximation to the 1-norm of f:

$$\|f\|_1 = \int_0^1 \int_0^1 |f(x, y)|\, dx\, dy \simeq h^2 \sum_{i=1}^n \sum_{j=1}^n |f_{ij}| = h^2 \|\text{vec}(F)\|_1$$

while numerical approximations to the 1-norms of the first and second partial derivatives are given by

$$\left\|\frac{\partial f}{\partial x}\right\|_1 \simeq h\|\mathrm{vec}(F\, L_1^T)\|_1 = h\|(L_1 \otimes I)\mathrm{vec}(F)\|_1$$

$$\left\|\frac{\partial f}{\partial y}\right\|_1 \simeq h\|\mathrm{vec}(L_1\, F)\|_1 = h\|(I \otimes L_1)\mathrm{vec}(F)\|_1$$

$$\left\|\frac{\partial^2 f}{\partial x^2}\right\|_1 \simeq \|\mathrm{vec}(F\, L_2^T)\|_1 = \|(L_2 \otimes I)\mathrm{vec}(F)\|_1$$

$$\left\|\frac{\partial^2 f}{\partial y^2}\right\|_1 \simeq \|\mathrm{vec}(L_2\, F)\|_1 = \|(I \otimes L_2)\mathrm{vec}(F)\|_1$$

$$\left\|\frac{\partial^2 f}{\partial x \partial y}\right\|_1 \simeq \|\mathrm{vec}(L_1\, F\, L_1^T)\|_1 = \|(L_1 \otimes L_1)\mathrm{vec}(F)\|_1.$$

Here and throughout the manuscript, I denotes the $n \times n$ identity matrix. Another useful result concerns the sum of 1-norms:

$$\left\|\frac{\partial f}{\partial x}\right\|_1 + \left\|\frac{\partial f}{\partial y}\right\|_1 \simeq h\left\|\begin{pmatrix} L_1 \otimes I \\ I \otimes L_1 \end{pmatrix} \mathrm{vec}(F)\right\|_1 \tag{10}$$

$$\left\|\frac{\partial^2 f}{\partial x^2}\right\|_1 + \left\|\frac{\partial^2 f}{\partial y^2}\right\|_1 \simeq \left\|\begin{pmatrix} L_2 \otimes I \\ I \otimes L_2 \end{pmatrix} \mathrm{vec}(F)\right\|_1. \tag{11}$$

These relations follow immediately from the definition of the vector 1-norm.

The matrices L_1 and L_2 can also be used to compute other quantities that involve derivatives. For example, to approximate $\nabla^2 f = \frac{\partial^2 f}{\partial x^2} + \frac{\partial^2 f}{\partial y^2}$ we can first compute approximations $L_2\, F$ and $F\, L_2^T$ to the second partial derivatives. However, we cannot immediately add these two matrices because their dimensions are incompatible (they are $(n-2) \times n$ and $n \times (n-2)$, respectively). One solution is to "peel off" the first and last columns of $L_2\, F$ and the first and last rows of $F\, L_2^T$, and then add the two $(n-2) \times (n-2)$ matrices.[1]

Another useful quantity that we can approximate is the total variation (TV) of f, defined as

$$J_{\mathrm{TV}}(f) = \int_0^1 \int_0^1 \left(\left(\frac{\partial f}{\partial x}\right)^2 + \left(\frac{\partial f}{\partial y}\right)^2 \right)^{1/2} dx\, dy. \tag{12}$$

The quantity $\left((\partial f/\partial x)^2 + (\partial f/\partial y)^2 \right)^{1/2}$ is known as the gradient magnitude in the image processing literature [14, §4.3.2]. The TV functional gives information about the discontinuities in the image. For example, if f takes on a constant value in a region Ω_0 and takes on another constant value on the complementary of Ω_0,

---

[1] The resulting matrix is identical, except for a scaling factor, to the inner points in the matrix computed by means of Matlab's del2 function.

then $J_{\text{TV}}(f)$ is the length of the boundary of Ω_0 multiplied by the magnitude of the jump [15]. Various aspects of the TV function and TV regularization are discussed in [4] and [15]. To approximate $J_{\text{TV}}(f)$, define the $(n-1) \times n$ matrix

$$M = \frac{1}{2}|L_1| = \frac{1}{2}\begin{pmatrix} 1 & 1 & & \\ & \ddots & \ddots & \\ & & 1 & 1 \end{pmatrix}$$

and the two $(n-1) \times (n-1)$ matrices

$$A_x = M\,F\,L_1^T, \qquad A_y = L_1\,F\,M^T.$$

Then we have found experimentally that a reasonable numerical approximation to the TV of f is given by the following sum of n^2 2-norms:

$$J_{\text{TV}}(F) = \sum_{i=1}^{n}\sum_{j=1}^{n}\left((A_x)_{ij}^2 + (A_y)_{ij}^2\right)^{1/2}.$$

We note that some authors choose to approximate $J_{\text{TV}}(f)$ by

$$\tilde{J}_{\text{TV}}(f) = \int_0^1\int_0^1\left(\left|\frac{\partial f}{\partial x}\right| + \left|\frac{\partial f}{\partial y}\right|\right)dx\,dy = \left\|\frac{\partial f}{\partial x}\right\|_1 + \left\|\frac{\partial f}{\partial y}\right\|_1 \tag{13}$$

which is linear in the two derivatives and which, in turn, is approximated numerically by the quantity in Eq. (10).

4   2-D PP-TSVD Regularization Algorithms

Having realized that discretized 2-D problems have the same general structure as 1-D problems, cf. (8), it is obvious that we can derive 2-D versions of all the 1-D regularization algorithms simply by replacing the coefficient matrix with the Kronecker product $\bar{A} \otimes A$, the right-hand side with $\text{vec}(G)$, and the solution with $\text{vec}(F)$. Moreover, in the constraints, we replace the solution's seminorm with one of the functionals defined in the previous section. Of particular interest here are the functionals that can be written in terms of a matrix times $\text{vec}(F)$. This excludes the TV function, but not the approximation in (10).

Thus, we define the 2-D version of the PP-TSVD method as follows:

$$\min \|\mathcal{L}\,\text{vec}(F)\|_1 \qquad \text{subject to} \qquad \|\mathcal{A}_k\,\text{vec}(F) - \text{vec}(G)\|_2 = \min, \tag{14}$$

where the matrix $\mathcal{A}_k$ is the truncated SVD of the Kronecker product $\bar{A} \otimes A$, and the matrix $\mathcal{L}$ is one of the matrices

$$I \otimes I, \qquad \begin{pmatrix} L_p \otimes I \\ I \otimes L_p \end{pmatrix}, \qquad p = 1, 2, \ldots. \tag{15}$$

To obtain more insight into the 2-D PP-TSVD method, we use the fact that the SVD of the Kronecker product $\bar{A} \otimes A$ can be conveniently expressed in terms of the SVDs of the two matrices. Assume that the SVDs of A and $\bar{A}$ are given by

$$A = U \, \Sigma \, V^T = \sum_{i=1}^{n} u_i \, \sigma_i \, v_i^T, \qquad \bar{A} = \overline{U} \, \overline{\Sigma} \, \overline{V}^T = \sum_{i=1}^{n} \overline{u}_i \, \overline{\sigma}_i \, \overline{v}_i^T.$$

Then it follows from the properties of Kronecker products that

$$\bar{A} \otimes A = \left(\overline{U} \otimes U \right) \left(\overline{\Sigma} \otimes \Sigma \right) \left(\overline{V} \otimes V \right)^T$$

$$= \sum_{i=1}^{n} \sum_{i=1}^{n} \left(\overline{u}_i \otimes u_i \right) \left(\overline{\sigma}_i \, \sigma_i \right) \left(\overline{v}_i \otimes v_i \right)^T, \tag{16}$$

which, except for the ordering, constitutes the SVD of $\bar{A} \otimes A$. Hence, the matrix $\mathcal{A}_k$ is given by

$$\mathcal{A}_k = \sum_{(i,j) \in \mathcal{K}} \left(\overline{u}_i \otimes u_j \right) \left(\overline{\sigma}_i \, \sigma_j \right) \left(\overline{v}_i \otimes v_j \right)^T,$$

where $\mathcal{K}$ denotes the set of indices (i, j) corresponding to the largest k values of $\overline{\sigma}_i \, \sigma_j$.

We can also use expression (16) to define a slightly different 2-D PP-TSVD method, namely, by replacing $\mathcal{A}_k$ with the matrix

$$\widehat{\mathcal{A}}_\ell = \sum_{i=1}^{\ell} \sum_{j=1}^{\ell} \left(\overline{u}_i \otimes u_j \right) \left(\overline{\sigma}_i \, \sigma_j \right) \left(\overline{v}_i \otimes v_j \right)^T = \bar{A}_\ell \otimes A_\ell,$$

obtained by including the largest ℓ singular values of A and $\bar{A}$. Using the relation $\|\text{vec}(F)\|_2 = \|F\|_F$, this variation of the PP-TSVD method takes the form

$$\min \|\mathcal{L} \, \text{vec}(F)\|_1 \qquad \text{subject to} \qquad \|A_\ell \, F \, \bar{A}_\ell^T - G\|_F = \min. \tag{17}$$

Both versions of the 2-D PP-TSVD method yield regularized solutions; they differ in the way that they select which SVD components to be included in the solution.

5 Implementation Issues

The original 1-D version of the PP-TSVD algorithm was based on the fact that the PP-TSVD solution $z_{L,k}$ can be written as

$$z_{L,k} = z_k - V_k^o w_k, \qquad z_k = \sum_{i=1}^{k} \frac{u_i^T b}{\sigma_i} v_i, \tag{18}$$

where z_k is the TSVD solution, i.e. the solution to (3). Moreover, the matrix $V_k^o = (v_{k+1}, \ldots, v_n)$ consists of the last $n - k$ right singular vectors, and the vector w_k is the solution to the linear ℓ_1-problem

$$\min \|(L \, V_k^o) \, w - (L \, z_k)\|_1. \tag{19}$$

This algorithm thus requires two main computations: the *full SVD* of the matrix A and the solution to the unconstrained ℓ_1-problem. Note that this formulation gives insight into the PP-TSVD solution: its first k SVD components are identical to the TSVD components while its last $n - k$ SVD components, which lie in the null space of A_k, are chosen so as to minimize the 1-norm of the vector $L z$.

Although convenient for small- and medium-size problems, the original algorithm is not suited for large-scale problems where it is inconvenient, or impossible, to compute the full SVD — even if the Kronecker product formulation is used. Large-scale SVD algorithms, such as those based on Lanczos bidiagonalization, compute only (approximations to) the principal SVD components. Thus, we need an algorithm that avoids the explicit use of the matrix V_k^o.

One approach is to replace the unconstrained ℓ_1-problem with a related linearly constrained ℓ_1-problem. Let $V_k = (v_1, \ldots, v_k)$ consist of the first k right singular vectors, and rewrite the PP-TSVD solution as

$$z_{L,k} = z_k - y_k, \tag{20}$$

where $y_k = V_k^o w_k$ and therefore lies in the range of V_k^o. Hence y_k is orthogonal to the range of V_k, and it follows that y_k can be computed as the solution to the linearly constrained ℓ_1-problem

$$\min \|L y - (L z_k)\|_1 \qquad \text{subject to} \qquad V_k^T y = 0. \tag{21}$$

We have implemented this approach in Matlab in the function pptsvd, which is available via the home page[2] for the REGULARIZATION TOOLS package [5]. The principal SVD components are computed by means of the built-in Matlab function svds (which uses Lanczos bidiagonalization with implicit restarts), and to solve the linearly constrained ℓ_1-problem we implemented the classical algorithm by Barrodale and Roberts [2].

Currently we explore the possibilities for using the Lanczos vectors of the bidiagonalization process without transforming them into approximate SVD vectors. Specifically, assume that the mth iteration of the Lanczos bidiagonalization process provides the three matrices $U^{(m)}$, $B^{(m)}$, and $V^{(m)}$ such that

$$A V^{(m)} = U^{(m)} B^{(m)},$$

where $U^{(m)}$ and $V^{(m)}$ have orthonormal columns and $B^{(m)}$ is bidiagonal. Then the iterative LSQR algorithm [13], which is based on Lanczos bidiagonalization, computes a solution which is formally given by

$$\tilde{z}_m = V^{(m)} \left(B^{(m)} \right)^{\dagger} \left(U^{(m)} \right)^T b,$$

and which can be considered as a (rough) approximation to a TSVD solution z_k for some k [6, §6.4]. As discussed by Kilmer and Stewart [11] we can expect that $\tilde{z}_m$ contains more than m SVD components, i.e., $\tilde{z}_m$ approximates a TSVD

---

[2] The URL is http://www.imm.dtu.dk/~pch/Regutools/regutools.html.

180

solution z_k with $k > m$. It is worth exploring the possibility of replacing z_k and V_k in (20) and (21) with $\tilde{z}_m$ and $V^{(m)}$.

The main difficulty with the approach in (21) is that the coefficient matrix V_k^T in the linear equality constraints is a large *dense* matrix (while the matrix L is very sparse and hence not a problem). This limits the size of the problems tha can be treated with the current method, independently of the particular algorithm used to solve the ℓ_1-problem.

6   Properties of the 2-D PP-TSVD Solutions

For 1-D problems it is easy to quantify the appearance of the PP-TSVD solutions $z_{L,k}$, see [9, §2.2] for details. A key observation is that the vector $L_p \, z_{L,k}$ is precisely the residual vector in the linear ℓ_1-problem (19), and this residual vector will have many zero elements. The maximum number of nonzero elements equals the difference between the row and column dimensions of the coefficient matrix in (19). Hence, if the $(n - p) \times n$ matrix L_p is an approximation to the $p$the derivative operator and if $k > p$, then $L_p \, z_{L,k}$ will have at most $k - p$ nonzero elements. And since $z_{L,k}$ can be considered as samples of the pth integral of the function represented by $L_p \, z_{L,k}$, it follows that $z_{L,k}$ itself represents a piecewise polynomial of degree $p - 1$ with at most $k - p$ break points. If $k < p$ then the ℓ_1-problem is underdetermined and $z_{L,k}$ is identical to the MTSVD solution [10] which is known to be continuous and smooth.

Consider now the 2-D problems and assume that $\mathcal{L}$ is one of the matrices in Eq. (15) with L_p being an approximation to the pth derivative operator. Then the coefficient matrix $\mathcal{L} \, V_k^o$ in (19) is either $n^2 \times (n^2 - k)$ for $p = 0$, or $2n(n - p) \times (n^2 - k)$ for $p > 0$, ensuring that the ℓ_1-problem is overdetermined, and the number of nonzero entries in the residual vector $\mathcal{L} \, z_{\mathcal{L},k}$ corresponding to the PP-TSVD solution $z_{\mathcal{L},k}$ is at most k for $p = 0$, and at most $n(n - 2p) + k$ for $p > 0$. Our experience is that the number of nonzeros in the ℓ_1-residual vector is much smaller.

In order to illustrate this, consider the case $p = 1$ and $\mathcal{L}$ given by

$$\mathcal{L} = \begin{pmatrix} L_1 \otimes I \\ I \otimes L_1 \end{pmatrix} \tag{22}$$

which leads to regularized solutions that minimize the *approximate* TV functional $\tilde{J}_{TV}(F)$, cf. (10) and (13). Clearly, horizontal or vertical edges in F lead to smaller values of

$$\|\mathcal{L} \operatorname{vec}(F)\|_1 = \|\operatorname{vec}(L_1 \, F)\|_1 + \|\operatorname{vec}(F \, L_1^T)\|_1$$

than skew edges; for example, if

$$F = \operatorname{triu}(\operatorname{ones}(n))$$

then $\|\operatorname{vec}(L_1 \, F)\|_1 = \|\operatorname{vec}(F \, L_1^T)\|_1 = n - 1$, while if

$$F = [\operatorname{zeros}(n, n/2), \operatorname{ones}(n, n/2)]$$

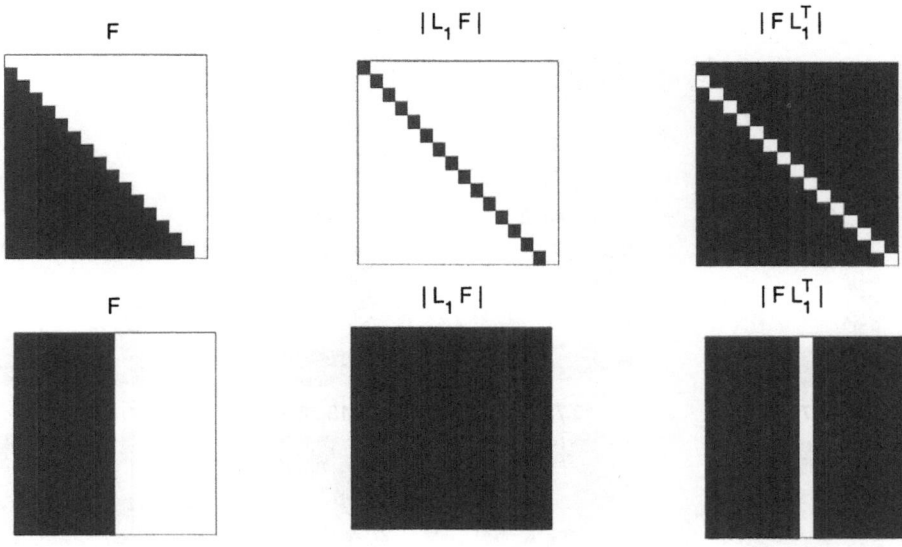

Fig. 1. A skew edge gives a higher value of $\|\mathcal{L}\operatorname{vec}(F)\|_1$ than a vertical (or horizontal) edge.

then $\|\operatorname{vec}(L_1 F)\|_1 = 0$ and $\|\operatorname{vec}(F L_1^T)\|_1 = n$; see Fig. 1. Hence, $\|\mathcal{L}\operatorname{vec}(F)\|_1$ will be made small by constructing F as a blocky image consisting of horizontal or vertical rectangles with the same intensity — and such images are precisely the ones that are produced by the 2-D PP-TSVD algorithm with regularization term $\|\mathcal{L}\operatorname{vec}(F)\|_1$. And since $\|\mathcal{L}\operatorname{vec}(F)\|_1$ can be considered as an approximation to the TV of the image, we see that with this regularization term the PP-TSVD produces regularized solutions which are related to the TV regularized images (which are also found to be blocky, cf. [4]).

To take the illustration further, consider an 8×8 image F consisting of a white 3×4 rectangle on a black background, with $\|\mathcal{L}\operatorname{vec}(F)\|_1 = 14$. Now add Gaussian blurring to F and compute regularized images F_k by means of the 2-D PP-TSVD algorithm with $k = 1, \ldots, 16$. These 16 solutions are shown in Fig. 2 along with the corresponding values of $\|\mathcal{L}\operatorname{vec}(F_k)\|_1$. We see that as k increases, the image becomes more complex and $\|\mathcal{L}\operatorname{vec}(F_k)\|_1$ increases until the original image F is recovered for $k = 16$. We stress that no noise was added in this experiment — the purpose is solely to illustrate the reconstruction capabilities of the 2-D PP-TSVD algorithms with $\mathcal{L}$ given by (22).

It is interesting to inspect the 2-D PP-TSVD solutions more closely for various values of p, in order to understand how "piecewise polynomial" should be interpreted in two dimensions. Figure 3 shows surface plots of four 16×16 reg-

182

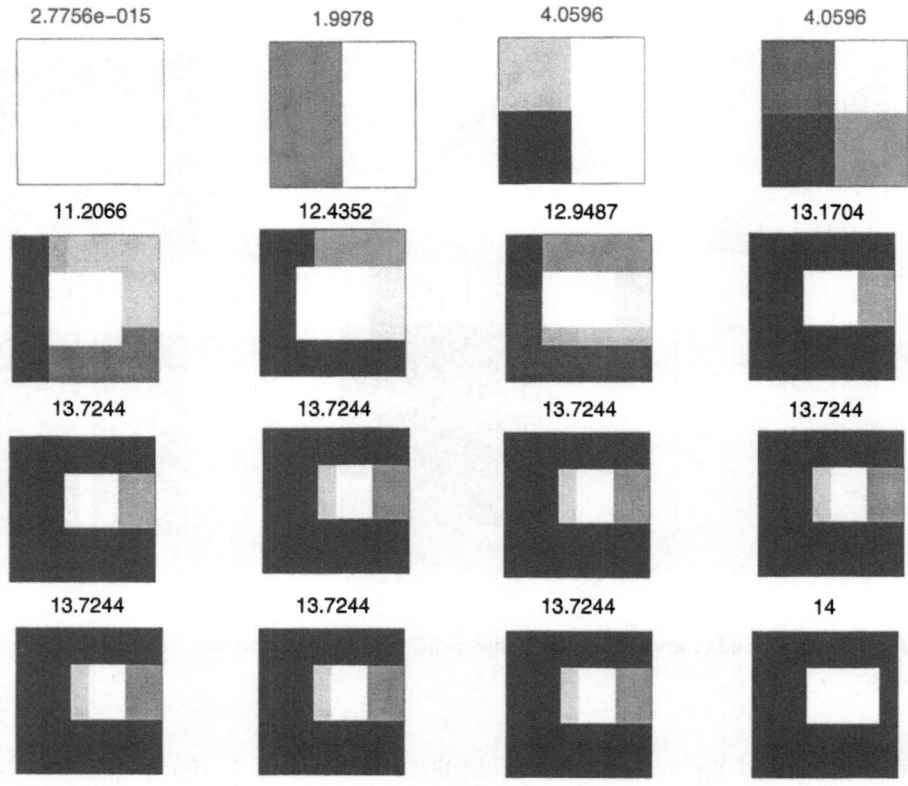

Fig. 2. PP-TSVD reconstructions F_k for $k = 1, \ldots, 16$. The numbers on top of each image are the corresponding values of $\|\mathcal{L}\,\mathrm{vec}(F_k)\|_1$ with $\mathcal{L}$ given by (22).

Matrix $\mathcal{L}$	Characterization of solutions
$I \otimes I$	Delta functions
$\begin{pmatrix} L_1 \otimes I \\ I \otimes L_1 \end{pmatrix}$	Piecewise constant functions
$\begin{pmatrix} L_2 \otimes I \\ I \otimes L_2 \end{pmatrix}$	Piecewise linear functions
$\begin{pmatrix} L_3 \otimes I \\ I \otimes L_3 \end{pmatrix}$	Piecewise quadratic functions

Table 1. The four choices of $\mathcal{L}$ used to illustrate 2-D PP-TSVD solutions.

183

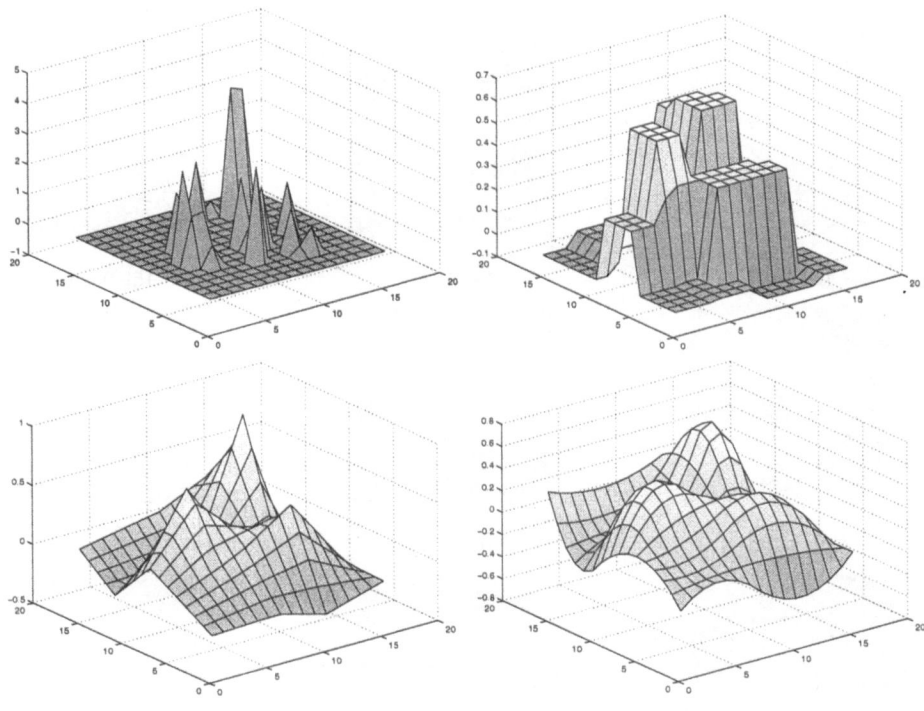

Fig. 3. Regularized 2-D PP-TSVD solutions for the four choices of $\mathcal{L}$ in Table 1.

ularized solutions F corresponding to the four choices of $\mathcal{L}$ listed in Table 1. We see that for $\mathcal{L}$ equal to the $n^2 \times n^2$ identity matrix, the regularized solution F consists of delta functions, and for $\mathcal{L} = L_i \otimes L_i$, $i = 1, 2, 3$ each row or column of F is a piecewise polynomial of degree $i - 1$.

7 Image Reconstructions

We conclude with a few illustrations of the reconstruction capabilities of the 2-D PP-TSVD algorithm. Our test problem is the image deblurring problem blur included in Version 3 of REGULARIZATION TOOLS [7], and the image size is 16×16 (which is similar to the test problem used in [6, §7.7.2]). The condition number of the matrix $\bar{A} \otimes A$ less than 100 (i.e., the system is well conditioned). The original image and the noisy blurred image are shown in Fig. 5.

First we use a noise-free test problem to demonstrate that the 2-D PP-TSVD algorithm is indeed capable of reconstructing images with sharp edges — in contrast to the classical regularization algorithms that produce smooth solutions. Figure 4 shows PP-TSVD reconstructions for $\mathcal{L}$ given by (22) and for $k = 10, 20, \ldots, 90$. We see that $\|\mathcal{L} \operatorname{vec}(F_k)\|_1$ increases monotonically with k until

184

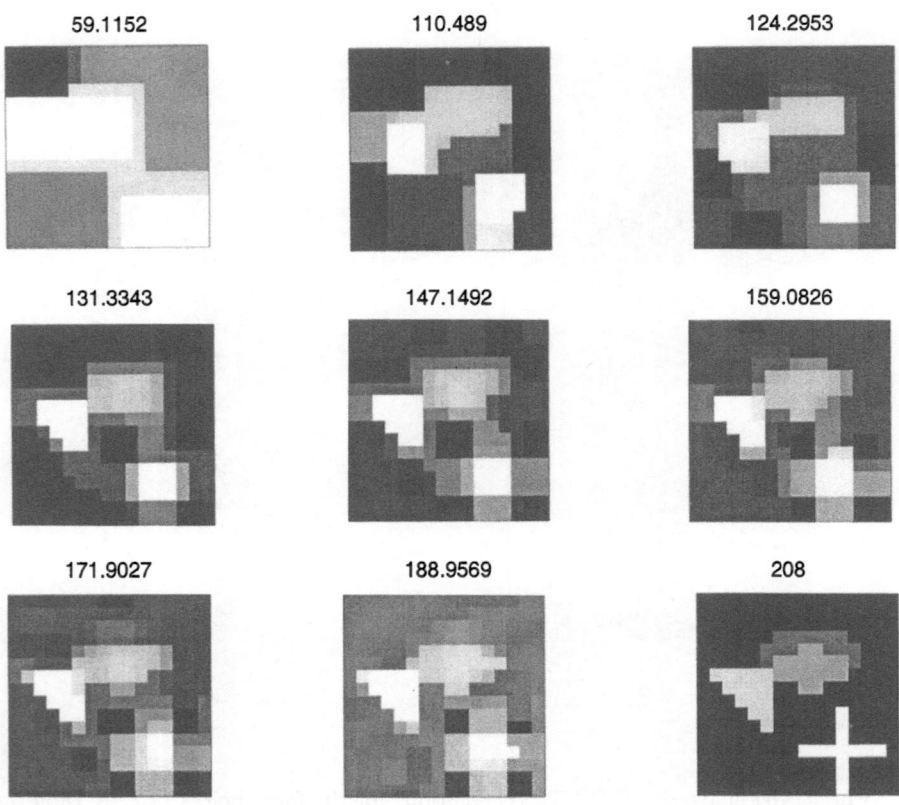

Fig. 4. Reconstructed 16×16 images F_k in the noise-free test problem for $k = 10, 20, \ldots, 90$. The numbers appearing above the plots are the quantities $\|\mathcal{L} \operatorname{vec}(F_k)\|_1$. Perfect reconstruction is achieved for $k = 90$.

we achieve perfect reconstruction for $k = 90$. All the reconstructions are blocky and the number of image blocks increases with k.

We now add Gaussian noise E to the blurred image G, and we choose a noise level such that the blurred signal-to-noise ratio is $\|G\|_F/\|E\|_F = 10$. Figure 5 shows the "optimal" PP-TSVD solution F_k (i.e., the one that minimizes $\|F^{\text{exact}} - F_k\|_F$), which is achieved for $k = 110$, along with the optimal TSVD solution achieved for $k = 180$. Clearly, the PP-TSVD algorithm is much better at reconstructing the sharp edges in the image than the TSVD algorithm.

The choice of the regularization parameter k is a complicated matter that lies outside the scope of this work.

Original image

Blurred noisy image

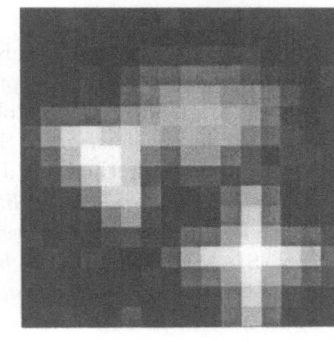

PP–TSVD, k = 110

TSVD, k = 180

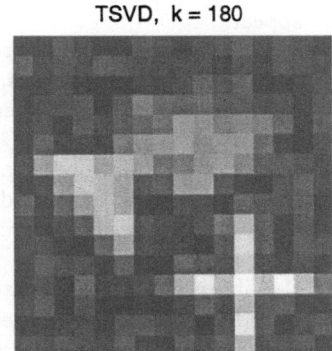

Fig. 5. Top: the original image F and the noisy blurred image $G + E$, with additive Gaussian noise E and blurred signal-to-noise ratio $\|G\|_{\mathrm{F}}/\|E\|_{\mathrm{F}} = 10$. Bottom: the optimal PP-TSVD and TSVD solutions to $A\,F\,\bar{A}^T = G + E$.

8 Conclusion

We have discussed the implementation of the PP-TSVD algorithm for 2-D problems, and we have characterized the 2-D PP-TSVD solutions as blocky images representing reconstructed functions $\widetilde{f}(x, y)$ in which both $\widetilde{f}(x_0, y)$ and $\widetilde{f}(x, y_0)$, with x_0 and y_0 being constant, are piecewise polynomials.

The main bottleneck of the current algorithm is the large dense system that constitutes the linear equality constraints in the ℓ_1-problem. We are currently seeking ways to overcome this problem.

References

1. M. R. Banham and A. K. Katsaggelos, *Digital image restoration*, IEEE Signal Proc. Magazine, 14 (1997), pp. 24–41.
2. I. Barrodale and F. D. K. Roberts, *An efficient algorithm for discrete approximation with linear constraints*, SIAM J. Numer. Anal., 15 (1978), pp. 603–611.
3. T. Corbard, G. Berthomieu, J. Provost, and P. Morel, *Inferring the equatorial solar tachocline from frequency splittings*, Astron. Astrophys., 330 (1998), pp. 1149–1159.
4. D. C. Dobson and F. Santosa, *Recovery of blocky images from noisy and blurred data*, SIAM J. Appl. Math., 56 (1996), pp. 1181–1198.
5. P. C. Hansen, *Regularizations Tools: A Matlab package for analysis and solution of discrete ill-posed problems*, Numer. Algo., 6 (1994), pp. 1–35.
6. P. C. Hansen, *Rank-Deficient and Discrete Ill-Posed Problems. Numerical Aspects of Linear Inversion*, SIAM, Philadelphia, 1998.
7. P. C. Hansen, *Regularization Tools, Version 3.0 for Matlab 5.2*, Numer. Algo., to appear.
8. P. C. Hansen and K. Mosegaard, *Piecewise polynomial solutions to linear inverse problems;* in B. H. Jacobsen, K. Mosegaard, and P. Sibani (Eds.), *Inverse Methods*, Lecture Notes in Earth Sciences 63, Springer, Berlin, 1996.
9. P. C. Hansen and K. Mosegaard, *Piecewise polynomial solutions without a priori break points*, Num. Lin. Alg. Appl., 3 (1996), pp. 513–524.
10. P. C. Hansen, T. Sekii, and H. Shibahashi, *The modified truncated SVD method for regularization in general form*, SIAM J. Sci. Stat. Comput., 13 (1992), pp. 1142–1150.
11. M. Kilmer and G. W. Stewart, *Iterative regularization and MINRES*, Report CMSC TR-3949, Dept. of Computer Science, Univ. of Maryland, 1999.
12. T. K. Nielsen, *Linear inversion of gravity and magnetic data using truncated SVD and piecewise polynomial TSVD;* in B. H. Jacobsen (Ed.), *Proc. Fifth Interdisciplinary Inversion Workshop*, Aarhus, 1997.
13. C. C. Paige and M. A. Saunders, *LSQR: An algorithm for sparse linear equations and sparse least squares*, ACM Trans. Math. Software, 8 (1982), pp. 195–209.
14. M. Sonka, V. Hlavac, and R. Boyle, *Image Processing, Analysis and Machine Vision*, Chapman & Hall, London, 1993.
15. C. R. Vogel, *Nonsmooth regularization;* in H. W. Engl, A. K. Louis, and W. Rundell (Eds.), *Inverse Problems in Geophysical Applications*, SIAM, Philadelphia, 1995.
16. C. R. Vogel and M. E. Oman, *Iterative methods for total variation denoising*, SIAM J. Sci. Comput., 17 (1996), pp. 227–238.

Smooth versus Sharp Frechet Kernels in Time-Distance Helioseismology

J.M. Jensen*, B.H. Jacobsen* and J. Christensen-Dalsgaard+

* Department of Earth sciences, University of Aarhus, Denmark.
+ Institute of Physics and Astronomy, University of Aarhus, Denmark
and Theoretical Astrophysics Center, Danish National Research Foundation

Abstract. Seismic tomography in geophysics has traditionally computed Fréchet kernels from the ray paths obtained in the ray approximation. This yields kernels that are constant along the ray path and zero elsewhere. Finite-wavelength effects imply that actual Fréchet kernels are smoother than the implicit deltaness of the ray kernels. One way to account for the finite-wavelength effects is to make use of Fréchet kernels based on the first Fresnel zone (Snieder and Lomax 1996).

In helioseismic tomography the finite-wavelength problem is aggravated because the ratio of wavelength to ray length is smaller in the Sun compared to the Earth. Therefore the need to go beyond the sharp ray-based Fréchet kernels is even more present in helioseismology than in 'terrestrial' seismology.

We present Fréchet kernels from the ray approximation and Fresnel-zone-based Fréchet kernels. The different kernels are compared by inverting a synthetic data set. The results show that inversion based on Fresnel zones is typically superior to inversion based on rays.

1. Introduction

In the past decades helioseismology has been very successful in investigating the solar interior using the solar global oscillation modes. Due to the similarity with 'terrestrial' seismology, methods from geophysics have been employed with great effect (Gough 1996). Recently a new approach for investigating the solar interior, based on local wave propagation, has emerged. This new approach has become known as Time-Distance helioseismology (Duvall et al. 1993) and is quite similar to explorational seismology performed on Earth. In Time-Distance helioseismology travel-time data are obtained by cross correlating oscillation patterns on the surface of the Sun, thus producing solar seismograms. From these data it is possible to do a tomographic imaging of local structures in the outer part of the Sun (Kosovichev and Duvall 1997).

We have used an inversion technique based on a Multi Channel Deconvolution (MCD) algorithm (e.g. Jacobsen et al. 1999) to perform the imaging. The MCD technique speeds up the calculations by doing the computational work in the Fourier domain. This means that inversions can be done in a matter of seconds or minutes, even for high-volume data sets. This technique has been developed recently within geophysics for inversion of electromagnetic data. It is based on the Born approximation, and utilises the horizontal translational invariance of the reference model.

The quality of inversion results depends on the accuracy of the Fréchet kernels employed. To test the different Fréchet kernels we computed a synthetic data set. These data were inverted using both conventional ray tomographic Fréchet kernels and smoother Fréchet kernels based on the first Fresnel zone. These results show that the smoother Fréchet kernels seem to perform better in the inversion than the ray-based kernels.

2. Helioseismic Time-Distance Data

The idea of Time-Distance helioseismology is to measure the travel time of sound waves propagating through the solar interior, between different points on the solar surface. These travel times can then be used to perform a tomographic imaging of the Sun's interior, using techniques known from terrestrial seismology. The data used are time series of Dopplergrams, which measure the movement of the Sun's surface. As seismic energy is continuously generated by the solar convection, it is not possible to measure travel times as in terrestrial seismology. Instead the vibration patterns of the surface are compared by calculation of the cross-correlation (Duvall et al. 1997)

$$\Psi(\mathbf{r_1}, \mathbf{r_2}) = \int_0^T f(t, \mathbf{r_1}) f(t - \tau, \mathbf{r_2}) \tag{1}$$

where $\mathbf{r_1}$ and $\mathbf{r_2}$ are position vectors for two points at the surface, T is the total time of observation, τ is the time delay and $f(t,\mathbf{r})$ is the observed oscillation pattern at $\mathbf{r}$. This procedure yields solar seismograms which are very much like terrestrial seismograms. The first arrivals in these seismograms are fitted with a Gabor wavelet. The arrival time of this wavelet is taken to be the travel time for a sound wave propagating from $\mathbf{r_1}$ to $\mathbf{r_2}$. To increase the signal-to-noise ratio the travel times are not calculated between points on the surface, but averaged over larger areas. One approach has been to calculate travel times from a central point to a surrounding annulus, averaging over the annulus. By measuring

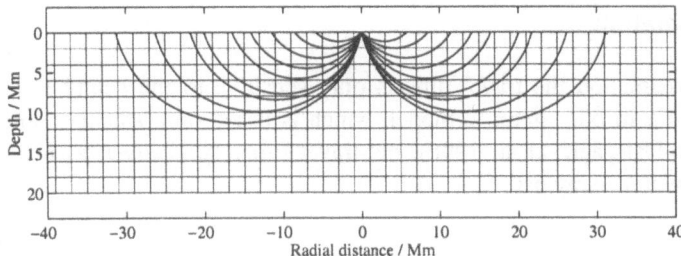

Fig. 1. Delimiting ray paths.

the travel times between the central point and several annuli a resolution with depth can be obtained, as the waves propagating to the different annuli reach different depths.

Figure 1 shows a cross section of ray paths for waves propagating to different annuli. The full ray pattern is rotationally symmetric around the central point, as we are averaging over the full annulus at the surface. The data can for example be averaged over the annuli delimited by the ray paths shown. In this case there would be eight different data annuli.

3. Forward Modeling

The Time-Distance helioseismic forward problem is not at present fully understood. Therefore one has to make use of various approximations to solve the forward problem, and different approximations may yield different results. This work tests two approaches to the forward problem. The first is the well-known and widely used ray approximation and the second is an approximation, based on the first Fresnel zone, suggested by Snieder and Lomax (1996).

The ray approximation or geometrical acoustics, which is valid at high frequencies, solves the wave-propagation problem in terms of rays. This approximation is computationally efficient and thus widely used in seismic tomography. In the case of the Sun we deal with very low frequencies compared to terrestrial seismology, but the ray approximation has still been used (Kosovichev, 1996; Kosovichev and Duvall, 1997). The question of validity in the case of the Sun has been addressed by Bogdan (1997), who showed that a wave packet propagated following the paths given by the ray approximation but sampled a large region of the subsurface surrounding the ray. The Fresnel-zone approximation takes this finite-wavelength effect into account and thus represents a step beyond the ray approximation.

3.1 Ray Approximation

In the ray approximation the travel time is only sensitive to the sound speed along the ray path Γ_i. In our reference model, which is based on a standard solar model (Christensen-Dalsgaard et al. 1996), the travel time, τ_0, is given as a line integral along the reference ray path $\Gamma_{0,i}$:

$$\tau_{0,i} = \int_{\Gamma_{0,i}} \frac{dl}{c_0(\mathbf{x})} \tag{2}$$

where c_0 is the reference sound speed and $\mathbf{x}$ is the position vector. In the Sun we must take differences from the reference model into account; these can be sound-speed perturbations or subsurface flows. In this case the travel time is given as (Munk et al. 1995)

$$\tau_i^{\pm} = \int_{\Gamma_i} \frac{dl}{c(\mathbf{x}) \pm \mathbf{v}(\mathbf{x}) \cdot \mathbf{n}} \tag{3}$$

where $c(\mathbf{x}) = \delta c(\mathbf{x}) + c_0(\mathbf{x})$ is the sound speed and $\delta c(\mathbf{x})$ is the sound-speed perturbation. Furthermore $\mathbf{v}(\mathbf{x})$ is the flow, $\mathbf{n}$ is a unit tangent vector to the ray and $\pm$ denotes waves traveling in opposite directions along the ray path. To linearize the equations we consider travel-time differences between the observed travel times and the travel times obtained from the reference model. As we consider both the flow and sound-speed perturbations to be small compared to the reference sound speed, we can use the ray paths from the reference medium in the perturbed medium, thus ignoring effects of ray bending by the perturbations. The travel-time difference $\delta \tau_i = \tau_i - \tau_{0,i}$ is then given as

$$\delta \tau_i^{\pm} = \tau_i^{\pm} - \tau_{0,i} = -\int_{\Gamma_{0,i}} \frac{\delta c(\mathbf{x}) \pm \mathbf{v}(\mathbf{x}) \cdot \mathbf{n}}{c_0^2} dl \tag{4}$$

By taking the average and difference of these two travel-time differences, we obtain:

$$\delta \tau_i^{mean} = \tfrac{1}{2}\left(\delta \tau_i^+ + \delta \tau_i^-\right) = -\int_{\Gamma_i} \frac{\delta c(\mathbf{x})}{c_0^2(\mathbf{x})} dl \tag{5}$$

$$\delta \tau_i^{diff} = \left(\delta \tau_i^+ - \delta \tau_i^-\right) = -2\int_{\Gamma_i} \frac{\mathbf{v}(\mathbf{x}) \cdot \mathbf{n}}{c_0^2(\mathbf{x})} dl \tag{6}$$

Here we see that the effect of the perturbations has been separated as $\delta \tau_i^{mean}$ is sensitive only to the perturbation in sound speed and $\delta \tau_i^{diff}$ is only sensitive to the flow. Henceforth we will only be concerned about variations in the sound speed, thus we need only work with the mean travel-time difference.

To make the equations simpler and more linear we formulate them in terms of slowness ($s = 1/c$) instead of sound speed. The sound-speed perturbation can then be written as

$$\delta c = -\delta s c_0^2 \tag{7}$$

where δs is the slowness perturbation. The equation for the mean travel-time difference then becomes

$$\delta \tau_i^{mean}(x,y) = \oint_{\Gamma_i} \delta s\, dl \tag{8}$$

where (x,y) is the central point used in the data analysis and the travel-time differences are taken between waves propagating between the central point and the i'th annulus. For use in the numerical treatment we discretized the subsurface in a regular grid with labels m, n and k. An example of such a grid is shown in Figure 1. The mean travel time can then be written as

$$\delta \tau_i^{mean}(x,y) = \sum_m \sum_n \sum_k B_{xyi}^{mnk} \delta s_{mnk} \tag{9}$$

where m, n and k are indices in the x, y and z directions and B_{xyi} is the Fréchet kernel for the waves propagating between the point (x,y) and the i'th annulus.

Since the line integral through a grid cell with constant slowness gives the length of the ray path times the slowness, we find that in the ray approximation the value of the Fréchet kernel in a grid cell is the length of the ray path in the grid cell. An example of a discretized Fréchet kernel is shown in Figure 2. To compute the ray paths we have taken a solar model (Christensen-Dalsgaard et. al. 1996) and assumed that the velocity increases linearly between the discrete layers. The ray path in a layer with constant velocity gradient becomes a circular arc. Gibson et al. (1979) give a convenient expression for the ray path in this case.

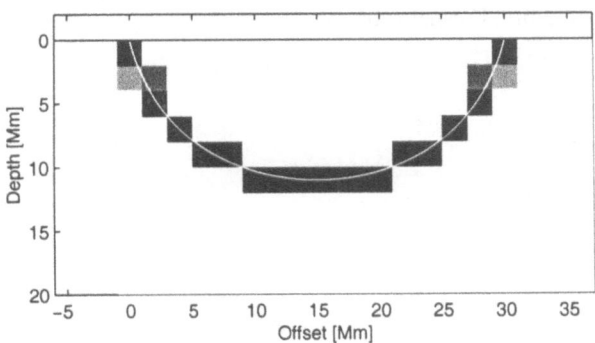

Fig. 2. Ray approximation based Fréchet kernel with corresponding ray path (in white)

3.2 Fresnel-Zone-Based Fréchet Kernels

The first Fresnel zone is the part of the refracting medium through which energy is transmitted from the source to the receiver less than a quarter of a period after the ray arrival; thus this energy interferes more or less constructively. Snieder and Lomax (1996) showed that the phase and amplitude perturbation can be written to first order as weighted averages over the first Fresnel zone. In the high-frequency limit, where the first Fresnel zone collapses to a curve, this corresponds to the ray approximation. The Fresnel-zone approximation, however, is not limited to high frequencies and is valid even if the velocity perturbation is not smooth over the first Fresnel zone, which is a criterion for the validity of the ray approximation.

A point P is taken to be in the first Fresnel zone, if the corresponding travel times satisfy the equation

$$\left| \tau(A, P) + \tau(P, B) - \tau_0 \right| \leq \frac{T}{4} \tag{10}$$

where $\tau(X, Y)$ is the travel time corresponding to the ray Γ_{XY} from point X to point Y (see Figure 3). Here τ_0 is the travel time corresponding to the reference ray path Γ_0 from A to B, and T is the dominant period of the wave.

We have calculated the first Fresnel zone for a distance between the reference points of 30 Mm, and a corresponding prevailing period of 5 minutes (Figure 4). For the ray approximation to be valid the medium has to be smooth over the first Fresnel zone as previously pointed out. This means that the sound-speed perturbations we are looking for must be smooth on a scale of 10 Mm for the ray approximation to be valid. We cannot expect the Sun to obey this degree of smoothness and therefore we have to go beyond the ray approximation to do a tomographic imaging of the solar interior.

As a first step beyond the ray approximation we define an approximate Fréchet kernel delimited by the first Fresnel zone as

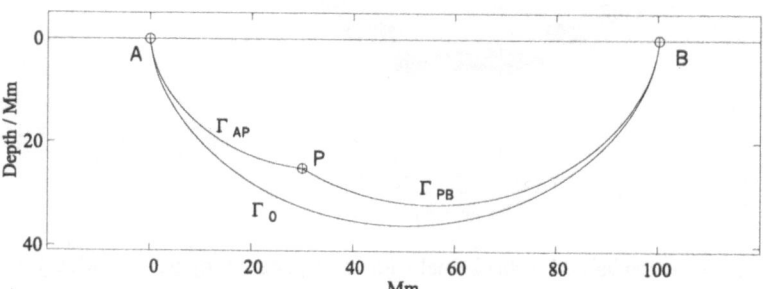

Fig. 3. Ray geometry of Fresnel-zone definition

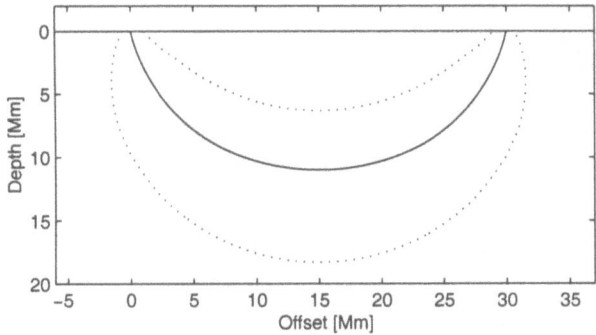

Fig. 4. First Fresnel zone (dotted) and corresponding ray path (solid).

$$S(P) \approx K \cdot amp(A, P) \cdot amp(P, B) \cdot \exp\left(-\left(\frac{\tau(A, P) + \tau(P, B) - \tau_0}{T/4}\right)^2\right) \tag{11}$$

where $S(P)$ is the sensitivity at P, $amp(X, Y)$ is the amplitude of the wave field propagating from X at Y, and K is a normalization constant. The amplitude factors have been approximated by the geometrical spreading in a spherical wave field. The normalization of the kernels is achieved by ensuring that their sum is equal to the length of the reference ray path. A kernel obtained by this procedure is shown in Figure 5 with the reference ray path plotted on top.

4. The MCD Technique

As the data sets involved may be huge (up to 10^6 data and 10^6 modelparameters), conventional inversion schemes cannot be used due to the heavy computations. One option is to use an iterative solver such as a conjugate gradient (LSQR) scheme (Paige and Saunders 1982). The number of iterations in the LSQR scheme plays the role of tradeoff parameter, but the regularizing properties of the LSQR are not yet fully understood (Hansen 1996). The LSQR technique has been used by Kosovichev (1996) to perform Time-Distance inversions. We have used a Fourier-domain-based technique developed recently within geophysics (Li and Oldenburg (1994), Jacobsen (1996) and Jacobsen et al. (1996)) This technique exploits the translational invariance of the problem to write the forward and inverse problems in terms of convolutions. The technique has been called the MultiChannel Deconvolution technique or MCD for short. As the convolutions involved can be evaluated as multiplications in the Fourier domain, we get a tremendous speed up of both the forward and inverse calculations, while still solving the exact inverse problem. This means a better understanding and more flexible control of the regularization imposed on the

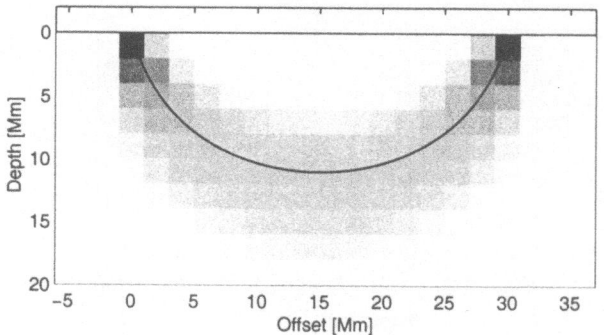

Figure 5: Fresnel-zone-based Fréchet kernel with ray path plotted on top.

inversion. To perform the Fourier transforms one has to make an assumption of periodic boundary conditions. In order to compensate for this assumption the data have been padded in a rim around the investigated area with a constant average value.

The MCD approach transforms the huge linear forward problem into a large number of small linear equations in the Fourier domain. Here I will only schetch the technique; further details can be found in Jensen et al. (1998), Jacobsen et al. (1999) and Møller (this volume). Because of the translational invariance of the forward problem it can be written as

$$\delta\tau_i(x, y) = \sum_m \sum_n \sum_k B_{xyi}^{mnk} \delta s_{mnk} = \sum_k B_{i,k}(x, y) * \delta s_k(x, y) \tag{12}$$

Here B_i is the translationally invariant Fréchet kernel which is independent of the central point (x,y). In the last equality the summations over x and y have been replaced by a 2D convolution denoted by $*$. The convolution can by evaluated very efficiently in the Fourier domain according to the convolution theorem (Bracewell 1965). In the Fourier domain the forward problem becomes a large number of small linear problem. For each pair of wavenumbers we get the following forward relation:

$$\mathbf{d} = \mathbf{G} \cdot \mathbf{m} + \mathbf{e} \tag{13}$$

Here $\mathbf{e}$ denotes the observational error and $\mathbf{d}$, $\mathbf{m}$ and $\mathbf{G}$ contain the Fourier transforms of data, model and Fréchet kernel at a given pair of wavenumbers. More explicitly they are given as

$$\begin{aligned} G_{i,k} &= \tilde{B}_{i,k}\left(\kappa_x, \kappa_y\right) \\ d_i &= \delta\tilde{\tau}_i\left(\kappa_x, \kappa_y\right) \\ m_k &= \delta\tilde{s}_k\left(\kappa_x, \kappa_y\right). \end{aligned} \tag{14}$$

Here the tilde denotes the Fourier-transformed quantities, k indicates the layer and (κ_x, κ_y) are the wave numbers in the x and y directions. The 2D Fourier transforms are done layer by layer and for all data corresponding to the same annulus.

The problem can then be inverted using conventional linear inversion as

$$\mathbf{m}_{est} = \tilde{\mathbf{H}} \cdot \mathbf{d}_{obs} \tag{15}$$

where $\tilde{\mathbf{H}}$ is an inverse operator. The inverse operator can be chosen in many different ways. We have used a Regularized Least Squares (RLS) (or Tikhonov) inverse operator with a depth-dependent regularization. The operator has the form

$$\tilde{\mathbf{H}} = \left(\mathbf{G}^H \mathbf{G} + \varepsilon^2 \mathbf{V}^2\right)^{-1} \mathbf{G}^H \tag{16}$$

Here ε is the tradeoff parameter, $\mathbf{G}^H$ is the conjugate transpose of $\mathbf{G}$ and

$$\underline{\underline{V}} = \begin{bmatrix} 1 & 0 & 0 & \cdots & 0 \\ 0 & \dfrac{s_{0,1}}{s_{0,2}} & 0 & & 0 \\ 0 & 0 & \dfrac{s_{0,1}}{s_{0,3}} & & 0 \\ \vdots & & & \ddots & \vdots \\ 0 & 0 & 0 & \cdots & \dfrac{s_{0,1}}{s_{0,N}} \end{bmatrix} \tag{17}$$

where $s_{0,k}$ is the slowness in the k'th layer of the reference model and N is the number of layers in the model. This inverse operator ensures that the relative sound-speed perturbation $\delta c/c_0$ has the same amplitude in all layers.

The model estimate in the space domain is then found by inverse Fourier transforming $\mathbf{m}_{est}$ layer by layer. This gives the estimated slowness perturbations at all gridpoints. In the space domain the model estimations are given as

$$\delta s_k(x,y) = \sum_l H_{l,k}(x,y) * \delta \tau_l(x,y) \tag{18}$$

This is the MultiChannel Deconvolution solution of the inverse problem.

To obtain the inverse operator used we have assumed that the measurement noise is uncorrelated and have the same variance, σ_e^2, for all data. Furthermore the model parameters are assumed uncorrelated and the relative variance of the sound-speed perturbations, $\sigma^2(\delta s/s_0)$, is assumed to be the same. Because of these assumptions we expect that the optimal choice of the tradeoff parameter should be approximately:

$$\varepsilon^2 = \frac{\sigma_e^2}{\sigma^2\left(\frac{\delta s}{s_0}\right) s_{0,1}^2} \tag{19}$$

5. Synthetic Test

To test the inversion method and the different Fréchet kernels, we calculated 2D synthetic data using a finite-difference forward modeling. A second-order finite difference scheme, as given by Alford et el. (1974), was used and the implementation was taken from the Colorado School of Mines Seismic Unix package (Cohen and Stockton 1998). The synthetic data we obtained from this modeling were then inverted. The synthetic data were first averaged in eight different data intervals according to the pattern shown in Figure 1, and then the mean travel time for each annulus was subtracted to yield the travel time differences. The mean travel-times were subtracted instead of the travel time for the reference model since this corresponds to the procedure used in actual helioseismic Time-Distance data processing (Tom Duvall, pers. comm.). We then added 10 per cent white noise to the data. These data were inverted using both ray-approximation kernels and Fresnel-zone-based Fréchet kernels. Inversions were done with different values of ε, the tradeoff parameter.

To find the 'right' value of the tradeoff parameter, ε, we plotted the L-curves of the inversions. The L-curve is a tool used in connection with the Tikhonov approach to help choose the value of ε. The L-curve is the L_2-norm of the model misfit plotted against the L_2-norm of the property of the model we wish to minimize, which in our case is the slowness perturbation $\delta s V$. The L-curve shows the tradeoff between fitting the data and obtaining a smooth model. The optimal value of ε is supposed to be one that gives a solution that corresponds to the bend of the L-curve (Hansen 1998). The L-curves for the inversions are plotted in Figure 6 a) and c).

Since the true model is known (Figure 7 a) it is possible to compute the model misfit between the estimated model and the true model. This model misfit is plotted as a function of ε in Figure 6 b) and d). We can then choose the solution that has the smallest model misfit. This is the *'minimum model estimation error'* (mmee) solution. These solutions are indicated in both model misfit curves and L-curves with a +.

According to eq. (19) we know from our prior knowledge of model and data error what value to expect for ε. The data error has mean of $\sigma_e \approx 4$ s. The perturbation of the model is around ten per cent, so $\sigma(\delta s/s_0) \approx 0.10$ and the slowness of the upper layer in the reference model is $s_{0,1} \approx 0.098$ s/km. With these values the estimated value of the trade-off parameter becomes $\varepsilon \approx 400$. The solution with this tradeoff is indicated with a o in Figure 6 and will be referred to as the *'prior ε solution'*. For the inversion using Fresnel-zone-based kernels the mmee solution has an ε of 900 and for the ray inversion $\varepsilon \approx 1600$. So our prior guess for ε was within a factor 2-4 of the mmee solution, which shows that our prior understanding of the data errors and model variability is consistent with our results. If it is necessary to vary ε much more than a factor of 10 to obtain a stable

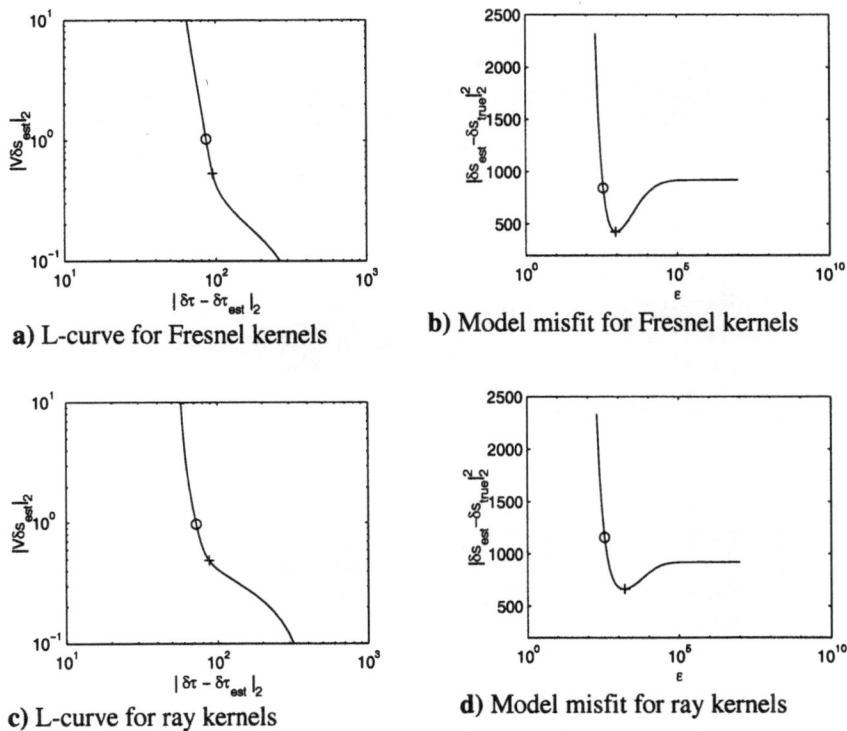

a) L-curve for Fresnel kernels

b) Model misfit for Fresnel kernels

c) L-curve for ray kernels

d) Model misfit for ray kernels

Fig. 6. L-curves and model misfit for different kernels. (+) denotes the mmee solution and (o) is our prior guess for ε (See text for details)

solution our prior knowledge is lacking in some way. In both cases we see that the mmee solution lies closer to the corner of the L-curve than the prior ε solution.

Figure 7 shows the test model and the inversion results. Figure 7 b) and 7 c) show the mmee solutions in the Fresnel- and ray inversions, while 7 d) and 7 e) show the prior ε solutions.

These plots clearly show the effect of the regularization. The mmee solutions which are more strongly regularized have smaller amplitudes, while the less regularized prior ε solutions have higher and more rapidly varying amplitudes. This illustrates the damping effect of the regularization. In all cases we see some evidence of the buried black feature being recovered. The white features are very difficult to find in the prior ε plots, while they are much more visible in the mmee solutions.

There are several differences between the type of solutions we get from the ray inversions and the ones obtained from the Fresnel inversion. First of all the ray inversion cannot recover anything below the turning point of the deepest

198

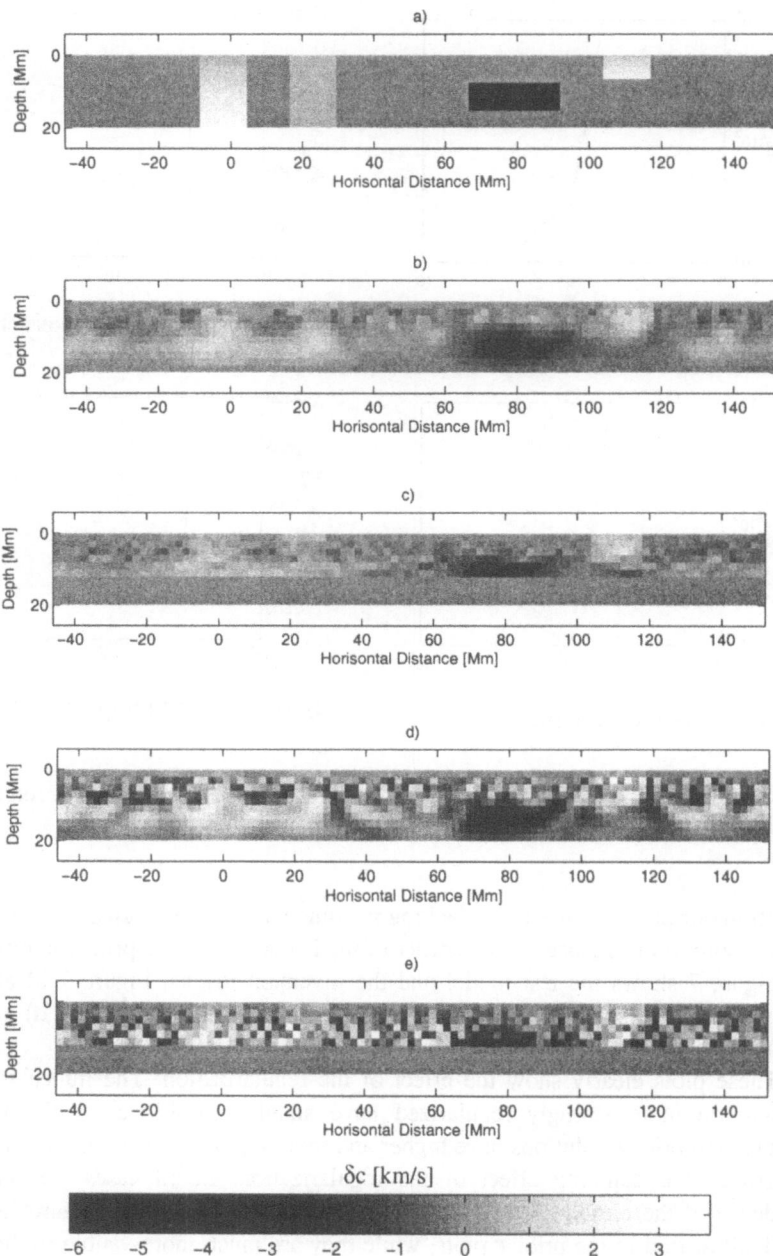

Fig. 7. a) Test model, b) Inversion using Fresnel-zone-based kernels (mmee) c) Inversion using ray kernels (mmee) d) Inversion using Fresnel-zone-based kernels (prior ε) and e) Inversion using ray kernels (prior ε)

penetrating ray, whereas the Fresnel inversions go much deeper. Possibly the bottom of the buried anomaly can be recovered but this is questionable, as everything at this depth appears to be smeared out. It seems that the white anomaly which increases with depth is cut off at this depth as well. This indicates that it is not the bottom of the buried anomaly but a general effect we are seeing. Another difference between the two types of inversion is that the Fresnel inversions seem to be more smeared out and coherent deeper in the model, whereas the ray inversions still vary rapidly. The smearing that occurs in the Fresnel kernel inversions is caused by the fact that when the sound speed increases with depth so does the wavelength, and as the wavelength increases the resolution power decreases. The highly varying model estimates in the lower layers of the ray inversion can be seen as reflecting an artificially increased resolution obtained by ignoring the finite wavelength of the waves.

To investigate the resolution of the inversion a localized sound-speed anomaly (2Mm by 2Mm) was placed and the forward response calculated using the previously mentioned finite-difference method. We then inverted the data to see how well this anomaly could be resolved. Figure 8 shows the results for several depths of the anomaly as grayscale plots. The tradeoff parameters were chosen to correspond to the mmee solutions, from the previous inversions. The grayscale is not the same from plot to plot so the maximum value of the resolution kernels differs even though the color is the same. In the ideal case one would just see a white spike at the location of the anomaly (indicated by + in the plots). Figure 8 a) and e) show that both the Fresnel inversion and ray inversion resolve the upper layers very well. As the anomaly is located deeper the resolution gets worse, which is what could be expected. It is seen that the Fresnel inversion is able to resolve the anomaly in the deepest layer much better than the ray inversion. The ray inversion distorts seriously all anomalies below the turning point of the deepest ray. The resolution kernels of the Fresnel inversion are smooth and localized around the target depth in all figures, although in Figure 8 b) and Figure 8 c) the maximum does not correspond to the anomaly depth. The resolution kernels for the ray inversion are much sharper, but has a strange triangular shape in the deeper cases. In Figure 8 f) we also see some sidelobes to the resolution kernel, and the resolution kernel is not very well localized. In Figure 8 g) the maximum is not at the depth of the anomaly. An interesting feature of the resolution kernels of the deepest case d) and h) is that we see a dark anomaly above the true white anomaly. This is an artifact of the inversion but could be interpreted as a low sound-speed anomaly lying above a fast anomaly in the inversions. This shows how one has to be cautious about interpreting the inversion results.

The resolution kernels show that the Fresnel inversion is able to localize features the ray inversion cannot find due to the limitations of the ray approximation. In general the Fresnel-zone approximation yields resolution

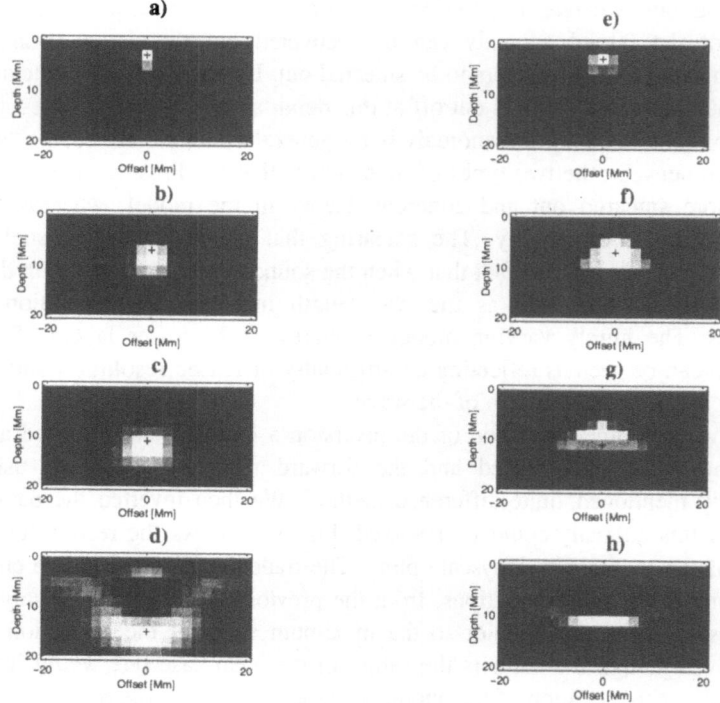

Fig. 8. Resolution kernels for Fresnel-zone-based kernel a)-d) and ray kernels e)-h). The depth of the localized anomaly is marked with a +. (The grayscale differs from plot to plot)

kernels which are at least comparable in localization to those of the ray approximation, even though the Fréchet kernels used in the inversion are much broader.

6. Conclusion

We have presented a first step beyond the ray approximation in helioseismic Time-Distance inversions. The Fréchet kernels used in the inversion were based on the Fresnel-zone approximation suggested by Snieder and Lomax (1996). To be able to perform the huge inverse calculations we have used the MCD technique (Jacobsen et al. 1999.

The methods presented have been tested using a synthic 2D example. The results seem very promising for the use of Fresnel-zone-based Fréchet kernels in helioseismic studies. The Fresnel-zone-based resolution kernels are more localized and less distorted compared to the ray-based resolution kernels.

7. Acknowledgements

The authors wishes to thank the Danish National Science Research Council for its support of the Danish Interdisciplinary Inversion Group. This work was supported in part by the Danish National Research Foundation through its establishment of the Theoretical Astrophysics Center.

8. References

Alford, R.M., Kelly, K.R. and Boore, D.M., 1974, Accuracy of finite-difference modeling of the acoustic wave equation. Geophysics, 39, 834-842.

Bogdan, T.J., 1997, A comment on the relationship between the model and time-distance formulations of local helioseismologi. Astrophys. J., 477, 475-484.

Bracewell, R.N., 1965, The Fourier transform and its applications. McGraw-Hill.

Christensen-Dalsgaard, J. et al, 1996, The current state of solar modelling. Science, 272, 1286-1291.

Cohen, J.K. and Stockton, Jr. J. W., 1998, CWP/SU: Seismic Unix Release 32: a free package for seismic research and processing, Center for Wave Phenomena, Colorado School of Mines.

Duvall Jr, T.L., Jefferies, S.M., Harvey, J.W. and Pomerantz, M.A., 1993, Time-Distance helioseismology. Nature, 362, 430-432.

Duvall Jr., T.L. et al, 1997, Time-Distance helioseismologi with the MDI instrument: initial results. Solar Phys., 170, 63-73.

Gibson, B.S., Odegard, M.E. and Sutton, G.H., 1979, Nonlinear least-squares inversion of traveltime data for a linear velocity-depth relationship. Geophysics, 44, 185-194.

Gough, D.O., 1996, The success story of the transfer and development of methods from geophysics to helioseismology. In B.H. Jacobsen, K. Mosegaard, and P. Sibani (eds.): "Inverse Methods - Interdisciplinary Elements of Methodology, Computing and Applications." Lecture Notes in Earth Sciences, 1-31, Springer-Verlag, Berlin-Heidelberg.

Hansen, P.C., 1998, Rank-Deficient and Discrete Ill-Posed Problems: Numerical Aspects of Linear Inversion, SIAM, Philadelphia.

Jacobsen, B.H., 1996, Rapid 2-D Inversion by Multichannel Deconvolution of Geophysical Profile Data. Proc. Symp. Application of Geophysics to Engineering and Environmental Problems. Keystone, Colorado. 659-668.

Jacobsen, B.H., Møller, I.M., Jensen, J.M. and Effersø, F., 1999, Multichannel deconvolution (MCD) in geophysics and helioseismology. Physics and Chemistry of the Earth, 24, 215-220.

Jensen, J.M., Jacobsen, B.H. and Christensen-Dalsgaard, J., 1998. MCD inverson for sound speed using time-distance data. In "Structure and dynamics of the interior of the Sun and Sun-like stars"; Proc. SOHO 6/GONG 98 Workshop, eds S.G. Korzennik and A. Wilson, ESA SP-418, ESA Publications Division, Noordwijk, The Netherlands, 635 - 640.

Kosovichev, A.G., 1996, Tomographic Imaging of the Sun's Interior. Astrophys. J., 462, L55-L57.

Kosovichev, A.G., and Duvall JR, T.L., 1997, Acoustic tomography of solar convective flows and structure, in: "Solar Convection and Oscillations and their Relationship", eds F.P. Pijpers, J. Christensen-Dalsgaard and C.S. Rosenthal, Proc. of SCORe'96 Workshop, Aarhus (Denmark), 241 - 260, Kluwer Acad. Publ

Li, Y., and Oldenburg, D.W., 1994, Inversion of 3-D DC resistivity data using an approximate inverse mapping. Geophys. J. Int., 116, 527-537.

Munk, W., Worchester, P. and Wunsch, C.,1995, Ocean Acoustic Tomography. Cambridge University Press.

Møller, I., Christensen, N.B., Jacobsen, B.H., 1996, Deconvolution of geophysical multioffset resistivity profile data. In B.H. Jacobsen, K. Mosegaard, and P. Sibani (eds.): Inverse Methods - Interdisciplinary Elements of Methodology, Computing and Applications. Lecture Notes in Earth Sciences, 197-204, Springer-Verlag, Berlin-Heidelberg.

Paige, C.C., Saunders, M.A., 1982, LSQR: an algorithm for sparse linear equations and sparse least squares. ACM Trans.Math.Software, 8, 43-71.

Snieder, R. and Lomax, A. 1996, Wavefield smoothing and the effect of rough velocity perturbations on arrival times and amplitudes. Geophys. J. Int., 125, 796-812.

Covariance Modelling for Merging of Multi-Sensor Ocean Surface Data

Olwijn Leeuwenburgh

National Survey and Cadastre
Rentemestervej 8, 2400 Copenhagen NV, Denmark
Email: ol@kms.dk
Home page: http://www.kms.dk

Abstract. The problem of estimating the sea surface height from observations made by remote sensing satellites is discussed. The potential for improving altimetric estimates by merging different types of surface data is explored. The focus in this paper is on the modelling of covariance functions of sea surface height and sea surface temperature and identification of the appropriate spatial and temporal correlation scales. It is shown that on the mesoscale, westward propagating planetary waves can be detected in both fields. The dependence of the phase speed on latitude and depth, as well as the local enhancement of variability near ocean fronts, renders the auto and cross covariance functions to be complicated functions of latitude, longitude and time. It is shown how propagating features can be accounted for in the covariance model for the purpose of optimal merging of sea surface height and temperature data.

1   Introduction

Sea surface height (SSH) is an important parameter in both the fields of physical oceanography and geodesy. The mean shape of the ocean surface is determined by the earth's gravity field and by the mean ocean circulation. Variations with respect to this mean contain information about changes in the ocean interior such as heat content and paths and velocities of currents. Satellite altimetry is the most widely used technique for determining sea surface height in a global sense. Estimates of the sea surface elevation are obtained every second at about 7 km intervals along a 'ground track', the projection of the satellite's position on the earth's surface. The altimeter thus works as a 'uniform' measuring system globally. Changes in the error of the height estimate from one place to another result primatily from the sea-state. All current altimetric satellites are positioned in repeating orbits. This means that the satellite returns to the same location at fixed time intervals. There is a trade-off between the length of the repeat interval and the spatial coverage (distance in longitude between neighbouring ground tracks). For TOPEX/POSEIDON a compromise has been found in a 10 day repeat resulting in a track spacing of about 300 km at the equator. Chelton and Schlax (1994) and Greenslade et al. (1997) investigated the frequency and wavenumber content of midlatitude gridded SSH fields and concluded that the spatial-temporal resolution that can uniformly be resolved from altimetry is limited to about 2.5° and 25 days. These scales are to large to resolve much of the mesoscale variability which dominates the energy spectrum outside of the annual frequency band. Some recent studies however have indicated the existence of correlations between sea surface temperature (SST) as obtained from satellite radiometers and SSH (Halliwell et al., 1991; Cipollini et al., 1997; Jones et al., 1998; Polito and Liu, 1998). Since SST is typically sampled with a more favourable resolution than SSH, these measurements can potentially be used to support the interpolation of SSH away from the satellite tracks. This new concept is explored here.

In section 2 the basic merging scheme for the two data types is outlined. In section 3 a method to determine the covariance function from irregularly distributed SSH observations is described and empirical covariance functions estimated from T/P altimetry are presented. The physical relation between SSH and SST through wave dynamics is discussed. In section 4 the empirical covariance functions are modelled by analytical expressions which are able to account for propagating waves. Section 5 concludes with a short discussion of current results.

2   Optimal Interpolation

For the purpose of interpolating the irregularly distributed SSH observations the selected method is minimum error variance, or Gauss-Markov, estimation, applied in a variety of disciplines under different names such as objective analysis, optimal interpolation, collocation and Kriging.

If it is assumed that the mean is constant and equal to zero (this can be arranged by removing an appropriate mean field prior to the interpolation), the single estimate $\hat{x}$, for example of SSH, can be obtained from a set of observations $\mathbf{x}$ as (see e.g. Moritz, 1980; Wackernagel, 1995)

$$\hat{x}(r) = C_{xx}(r, \mathbf{r}) \, C_{xx}^{-1}(\mathbf{r}, \mathbf{r}) \, \mathbf{x}(\mathbf{r}) \tag{1}$$

and its error is

$$\sigma_{\hat{x}}^2 = \sigma_x^2 - C_{xx}(r, \mathbf{r}) C_{xx}^{-1}(\mathbf{r}, \mathbf{r}) C_{xx}(r, \mathbf{r})^T \tag{2}$$

where C_{xx} is the autocovariance matrix of the variable x, and r and $\mathbf{r}$ denote the positions of the estimate $\hat{x}$ and the observations $\mathbf{x}$, for example the latitude and longitude (and, for time-dependent phenomena, time). This procedure is used extensively in oceanography for estimating grids of sea surface height at regular time intervals which are easier to analyse than the irregularly sampled along track data. Expression (1) gives the estimate at a single location, but can easily be extended into an expression for an array of estimates located in a regular grid. It can furthermore be adapted to include observations $\mathbf{y}$ of a different type y, e.g. SST,

$$\hat{x}(r) = \begin{bmatrix} C_{xx}(r, \mathbf{r}) & C_{xy}(r, \mathbf{r}') \end{bmatrix} \begin{bmatrix} C_{xx}(\mathbf{r}, \mathbf{r}) & C_{xy}(\mathbf{r}, \mathbf{r}') \\ C_{yx}(\mathbf{r}', \mathbf{r}) & C_{yy}(\mathbf{r}', \mathbf{r}') \end{bmatrix}^{-1} \begin{bmatrix} \mathbf{x}(\mathbf{r}) \\ \mathbf{y}(\mathbf{r}') \end{bmatrix} \tag{3}$$

where C_{xy} is the cross-covariance matrix. The expression for the error variance now becomes

$$\sigma_{\hat{x}}^2 = \sigma_x^2 - \left[\, C_{xx}(r, \mathbf{r})\ C_{xy}(r, \mathbf{r}')\,\right] \begin{bmatrix} C_{xx}(\mathbf{r}, \mathbf{r}) & C_{xy}(\mathbf{r}, \mathbf{r}') \\ C_{yx}(\mathbf{r}', \mathbf{r}) & C_{yy}(\mathbf{r}', \mathbf{r}') \end{bmatrix}^{-1} \begin{bmatrix} C_{xx}(\mathbf{r}, r) \\ C_{yx}(\mathbf{r}', r) \end{bmatrix} \quad (4)$$

Here it has been assumed that the observations **y** were made at different positions (or epochs) denoted by $\mathbf{r}'$. In geological applications, this scheme is referred to as 'simple Kriging'. There exists a variation known as 'ordinary Kriging' in which the mean is assumed to be constant only locally for the points that enter the estimation. This procedure may be useful if there is low wavenumber variability present that is not adequately accounted for in the procedure to remove the mean prior to the estimation.

3 Covariances from Observations

3.1 Sea Wurface Height

As pointed out in the previous section, minimum error variance estimation requires the use of the covariance of the sampled field. Here it will be assumed that the frequency-wavenumber content of the observed field remains constant in time locally, while allowing for spatial variation. In other words, it is assumed that the local dominant physical processes at work remain unchanged over time, thus imposing local spatial and temporal stationarity on the SSH field (even this loose assumption is not always strictly valid as will be seen later on). With these assumptions the covariance function can be written as

$$C_{xx}(r, r') = C_{xx}(r - r') \quad (5)$$

The covariance only depends on the seperation in space (and time) between two observations and not on the precise location and epoch (within a certain local neighbourhood). Often covariances are determined from gridded fields where observations are first averaged into equally sized bins. The coarseness of the sampling limits this averaging to binsizes of about 2 spherical degrees at lower latitudes. This strongly limits the resolution of the covariance function and inherently filters out much of the mesoscale structure of interest here. Some authors have presented covariance functions from isotropic frequency-wavenumber power spectra obtained from along-track observations (see e.g. Fu, 1983; Le Traon et al., 1990; Stammer, 1997). The approach adopted here was proposed by Glazman et al. (1996) and makes use of the ungridded along-track observations by averaging cross-products of the data over discrete time and range bins, thus extending procedures for purely spatial applications (see e.g. Tscherning et al., 1992). No prior gridding or spectral filtering is required. The major problem in this approach is the enormous amount of observations that go into the modelling process. It may be worth reducing the number of data by retaining for example only every fifth observation. This reduction scheme should be applied carefully in accordance with the desired resolution of the covariance function.

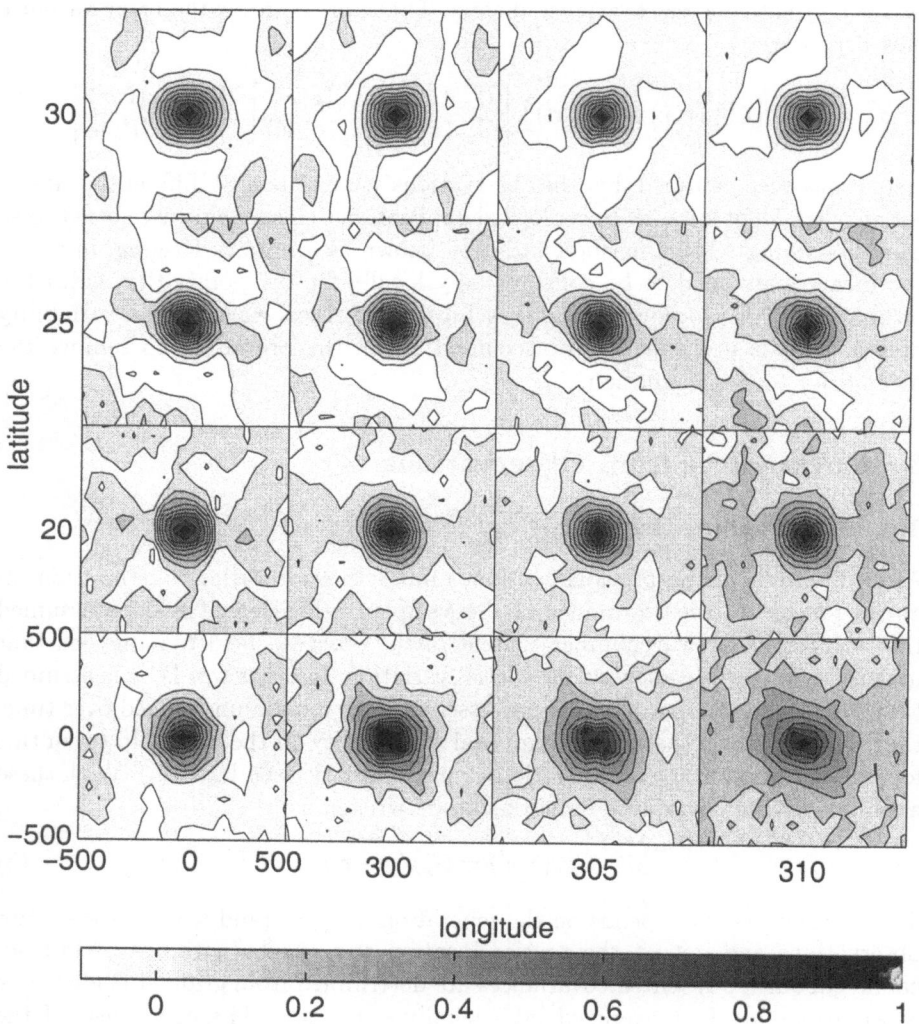

Fig. 1. Zonal-meridional SSH autocorrelation functions for 5°by 5°boxes centered at 5°intervals from 290°E to 305°E and 15°N to 30°N in the North Atlantic. The dimensions of each box are shown with the bottom left box with the zonal and meridional lags, along the horizontal and vertical axes respectively, ranging from −500 km to +500 km. The contour interval is 0.1.

The oceanic mesoscale variability is characterized by fluctuations which lie roughly on scales between 20 and 150 days and between 50 and 500 km (Stammer, 1997). The spatial bin size should be small enough to properly reflect the influence of these scales, and at the same time be large enough to result in statistically reliable covariances. In order to filter out variability on very short time scales, products of observations along the same track are excluded from the estimation procedure adopted here. Empirical experiments showed that significant biases are introduced by using the cross products of observations on parallel tracks, so it was decided only to use products of observations from crossing tracks.

Using the approach described above, spatial-temporal covariance functions were computed for boxes of size 5 by 5 degrees in latitude and longitude in the North Atlantic Ocean. Figure 1 shows the zonal-meridional autocorrelation functions of SSH (for a zero time lag) for the south-western part of the North Atlantic. The correlations are nearly isotropic for short ranges, but become more anisotropic for large ranges, indicating that large scale anomalies may have a preferred shape and orientation. Figure 2 shows the zonal-time autocorrelations of SSH (for a zero meridional lag) for the same areas. It can be seen that positive correlations exist for observations seperated both in zonal distance and time, thus representative of zonally (in this case westward) propagating features. It can also be seen that the speed with which these features propagate decreases with increasing latitude. These speeds are roughly in agreement with the dispersion relationship for oceanic Rossby waves

$$\sigma_n = \frac{-\beta\, k}{k^2 + l^2 + 1/R_n^2} \tag{6}$$

where σ is the frequency, k and l are the zonal and meridional wavenumbers, β is the meridional gradient of the Coriolis parameter, and R_n is the Rossby deformation radius, where $n = 0, 1, 2, \ldots$ denotes the wave mode. The phase speed is defined as σ/k and is related in Figure 2 to the slope of the structure of high correlation. In principle all modes are present, but the resolution of the correlation estimate may allow only the estimation of a single lumped phase speed. Such estimates are generally compared with the first baroclinic mode ($n = 1$), but Zhang and Wunsch (1996) point out that for sea surface elevation this may not be appropriate everywhere (see also Wunsch, 1997). Similar figures were made for the meridional-temporal autocorrelation (not shown). For 25°N and 30°N some southward propagation was detected. From Figures 1 and 2 it can be concluded that the dominant spatial scales (defined, for example, as the e-folding length) of SSH variability for this region are on the order of 100–200 km and with periods of about 200–400 days. Furthermore, the correlation structure exhibits large changes from one location to another reflecting the variation in spatial and temporal scales set by the physical processes at work. It is therefore necessary to model the covariance function over the complete area of investigation and allow for anisotropy, variation in the spatial and temporal correlation scales, and for variation in the phase speed of propagating features.

3.2 Sea Surface Temperature

SSH is directly related to the subsurface temperature through the expansion (contraction) of the water column subject to heating (cooling). The heat flux forcing of the ocean surface is due to atmospheric variability on time scales from

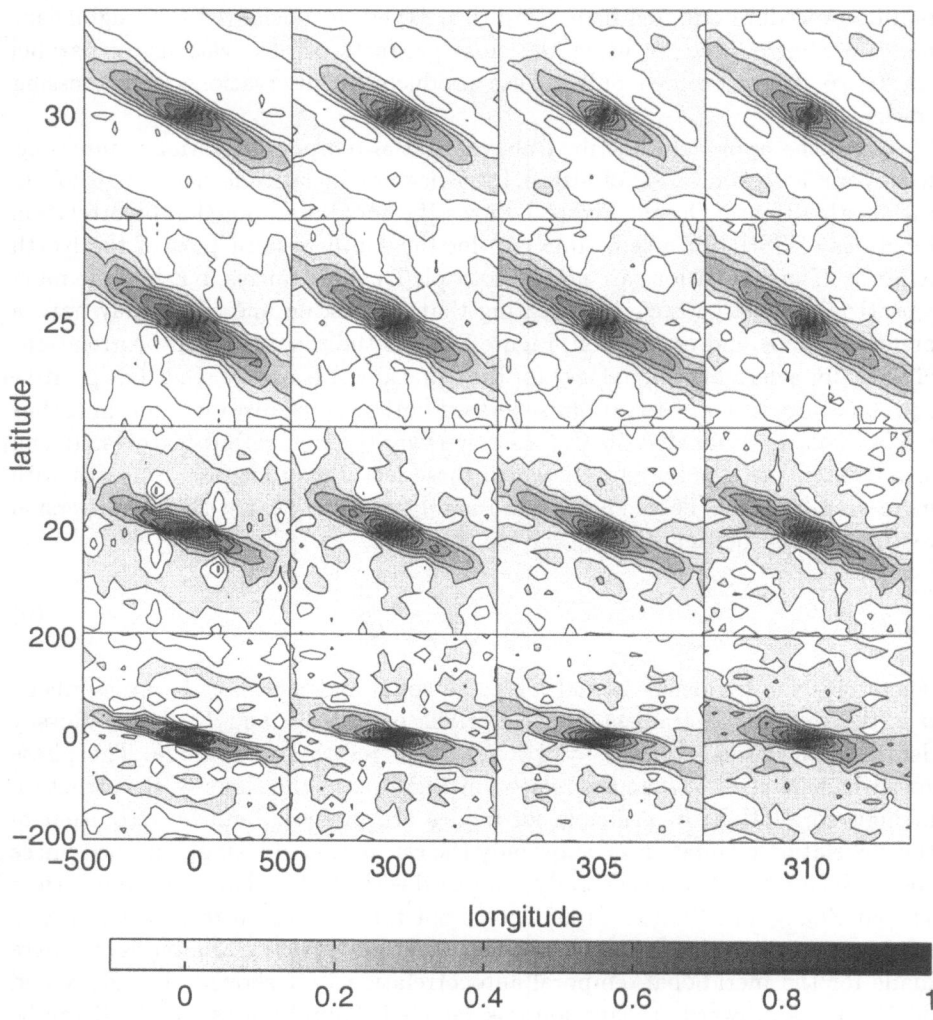

Fig. 2. Zonal-time SSH autocorrelation functions for the same area as figure 1. The dimensions of each subplot now range from −200 days to +200 days (time lag) and from −500 km to +500 km (zonal lag). The contour interval is 0.1.

days to years or longer and on spatial scales of the order of 1000 km (the synoptic scale of the atmosphere). Heat may also be advected internally by anomalous ocean currents. These currents may be due to wind forcing at the surface or to instability of the mean flow.

Water columns in the ocean tend to maintain their absolute vorticity (angular momentum). Columns which are displaced northwards for example will gain planetary vorticity which is balanced by a gain in negative relative vorticity, thus creating anomalous currents. Once these eddie currents are formed and

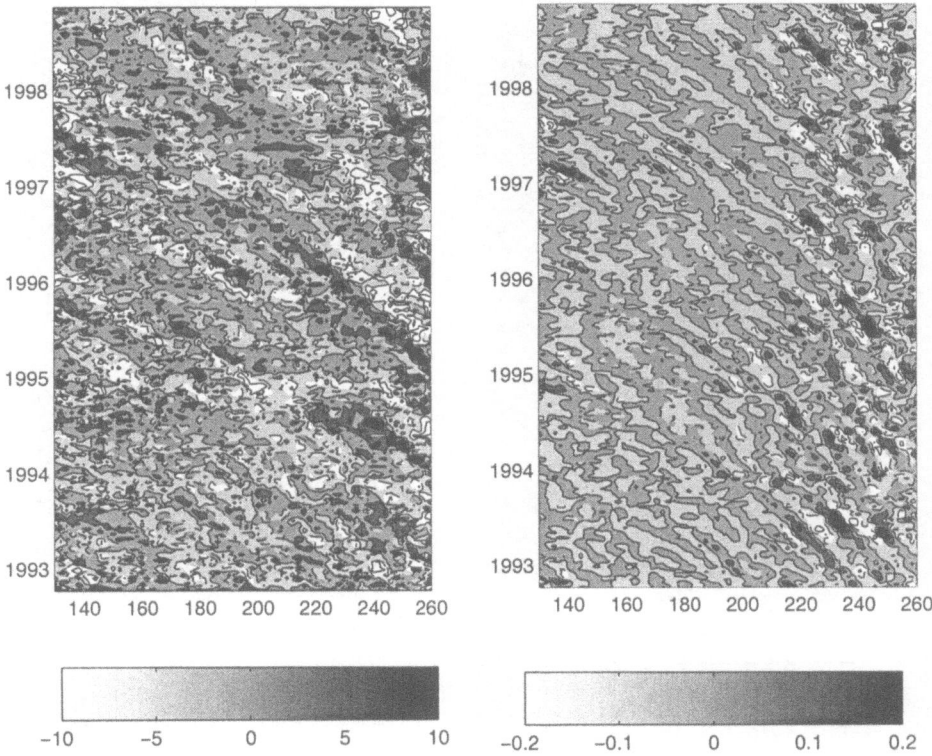

Fig. 3. Longitude-time plot for SSH (left) and SST (right) along 14° N in the North Pacific. Longitude is depicted along the horizontal axis and time is depicted along the vertical axis. The contour intervals are 5 cm and 0.1°C respectively.

reach a certain spatial dimension they may propagate as free waves subject to the dispersion relationship (6). These eddy currents are in near-geostrophic balance (Gill et al., 1974). At the sea surface this means that to first order

$$v_g = \frac{g}{f}\frac{\partial \eta}{\partial x} \tag{7}$$

$$u_g = -\frac{g}{f}\frac{\partial \eta}{\partial y} \tag{8}$$

where u_g and v_g are the surface geostrophic velocities, g is the local gravity, f is the Coriolis parameter, and η is the surface elevation. The mean sea surface temperature field is characterized by a large north-south gradient, which is especially strong during the winter months (in the summer a warm surface layer is formed which largely eclipses the underlying temperature field). Anomalous surface currents tend to distort this temperature gradient and transport relatively warm surface water northwards and colder water southwards. In terms of

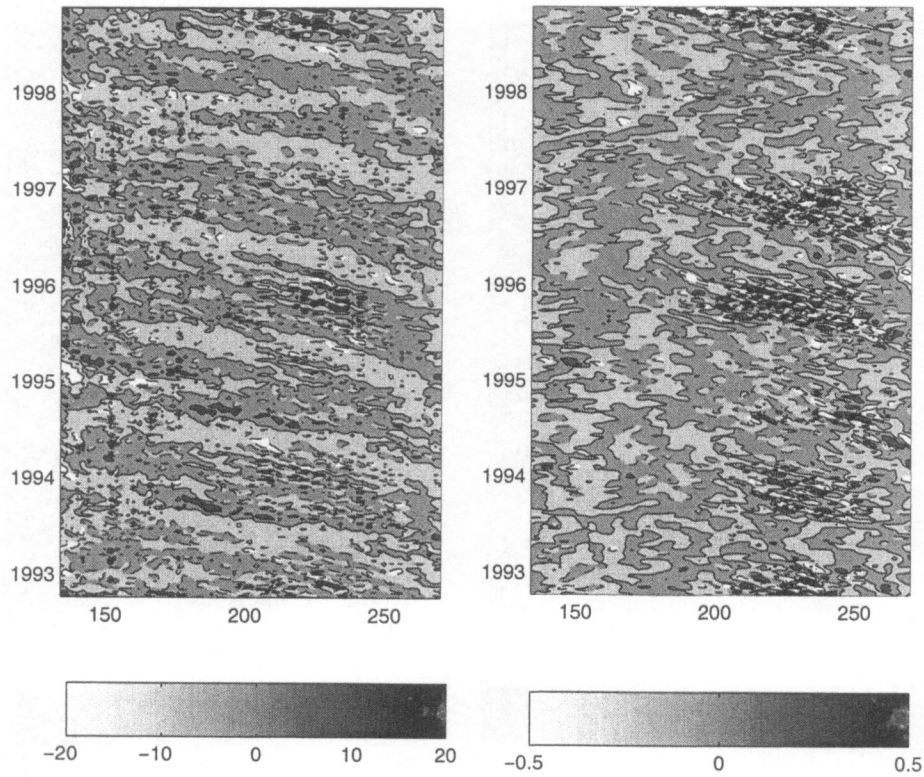

Fig. 4. Longitude-time plot for SSH (left) and SST (right) along 6°N in the North Pacific. Longitude is set along the horizontal axis and time is set along the vertical axis. The contour intervals are 10 cm and 0.25°C respectively.

the local heat balance (see e.g. Frankignoul, 1985)

$$\frac{\partial T'}{\partial t} \approx -\mathbf{u}'_g \, \nabla T \approx -v_g \, \frac{\partial \overline{T}}{\partial y} \tag{9}$$

where T and T' are the total and anomalous temperature, ∇ is the horizontal gradient operator, and $\mathbf{u}_g$ is the horizontal geostrophic velocity vector.

From combining (7) and (9) it can be seen that the temperature anomaly tendency is proportional to the zonal gradient of SSH

$$\frac{\partial T'}{\partial t} \approx -\frac{g}{f} \, \frac{\partial \eta}{\partial x} \, \frac{\partial \overline{T}}{\partial y} \tag{10}$$

This relation was investigated by Halliwell et al. (1991) in the Sargasso Sea and by Polito and Liu (1998) along the 15°N parallel in the North Pacific. Figure 3 shows a longitude-time plot of both (filtered) SSH and SST along 14°N in the North Pacific. Westward propagating Rossby waves can be detected in both fields. According to (10) the amplitude of the waves in the SST depends on both the magnitudes of the eddy current velocity and the mean temperature gradient. Furthermore, a more complete version of (9) will include the contributions of horizontal mixing and mixed layer depth variability, which might have a significant effect on the SST on the scales considered here.

Figure 4 shows longitude-time plots for SSH and SST for the 6°N parallel in the North Pacific. In both the SSH and SST fields one can detect series of high frequency waves which are most prominent in the latter part of the year and virtually nonexistent in spring and early summer. These waves are known as Tropical Instability Waves and result from the instability of the equatorial current system. This instability is due to the velocity shear between the South Equatorial Current (SEC) and the North Equtorial Counter Current. The SEC, and thus the shear, undergoes a seasonal cycle in response to variation in the trade wind forcing (Chelton et al., 1999). For these latitudes the SSH and SST covariance will thus be a function of the time of year.

4 Covariance Models

For interpolation puposes it is necessary to approximate the empirical covariance estimate with a positive definite analytical function. Several options are available here, but based on experimental results a Gaussian function was chosen. The correlation functions obtained from altimetric SSH anomalies tend to become negative for large lags which is not modelled by a Gaussian, but the main positive lobe appears to be better modelled by a second order Gaussian then by many alternative functions that were tested. Although other analytical models should be explored in future work, a Gaussian covariance function and its associated spectrum appear to model the main features of the observed covariance and spectrum reasonably well, and is well suited to demonstrate the current ideas. The covariance is made to be a function of zonal lag λ, meridional lag ϕ and time lag τ:

$$C_{\eta\eta}(\lambda, \phi, \tau) = C_0 \, e^{-\left((\lambda/L)^2 + (\phi/F)^2 + (\tau/T)^2\right)} \tag{11}$$

where L, F and T are the zonal, meridional and temporal (e-folding) decorrelation scales. In order to allow correct representation of zonal propagation of SSH anomalies η, the zonal-time covariance (i.e. F is arbitrary) is represented in a rotated coordinate frame. If two new coordinates λ' and τ' are introduced along the main axes of the zonal-temporal covariance structure, which has a slope α, then

$$\begin{bmatrix} \lambda' \\ \tau' \end{bmatrix} = \begin{bmatrix} \cos\alpha & \sin\alpha \\ -\sin\alpha & \cos\alpha \end{bmatrix} \begin{bmatrix} \lambda \\ \tau \end{bmatrix} \tag{12}$$

Structures as in Figure 2 can thus be modelled as

$$C_{\eta\eta}(\lambda', \phi = 0, \tau') = C_0 \, e^{-\left((\lambda'/L')^2 + (\tau'/T')^2\right)} \tag{13}$$

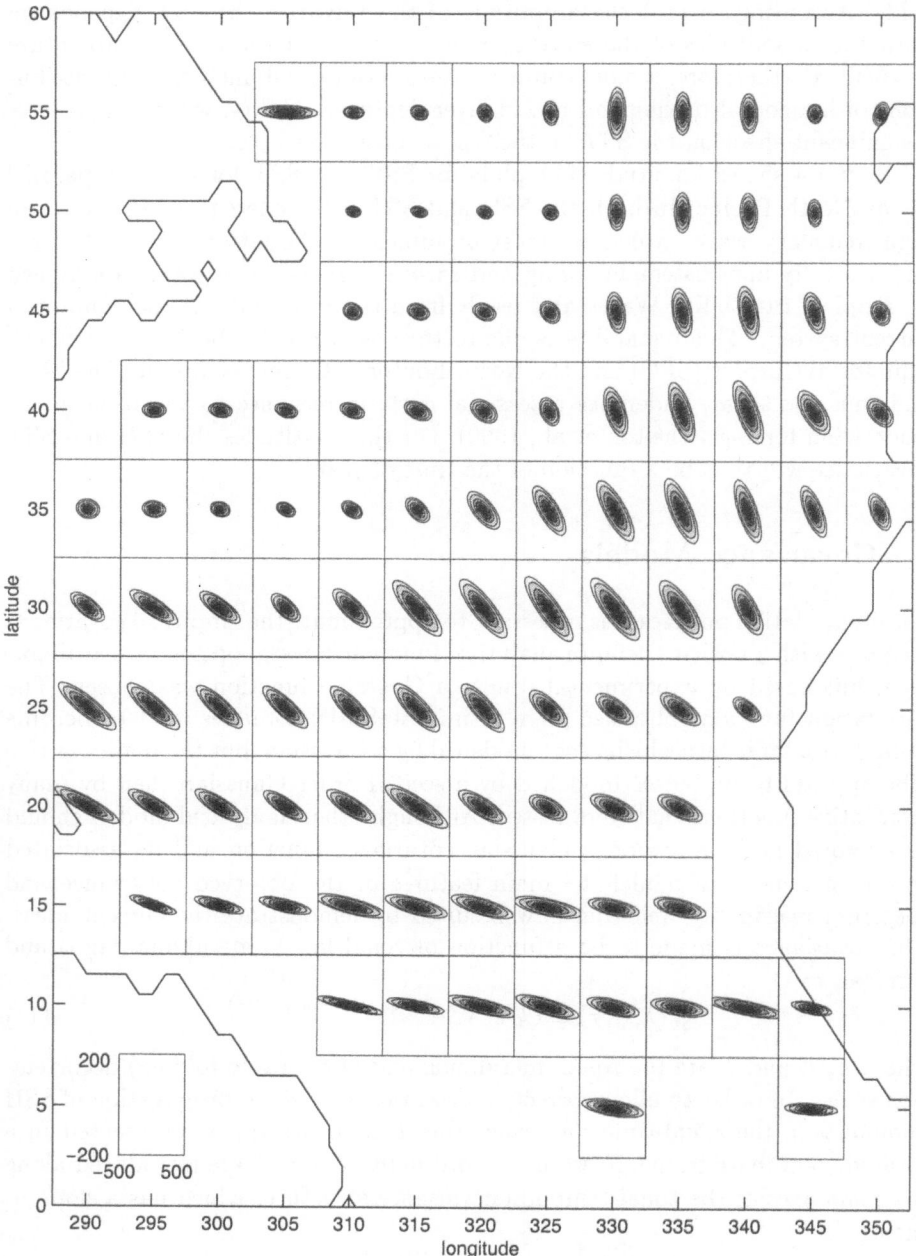

Fig. 5. Zonal-time SSH autocorrelation models for the North Atlantic. The zonal lag ranges from −500 km to +500 km (horizontal axis of each subplot) and the time lag ranges from −200 days to +200 days (vertical axis). Contour lines are drawn at 0.1, 0.2, 0.4, 0.6 and 0.8.

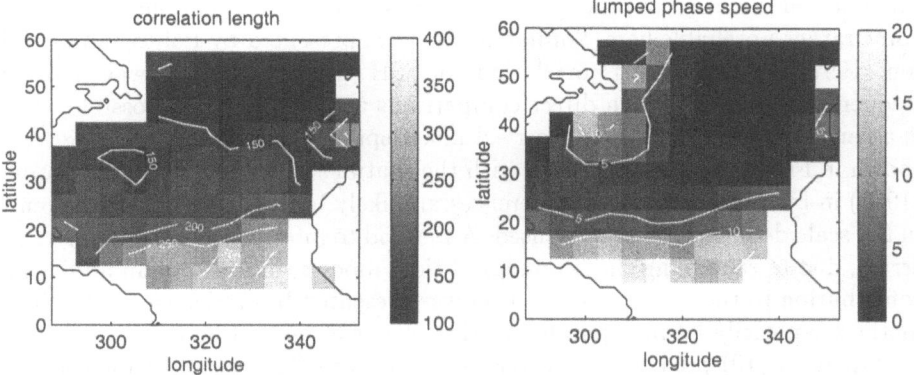

Fig. 6. (left) Estimates of the isotropic spatial correlation scale (km) estimated as $\sqrt{L^2 + F^2}$, with L and F as in (11), and (right) zonal phase speed (km/day) using (16).

Using (12) the exponent can be written in terms of the original lags λ and τ

$$\left(\frac{\lambda'}{L'}\right)^2 + \left(\frac{\tau'}{T'}\right)^2 = \left(\frac{\cos^2\alpha}{L'^2} + \frac{\sin^2\alpha}{T'^2}\right)\lambda^2 + \left(\frac{\sin^2\alpha}{L'^2} + \frac{\cos^2\alpha}{T'^2}\right)\tau^2$$
$$+ 2\cos\alpha\,\sin\alpha\,\left(\frac{1}{L'^2} - \frac{1}{T'^2}\right)\lambda\,\tau \tag{14}$$

The three-dimenional covariance function (11) is thus modified by a fourth term containing the product of the zonal and temporal lags. The new model is

$$C_{\eta\eta}(\lambda, \phi, \tau) = C_0\, e^{-(A\lambda^2 + B\phi^2 + D\tau^2 + E\lambda\tau)} \tag{15}$$

From equation (14) expressions can be found for the lumped phase speed parameter α and the correlation scales L' and T'

$$\alpha = \frac{1}{2}\tan^{-1}\frac{E}{A - D} \tag{16}$$

$$L' = \sqrt{\frac{2}{E/\sin 2\alpha + A + D}} \tag{17}$$

$$T' = \sqrt{\frac{1}{A + D - 1/L'^2}} \tag{18}$$

The above model (15) was used to model SSH covariances over the entire North Atlantic (Figure 5). Large variation exists in all scales involved (Figure 6). The Rossby waves have maximum zonal phase speeds at low latitudes and dissappear almost entirely beyond 50 degrees north (Figure 6). Along the western boundary the SSH is dominated by relatively small anomalies with apparently no preferred direction of propagation and the estimates of zonal phase speed become rather unreliable here. Similar figures were presented by Polito and Cornillon (1997) who computed correlations of SSH grids for specific wavenumber-frequency bands. Although direct comparisons are therefore not possible, there is a general agreement in the observed phase speeds. The correlation lengths as shown in Figure 6 agree fairly well with the spatial scales presented by Stammer (1997) in his Figure 21. Small differences are likely to exist due to the difference in the scale definitions that were used. A method to model correlated along-track errors, due to remaining shortcomings of tidal models, in terms of an additional contribution to the covariance function was presented by Andersen (1999). This method can easily be incorporated in the scheme suggested here.

Equation (10) predicts a zero phase-lag between SST and SSH for negative mean temperature gradients. In the presence of feedback however, the phase difference beteen SST and SSH becomes a function of the feedback coefficient and the frequency of the wave (see e.g. Halliwell et al., 1991). For a single wave frequency this may be represented in the cross-covariance function as an additional coordinate shift (on top of a rotation of the coordinate frame as above). Real ocean variability, however, consists of contributions from all possible wavelengths and frequencies, which would make this modelling procedure rather dubious. Only in regions where a single mode is dominant (which is the case for most of the mid-latitude ocean), and where waves are forced at a steady frequency, such an approach might be reasonable. Such requirements are too strict however to apply to all parts of the North Atlantic. Some authors have chosen to use the observed cross-covariance function instead of an analytical model, but this would bring the risk of obtaining negative error covariances and is not advised. (The observed phase-relationship between SST and SSH is the subject of a manuscript currently in preparation.)

5 Conclusions

The results presented here show that it is necessary to account for propagating waves when estimating gridded SSH fields from altimetric observations. It was shown that this could be done by introducing an extra term in the spatial-temporal covariance function. Covariance models for SSH computed over the entire North Atlantic show a decrease in planetary wave phase-speed with latitude, in agreement with the decrease in the deformation radius. Longitude-time plots of SSH and SST show that propagating waves can also be detected in the temperature field (after appropriate filtering in space and time). These plots furthermore show that SSH and SST can not be considered stationary (in a statistical sense) in time everywhere. Seasonal modulation of the equatorioal current system and of the mean meridional temperature gradient introduce temporal dependence in the local statistics. Phase relationships between SST and SSH in real ocean conditions depend on many factors and are not easily modelled in terms of a cross-covariance function.

Acknowledgements

I wish to thank Detlef Stammer, Ole Andersen and an anonymous reviewer for useful comments on the manuscript. Funding was supplied through a grant of the Danish Research Council, Earth Observation Program. This is a contribution to GEOSONAR.

References

[Andersen (1999)] Andersen, O. B.: Shallow water tides in the northwest European shelf region from TOPEX/POSEIDON altimetry. J. Geophys. Res. **104** (1999) 7729-7741

[Chelton and Schlax (1994)] Chelton, D. B., Schlax, M. G.: The resolution capability of an irregularly sampled dataset: with application to Geosat altimeter data. Atmos. Oceanic Technol. **11** (1994) 534-550

[Chelton et al. (1999)] Chelton, D. B., Schlax, M. G., Lyman, J. M., de Szoeke, R. A.: The latitudinal structure of monthly variability in the tropical Pacific. submitted to J. Phys. Oceanogr.

[Cipollini et al. (1997)] Cipollini, P., Cromwell, D., Jones, M. S., Quartly, G. D., Challenor, P. G.: Concurrent altimeter and infrared observations of Rossby wave propagation near 34° N in the Northeast Atlantic. Geophys. Res. Letters **24** (1997) 889-892

[Frankignoul (1985)] Frankignoul, C.: Sea surface temperature anomalies, planetary waves, and air-sea feedback in the middle latitudes. Rev. Geophys. **23** (1985) 357-389

[Fu (1983)] Fu, L.-L.: On the wave number spectrum of oceanic mesoscale variability observed by the SEASAT altimeter. J. Geophys. Res. **88** (1983) 4331-4341

[Glazman et al. (1996)] Glazman, R. E., Fabrikant, A., Greysukh, A.: Statistics of spatial-temporal variations of sea surface height based on Topex altimeter measurements. Int. J. Remote Sensing **17** (1996) 2647-2666

[Greenslade et al. (1997)] Greenslade, D. J., Chelton, D. B., Schlax, M. G.: The mid-latitude resolution capability of sea level fields constructed from single and multiple satellite altimeter datasets. J. Atmos. Oceanic Technol. **14** (1997) 849-870

[Gill et al. (1974)] Gill, A. E., Green, J. S. A., Simmons, A. J.: Energy partition in the large-scale ocean circulation and the production of mid-ocean eddies. Deep Sea Res. **21**(7) (1974) 499-528

[Halliwell et al. (1991)] Halliwell, G. R., Ro, Y. J., Cornillon, P.: Westward-propagating SST anomalies and baroclinic eddies in the Sargasso Sea. J. Phys. Oceanogr. **21** (1991) 1664-1680

[Jones et al. (1998)] Jones, M. S., Allen, M., Guymer, T., Saunders, M.: Correlations between altimetric sea surface height and radiometric sea surface temperature in the South Atlantic. J. Geophys. Res. **103** (1998) 8073-8087

[Le Traon et al. (1990)] Le Traon, P. Y., Rouquet, M. C., Boissier, C.: Spatial scales of mesoscale variability in the North Atlantic as deduced from Geosat data. J. Geophys. Res. **95** (1990) 20267-20285

[Moritz (1980)] Moritz, H.: Advanced physical geodesy. Herbert Wichmann Verlag, Karlsruhe, (1980)

[Polito and Cornillon (1997)] Polito, P. S., Cornillon, P.: Long baroclinic Rossby waves detected by TOPEX/POSEIDON. J. Geophys. Res., **102** (1997) 3215-3235

[Polito and Liu (1998)] Polito, P. S., Liu, W. T.: The air-sea interaction of semi-annual signals at 15° N in the Pacific. Joint TOPEX/POSEIDON Jason-1 science working team meeting, October 13-15, Keystone, Colorado (1998)

[Stammer (1997)] Stammer, D.: Global characteristics of ocean variability estimated from regional TOPEX/POSEISON altimeter measurements. J. Phys. Oceanogr. **27** (1997) 1743-1769

[Tscherning et al. (1992)] Tscherning, C. C., Forsberg, R., Knudsen, P.: The GRAV-SOFT package for geoid determination. Proc. 1st continental workshop on geoid in Europe, Prague (1992) 327-334

[Wackernagel (1995)] Wackernagel, H.: Multivariate Geostatistics -an introduction with applications. Springer Verlag (1995)

[Wunsch (1997)] Wunsch, C.: The vertical partition of oceanic horizontal kinetic energy. J. Phys. Oceanogr. **27** (1997) 1770-1794

[Zang and Wunsch (1999)] Zang, X., Wunsch, C.: The observed dispersion relationship for North Pacific Rossby wave motions. to be published in J. Phys. Oceanogr.

Inversion of 2D Geoelectrical Data by Iterated Multichannel Deconvolution

Ingelise Møller

Department of Earth Sciences, University of Aarhus,
Finlandgade 8, 8200 Aarhus N, Denmark

Abstract. Inversion of large amounts of geoelectrical data by an iterative linearised Gauss Newton method is computationally heavy. Therefore, shortcuts in the inversion procedure are useful. This paper presents a 2D inversion procedure that uses the 2D Fréchet derivative of the homogeneous halfspace as approximate partial derivatives. The inversion at each iteration step is carried out as a multichannel deconvolution that solves the inverse problem in the wavenumber domain. Hence, the iterative inversion procedure spends almost all computation time on the forward modellings. The inversion scheme is stabilised through covariance matrices reflecting the stochastic properties of the earth resistivity and data errors. The iterative inversion is stable and warrants convergence. Simple modifications allow the inversion procedure developed here, which is periodic in nature, to be used with non-periodic field data. A good correspondence is demonstrated with electrical resistivity logs that reflect the in situ resistivity of the ground.

1 Introduction

Geoelectrical methods have many applications in applied geophysics. The development of fast data acquisition systems [27, 23] that can collect 10 – 15 kilometres of densely sampled profile data per field day has made DC-resistivity methods practicable in large-scale mapping [16]. Interpretation of such amounts of data is a computationally heavy task, where shortcuts in the inversion procedures are useful.

Geoelectrical data are nonlinear functions of the ground resistivity. Therefore, the inverse problem is usually solved iteratively using a linearised Gauss Newton method [24] where the Jacobian matrix (the partial derivatives of the forward solution with respect to the model parameters) has to be calculated at each iteration. When large data volumes are to be inverted, this procedure becomes very time consuming because partial derivatives must be calculated and large linear systems must be solved. Sasaki [25] approximated the partial derivatives by those of the homogeneous halfspace for a 3D inversion. Loke and Barker [13] presented a quasi-Newton least squares inversion method, by which the partial derivatives were updated from those of the homogeneous halfspace. A general approach of Approximate Inverse Mapping inversion [21] made use of an accurate forward mapping, an approximate inverse mapping and updating in model space as well as in data space. Li and Olderburg [11] used the method for 3D DC-resistivity pole-pole data. Solution of large matrices can be speeded up using conjugate gradients, as presented for 3D inverse cases by Ellis and Oldenburg [5] Zhang, Mackie, and Madden or [31], or avoided using subspace methods [22].

This paper presents a 2D inversion procedure that uses the partial derivatives of the homogeneous halfspace as Jacobian matrix. Because the partial derivative of the homogeneous halfspace is translationally invariant, the inversion at each iteration can be solved as a multichannel deconvolution [17]. A multichannel deconvolution formulation leads to an inverse solution in the wavenumber domain, where the 2D problem decouples into many 1D problems. The regularization follows a pragmatic stochastic approach, where measurement errors as well as correlated nonlinearity errors are taken into account. Furthermore, ground resistivity amplitude and correlation properties estimated directly from data determine the model covariance matrix and the nonlinearity error covariance matrix used.

This paper extends the multichannel deconvolution by Møller, Jacobsen, and Christensen [17], which is a one-pass inversion, to be part of an iterated inversion procedure.

2 Theory

Geoelectrical profile data expressed as apparent resistivities, ρ_a, are nonlinear functions of electrical resistivities, ρ, so with $d_{obs} = \log \rho_a$ and $m = \log \rho$

$$d_{obs}(x, s) = g(\mathbf{m}, s) \quad s = s_1, s_2, \ldots, s_S \tag{1}$$

where g contains the forward coordinate functionals, s is one out of S possible electrode configurations, and $m(x', z')$ is an element in $\mathbf{m}$. The x-axis is assumed to be perpendicular to the strike direction and the z-axis points downwards. The logarithmical transformation of data and resistivity parameters makes the problem more linear because it prevents non-physical negative resistivities and equalizes large and small resistivity levels.

Linearisation of (1) by Taylor series expansion around $\mathbf{m}_0$ leads to

$$d_{obs}(x, s) = g(\mathbf{m}_0, s) + \int_0^\infty dz' \int_{-\infty}^\infty \Phi_{\mathbf{m}_0}(x', z', x, s) \, [m(x', z') - m_0(x', z')] dx' + R, \tag{2}$$

where $\Phi_{\mathbf{m}_0}$ is the 2D Fréchet derivative evaluated at $\mathbf{m}_0$ and R is a remainder term that is discarded in the Born approximation. $\mathbf{m}_0$ is chosen to be a homogeneous halfspace, which is translationally invariant along the profile. Because $\Phi_{\mathbf{m}_0}$ inherits translational invariance from $\mathbf{m}_0$, the integral in (2) is actually a convolution which can be solved for $\mathbf{m}$ through a multichannel deconvolution [17].

Equation (2) forms the base in an iterative scheme obtained by replacing $\mathbf{m}$ with $\mathbf{m}_{n+1}$ and $\mathbf{m}_0$ with $\mathbf{m}_n$, respectively. A model perturbation

$$\delta \mathbf{m} = \mathbf{m}_{n+1} - \mathbf{m}_n \tag{3}$$

is formed. Then the model estimate $\mathbf{m}_{n+1}$ is found iteratively by solving for $\delta\mathbf{m}$ in

$$d_{\text{obs}}(x, s) - \mathbf{g}(\mathbf{m}_n, s) \simeq \int_0^\infty dz' \int_{-\infty}^\infty \Phi_{\mathbf{m}_n}(x', z', x, s) \, \delta m(x', z') dx' \qquad (4)$$

and updating

$$\mathbf{m}_{n+1} = \mathbf{m}_n + \delta\mathbf{m}. \qquad (5)$$

When evaluated at an arbitrary $\mathbf{m}_n$ the Fréchet derivative $\Phi_{\mathbf{m}_n}$ loses the translational invariance. In order to maintain the translational invariance, $\Phi_{\mathbf{m}_n}$ is approximated by $\Phi_{\mathbf{m}_0}$. Then the iterated solutions for $\delta\mathbf{m}$ in (5) remain multichannel deconvolutions. The 2D Fréchet derivative, $\Phi_{\mathbf{m}_0}$, evaluated for the homogeneous halfspace is given analytically; derivations can be found in [12]; note that there is a misprint in [12] which corrected in [17].

Substitution of $\Phi_{\mathbf{m}_0}$ and the data perturbation

$$\delta d(x, s) = d_{\text{obs}}(x, s) - \mathbf{g}(\mathbf{m}_n, s) \qquad (6)$$

in (4) yields

$$\delta d(x, s) \simeq \int_0^\infty dz' \int_{-\infty}^\infty \Phi_{\mathbf{m}_0}(0, z', x - x', s) \, \delta m(x', z') dx', \qquad (7)$$

where the inner integral is a convolution between $\Phi_{\mathbf{m}_0}(0, z', \cdot, s)$ and $\delta m(\cdot, z')$. The model, $\mathbf{m}$, is discretised into M layers; hence the resistivities $\rho(x, z)$ are approximated by functions $\rho(x, l)$ describing the lateral variation of each individual layer. Insertion of $\delta m(x, l)$ in (7) leads to

$$\delta d(x, s) \simeq \sum_{l=1}^M \int_{-\infty}^\infty \overline{\Phi}_{\mathbf{m}_0}(x - x', l, s) \, \delta m(x', l) \, dx', \qquad (8)$$

where $\overline{\Phi}_{\mathbf{m}_0}$ is the vertical average of the Fréchet derivative defined as

$$\overline{\Phi}_{\mathbf{m}_0}(x, l, s) = \int_{z_{l-1}}^{z_l} \Phi_{\mathbf{m}_0}(0, z', x, s) \, dz'. \qquad (9)$$

Equation (8) is a multichannel convolution, which requires data to be sampled regularly along the profile in all electrode configurations and the model to be partitioned laterally according to the data sampling. A Fourier transformation of (8) leads to

$$\widetilde{\delta d}(k, s) \simeq \sum_{l=1}^M \widetilde{\overline{\Phi}}_{\mathbf{m}_0}(k, l, s) \, \widetilde{\delta m}(k, l), \qquad (10)$$

where the wavenumber $k = 1/\lambda$ is the reciprocal wavelength and $\sim$ denotes Fourier transformation.

For each k, (10) forms a linear system

$$\widetilde{\delta\mathbf{d}}_k \simeq \widetilde{\mathbf{G}}_k \widetilde{\delta\mathbf{m}}_k + \widetilde{\mathbf{e}}_k, \qquad (11)$$

where the element $\widetilde{\delta \mathbf{d}}_k(s) = \widetilde{\delta d}(k,s)$, $\widetilde{\mathbf{G}}_k(s,l) = \widetilde{\widetilde{\mathbf{\Phi}}}_{\mathbf{m}_0}(k,l,s)$, $\widetilde{\delta \mathbf{m}}_k(l) = \widetilde{\delta m}(k,l)$, and $\widetilde{\mathbf{e}}_k$ contains the Fourier transforms of the data errors.

A solution to (11) is [10, 29]

$$\widetilde{\delta \mathbf{m}}_k = \left(\widetilde{\mathbf{G}}_k^{\mathrm{H}} \widetilde{\mathbf{C}}_{\mathrm{d}_k}^{-1} \widetilde{\mathbf{G}}_k + \widetilde{\mathbf{C}}_{\mathrm{m}_k}^{-1} \right)^{-1} \left(\widetilde{\mathbf{G}}_k^{\mathrm{H}} \widetilde{\mathbf{C}}_{\mathrm{d}_k}^{-1} \widetilde{\delta \mathbf{d}}_k + \widetilde{\mathbf{C}}_{\mathrm{m}_k}^{-1} (\widetilde{\mathbf{m}}_{\mathrm{prior}_k} - \widetilde{\mathbf{m}}_{n_k}) \right), \quad (12)$$

where $\widetilde{\mathbf{G}}_k^{\mathrm{H}}$ is the hermitian i.e. transpose and complex conjugate of $\widetilde{\mathbf{G}}_k$, $\widetilde{\mathbf{C}}_{\mathrm{d}_k}$ is the data error covariance matrix describing the statistical variability of the data errors at the wavenumber k, and $\widetilde{\mathbf{C}}_{\mathrm{m}_k}$ is the model covariance matrix describing the statistical variability of the difference, $(\widetilde{\mathbf{m}}_{\mathrm{prior}_k} - \widetilde{\mathbf{m}}_{n_k})$, where $\widetilde{\mathbf{m}}_{\mathrm{prior}_k}$ is an a priori model. Pragmatically, a regularization term, $\mathbf{W}$, is added to (12), so

$$\widetilde{\delta \mathbf{m}}_k = \left(\widetilde{\mathbf{G}}_k^{\mathrm{H}} \widetilde{\mathbf{C}}_{\mathrm{d}_k}^{-1} \widetilde{\mathbf{G}}_k + \widetilde{\mathbf{C}}_{\mathrm{m}_k}^{-1} + \mathbf{W} \right)^{-1} \left(\widetilde{\mathbf{G}}_k^{\mathrm{H}} \widetilde{\mathbf{C}}_{\mathrm{d}_k}^{-1} \widetilde{\delta \mathbf{d}}_k + \widetilde{\mathbf{C}}_{\mathrm{m}_k}^{-1} (\widetilde{\mathbf{m}}_{\mathrm{prior}_k} - \widetilde{\mathbf{m}}_{n_k}) \right). \quad (13)$$

$\mathbf{W}$ is made to dominate at the first iteration and is turned off over the next few iterations. The model perturbation is now

$$\delta \mathbf{m} = \mathcal{F}^{-1}(\widetilde{\delta \mathbf{m}}) \quad (14)$$

where $\mathcal{F}^{-1}$ denotes inverse Fourier transformation. The updating in (5) can be carried out resulting in the model estimate $\mathbf{m}_{n+1}$. This procedure is called iterated multichannel deconvolution, IMCD and the first iteration without the extra stabilization term $\mathbf{W}$ in (13) will be referred to as a multichannel deconvolution, MCD.

2.1 Covariance Matrices and Adaptive Regularisation

The specification of covariance matrices follows [17], where the covariance matrices were tuned to be used in inversion of data with focus depth between 1 m and 16 m and sampled every one metre like data collected with the PACES method [28, 18].

The random character of $\log \rho(x, z)$ is described in the framework of stationary stochastic processes. The spatial correlation is expressed as a self-affine process which is characterized by the von Kármán covariance functions [7, 14]

$$C(x, z, A, \nu) = A^2 C_0 \left(\frac{r}{L} \right)^\nu \mathrm{K}_\nu(r/L) \quad (15)$$

where A is the amplitude, C_0 a constant, L the correlation length, $r = \sqrt{x^2 + y^2}$ spatial distance, and K_ν the modified Bessel function of second kind and order ν. C_0 is tuned so that A is the rms amplitude in the band 4 m $\leq \lambda \leq$ 64 m. A is referred to as the amplitude of the resistivity and ν as the shape parameter. A small ν reflects less correlation than a larger ν. The 1-D auto and cross spectra, which are analytically given cosine transforms of (15) and required for equation

(13), are [6, eqs. 1.13(45) and 1.12(41)]

$$\widetilde{C}(k,z,A,\nu) = \frac{A^2 \, C_0 \, \sqrt{\pi} \, |z|^{\nu+1/2} \mathrm{K}_{-\nu-1/2}\left[|z|\sqrt{1/L^2 + (2\pi k)^2}\right]}{\sqrt{2} \, L^{2\nu} \, [1/L^2 + (2\pi k)^2]^{\nu/2+1/4}} \qquad z > 0, (16)$$

$$\widetilde{C}(k,0,A,\nu) = \frac{A^2 \, C_0 \, \sqrt{\pi} \, 2^\nu \, \Gamma(\nu+1/2)}{2 \, L^{2\nu}[1/L^2 + (2\pi k)^2]^{\nu+1/2}} \qquad \nu > -1/2. \qquad (17)$$

Element (i,j) in the model covariance matrix for the wavenumber k is given as $\widetilde{C}_{\mathrm{m}_k}(i,j) = \widetilde{C}(k, z_i - z_j, A, \nu)$. All examples use a model covariance matrix with correlation length $L = 1000$ m, but varied ν and A.

The multichannel deconvolution and the first iteration of the IMCD is a linear inversion procedure carried out on nonlinear data. Hence, the nonlinear part of the data should be treated as errors and taken into account as part of the data errors [30]. Møller, Jacobsen, and Christensen [17] analyze the stochastic properties of the nonlinearity error. Simulation experiments using model responses of resistivity distributions, obtained by realizations of the von Kármán covariance functions for various values of the shape parameter and the resistivity amplitude, lead to an empirical model for nonlinearity error autopower. The empirical model for nonlinearity error autopower depending on wavenumber k, shape parameter ν, resistivity amplitude A, and electrode configuration is

$$P_{\mathrm{non}}(k,\nu,A,z_{\mathrm{f}}) = \frac{c_{\mathrm{f}} \, z_{\mathrm{f}}^{2(\nu-1)} \, A^4}{16^{2\nu} \, [k^2 + 1/(4z_{\mathrm{f}})^2]^{(2\nu+3)/2}} \qquad (18)$$

where z_{f} is the vertical focus depth (defined as [4] median depth of investigations) for the electrode configuration used, and c_{f} is a scaling factor depending on the sampling distance and the Fast Fourier Transforms. The nonlinearity error autopower enters into the total error covariance matrix as

$$\widetilde{C}_{\mathrm{d}_k} = \widetilde{C}_{\mathrm{d}_k}^{\mathrm{meas}} + \widetilde{C}_{\mathrm{d}_k}^{\mathrm{non}} \qquad (19)$$

where a diagonal element in $\widetilde{C}_{\mathrm{d}_k}^{\mathrm{non}}$ is $P_{\mathrm{non}}(k,\nu,A,z_{\mathrm{f}})$. Because the measurement error is assumed to be white and equal in all electrode configurations, $\widetilde{C}_{\mathrm{d}_k}^{\mathrm{meas}}$ is a diagonal matrix with constant elements, $\sigma(\tilde{e}_{\mathrm{meas}})^2$.

Values for A and ν to be used in specification of the model covariance matrices and the nonlinearity error covariance matrices can be estimated from the data directly. Møller, Jacobsen, and Christensen [17] find simplified empirical equations for apparent ν

$$\nu_{\mathrm{app.}} = 1 - 0.9 \frac{\mathrm{RMS}_{\mathrm{short}}}{\mathrm{RMS}_{\mathrm{long}}} \qquad (20)$$

and for apparent A

$$A_{\mathrm{app.}} = 14(\mathrm{RMS}_{\mathrm{short}} + \mathrm{RMS}_{\mathrm{long}}) \qquad (21)$$

where $\mathrm{RMS}_{\mathrm{short}}$ and $\mathrm{RMS}_{\mathrm{long}}$ are the running rms amplitude for signals in bands $\lambda \in [4, 16]$ m and $\lambda \in [16, 64]$ m stacked from four intermediate electrode configurations.

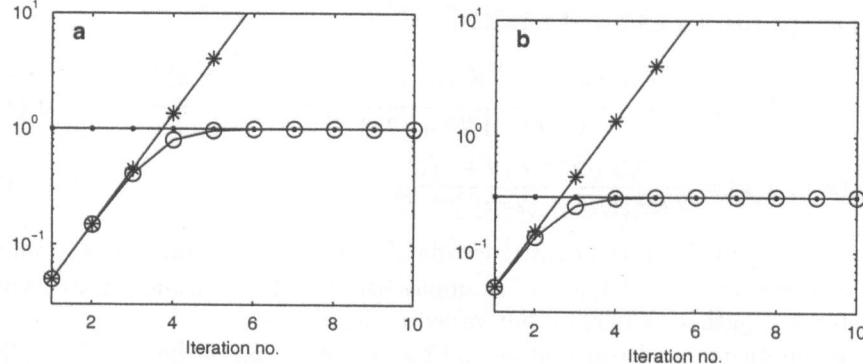

Fig. 1. The iterative scheme for $A = 1$ (a) and $A = 0.3$ (b). The resistivity amplitude A is displayed as $\cdots$, A_{W} is displayed as $***$, and A_{eff} as $\mathrm{o\ o\ o}$.

3 The Iterative Inversion Scheme

The iterative inversion scheme consists of (6), (13), (14), and (5). The forward modelling in (6) is carried out by a finite difference code [15]. The starting model $\mathbf{m}_0$ is usually chosen to be the homogeneous halfspace with a resistivity equal to the geometric mean of the measured apparent resistivities. The specific model covariance matrices $\widetilde{\mathbf{C}}_{\mathrm{m}_k}$ and the nonlinearity error covariance matrices $\widetilde{\mathbf{C}}_{\mathrm{d}_k}^{\mathrm{non}}$ used in (13) are chosen after estimation of ν and A from the data using (20) and (21).

$\mathbf{W}$ used in (13) stabilizing the iterative inversion scheme can be chosen in various ways. The choice

$$\mathbf{W} = (A/A_{\mathrm{W}})^2 \widetilde{\mathbf{C}}_{\mathrm{m}_k}^{-1}, \tag{22}$$

where A_{W} controls the strength of the additional regularisation, gives an effective amplitude A_{eff}

$$A_{\mathrm{eff}} = A/\sqrt{1 + (A/A_{\mathrm{W}})^2}. \tag{23}$$

Figure 1 shows how A_{W} and A_{eff} change depending on the iteration number in a rather conservative inversion scheme for two different values of A, where A_{W} has been chosen to increase exponentially with the iteration number.

The following examples use synthetic data computed for the PACES array. Figure 2 displays the electrode configurations used for the PACES array. All 8 electrode configurations use the same set of current electrodes separated by 30 m. Data collected with this array contain information about the upper 20–30 m of the ground. Data are collected continuously and simultaneously in all electrode configurations resulting in about one sample per metre [27, 18]. So, the synthetic data are also sampled with a sample rate of one metre.

Figure 3a displays a theoretical model consisting of resistive blocks buried in a conductive halfspace and covered with an overburden with an intermediate resistivity of 50 Ωm. The resistivity contrasts vary up to a factor of 50. Model

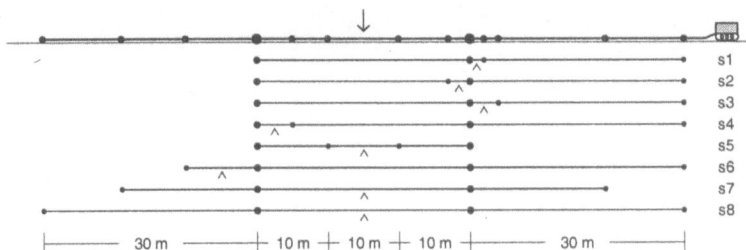

Fig. 2. PACES array. Sliding electrodes are mounted on a cable and towed by a small vehicle. The array has one set of current electrodes (larger ●) separated by 30 m. S1, S2, S3, S4, and S6 are pole-pole like electrode configurations with pole-pole separations of 2, 3, 4, 5, and 10 m. S5 and S8 are Wenner configurations with 10 m and 30 m electrode spacings. S7 is a symmetric configuration with 19 m between current and potential electrodes. The ↓ indicates the common lateral reference point and the ∧ individual lateral focus point.

responses computed for the PACES array are used as data in the inversions, which are carried out on noise-free data, although a data error of 3% has been assumed, so $\sigma_{\text{meas}} = 0.03$. The estimation of ν and A (Figure 3b) leads to an A around 0.7 and a ν around 0.5, which are the values used in inversions resulting in the model estimates in Figures 3c and d.

Figure 3c displays a model estimate after the first iteration using an iterative scheme, where A_{W} is chosen sufficient high that $A_{\text{eff}} = A$, so this is the MCD model estimate. Using an iterative scheme with an initial $A_{\text{W}} = 0.05$, the inversion is converged after 7 iterations and has resulted in the model estimate displayed in Figure 3d. A very similar model estimate with equal misfits is reached using the iterative scheme where A_{eff} is constantly equal to A. The iterated multichannel deconvolution leads to a model estimate with an enhanced vertical resolution compared to the model estimate obtained by multichannel deconvolution.

Figures 3e and f display model estimates obtained using covariance matrices with a ν changed to 0 and an A kept at 0.7. The MCD model estimate (Figure 3e) corresponding to the first IMCD iteration, where A_{eff} is constantly equal to A, is understabilized. Nevertheless, the IMCD inversion converged to a model estimate equal to the model estimate (Figure 3f) obtained using the iterative scheme, where the initial $A_{\text{W}} = 0.05$. The model estimate obtained using $\nu = 0$ (Figure 3f) has a better resolution than the IMCD model estimate (Figure 3d) obtained using a $\nu = 0.5$.

Misfits obtained for the four iterative inversion results are shown in Figure 4. Two data misfits are obtained. One that takes only the measuring error into account, i.e

$$\phi_{\mathrm{d_m}} = \delta\mathbf{d}^{\mathrm{T}}\mathbf{C}_{\mathrm{d_{meas}}}^{-1}\delta\mathbf{d} \tag{24}$$

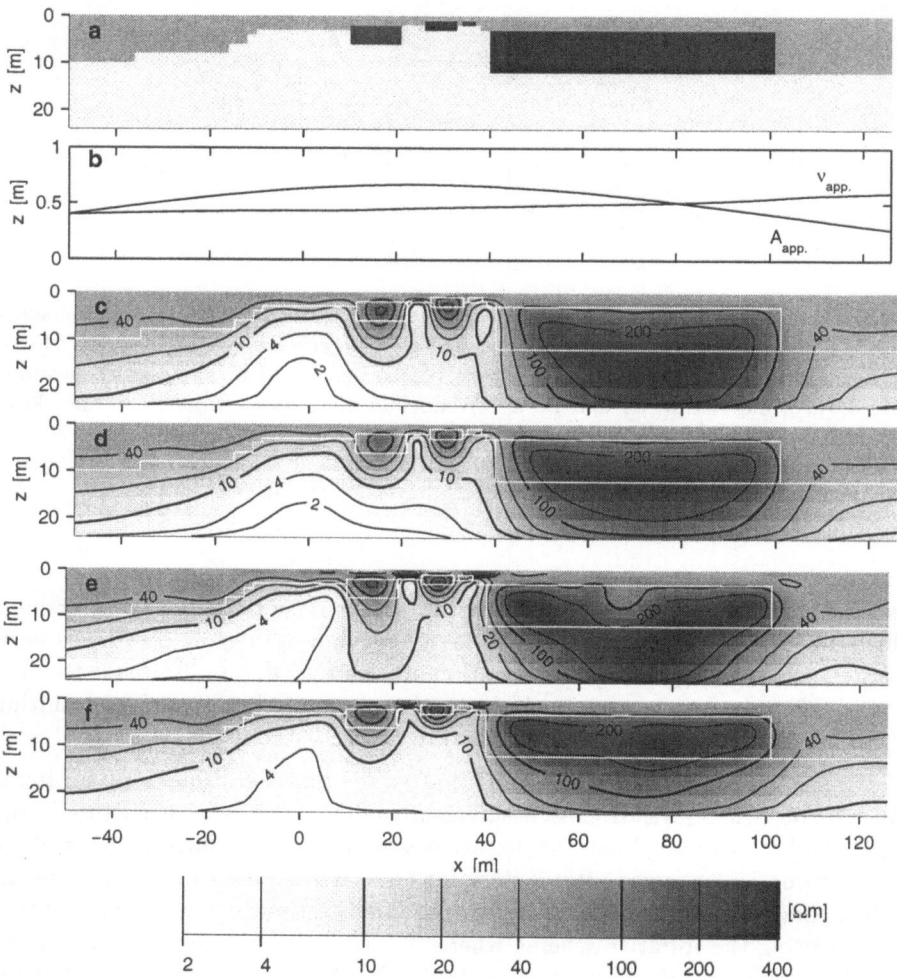

Fig. 3. Theoretical model (a) with resistive blocks (250–370 Ωm) buried in a conductive halfspace of 6.8 Ωm and an overburden of 50 Ωm. (b) Apparent ν and A estimated from data (Figure 6) computed for the PACES array. MCD model estimate (c) obtained using a $\nu = 0.5$. Model estimate (d) after 7 iterations with initial $A_\mathrm{W} = 0.05$ and using a $\nu = 0.5$. (e) is the same as (c) except that a $\nu = 0$ is used. (f) is the same as (d) except that a $\nu = 0$ is used. The contour interval is marked on the colour bar.

and another that takes the total data error, including the nonlinearity error, into account, i.e

$$\phi_\mathrm{d} = \delta\mathbf{d}^\mathrm{T}\mathbf{C}_\mathrm{d}^{-1}\delta\mathbf{d} \tag{25}$$

or expressed as a sum of the Fourier coefficients

$$\phi_\mathrm{d} = \sum_{k=0}^{N_\mathrm{x}} \widetilde{\delta\mathbf{d}}_k^\mathrm{T} \widetilde{\mathbf{C}}_{\mathrm{d}_k}^{-1} \widetilde{\delta\mathbf{d}}_k \tag{26}$$

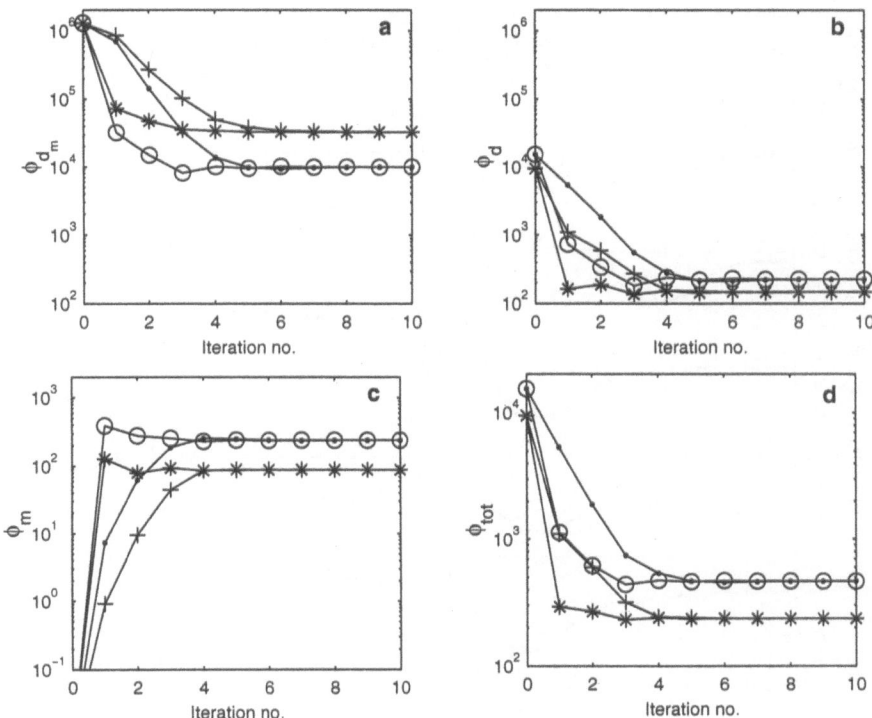

Fig. 4. (a) Data misfit, ϕ_{d_m}, assuming 3% uncorrelated measuring errors, equation (24). (b) Total data misfit ϕ_d, equation (25). (c) Model misfit ϕ_m relative to a priori model, equation (27). (d) Total misfit ϕ_{tot}, equation (29). The $*\,*\,*$ curve displays misfits where $\nu = 0.5$ and $A_{eff} = A$ is used (model estimate after first iteration in Figure 3c). The $+\,+\,+$ curve displays misfits, where $\nu = 0.5$ and an initial $A_W = 0.05$ is used (model estimate after convergence in Figure 3d). The $\circ\circ\circ$ curve displays misfits, where $\nu = 0$ and $A_{eff} = A$ is used (model estimate after first iteration in Figure 3e). The $\cdots$ curve displays misfits, where $\nu = 0$ and an initial $A_W = 0.05$ is used (model estimate after convergence in Figure 3f).

where N_x is the number of samples along the profile.

The model misfit relative to the a priori model is

$$\phi_m = (\mathbf{m}_{prior} - \mathbf{m}_n)^T \mathbf{C}_m^{-1}(\mathbf{m}_{prior} - \mathbf{m}_n) \tag{27}$$

or expressed as a sum of the Fourier coefficients

$$\phi_m = \sum_{k=0}^{N_x} (\widetilde{\mathbf{m}}_{prior_k} - \widetilde{\mathbf{m}}_{n_k})^T \widetilde{\mathbf{C}}_{m_k}^{-1}(\widetilde{\mathbf{m}}_{prior_k} - \widetilde{\mathbf{m}}_{n_k}) \tag{28}$$

and the total misfit is

$$\phi_{tot} = \phi_d + \phi_m. \tag{29}$$

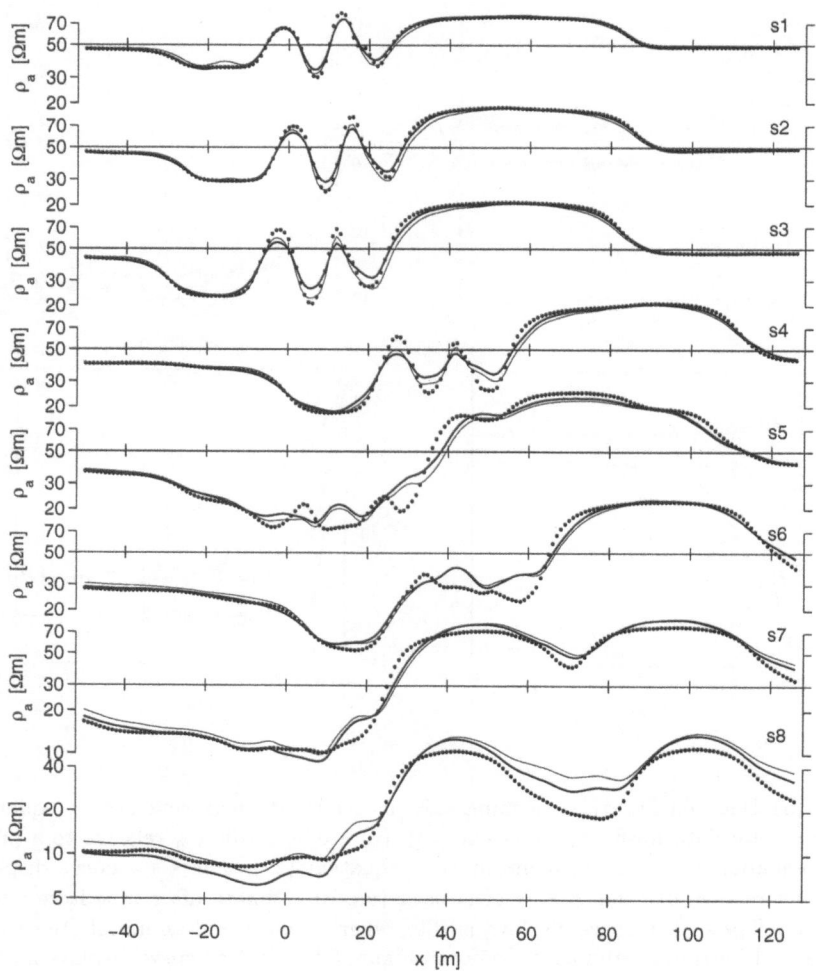

Fig. 5. Model responses computed for the model in Figure 3a using the PACES array are displayed as dotted curves for all electrode configurations (Figure 2). Thin curves show the response of the MCD model estimate (Figure 3c) obtained using $\nu = 0.5$ and $A_{\text{eff}} = A$. Heavy curves show response of the model estimate (Figure 3d) obtained using $\nu = 0.5$ and an initial $A_{\text{W}} = 0.05$.

The misfits can be computed very efficiently in the wavenumber domain.

Figure 4 shows that the misfits for the two iterative schemes follow different routes towards convergence. ϕ_{d_m} (Figure 4a) indicates that parts of the data are not very well fitted from the point of view of uncorrelated errors as ϕ_{d_m} reaches a level above 4096 which is the total number of data.

Figure 5 displays true model responses, responses of the model estimate after convergence (Figure 3d), and responses of the MCD model estimate (Figure 3c).

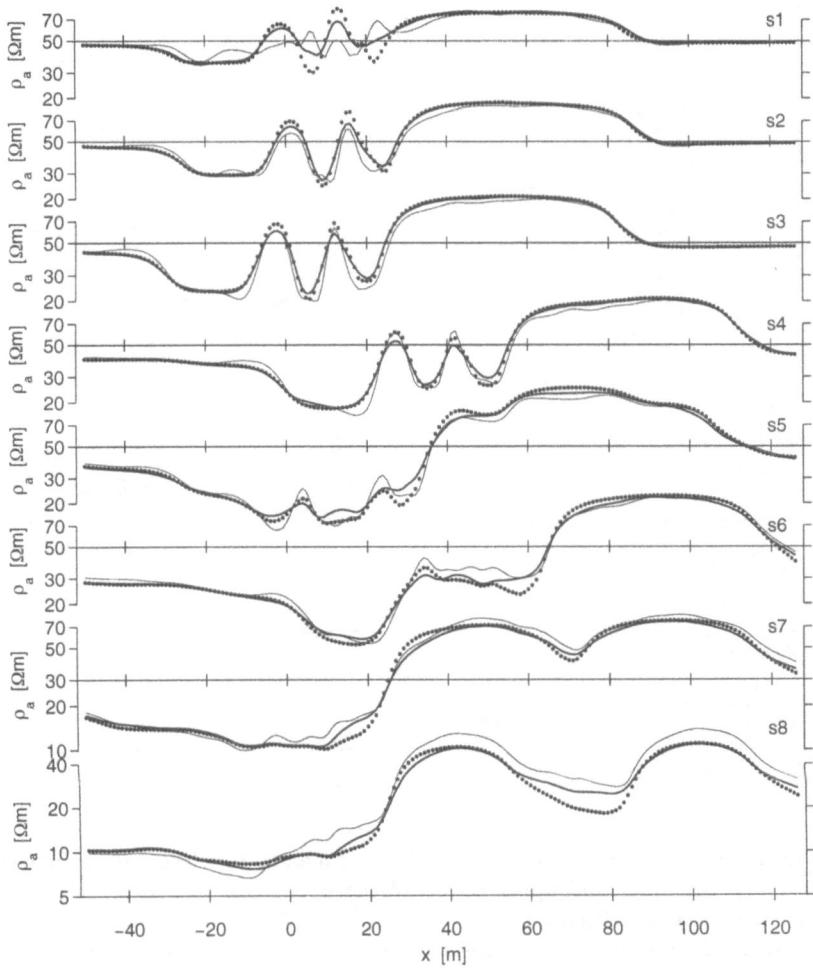

Fig. 6. Same as Figure 5 except that a $\nu = 0$ is used, so thin curves display the response of the MCD model estimate in Figure 3e and heavy curves show the response of the model estimate in Figure 3f.

Similarly, the responses of the model estimate shown in Figures 3e and f, obtained using covariance matrices with $\nu = 0$, are displayed in Figure 6. Both Figures 5 and 6 demonstrate that the first iteration yields a good data fit but that iterative inversion does lead to an improvement in data fit.

In general, the data fit for the model estimates obtained using a $\nu = 0$ (Figure 6) is better than the data fit for the model estimates obtained using a $\nu = 0.5$ (Figure 5). An exception is the model response of the shortest electrode configuration s1 (Figure 2) in the interval between -10 m and 30 m (upper pairs of curves in Figure 7a), where three shallow resistive blocks occur (see Figure

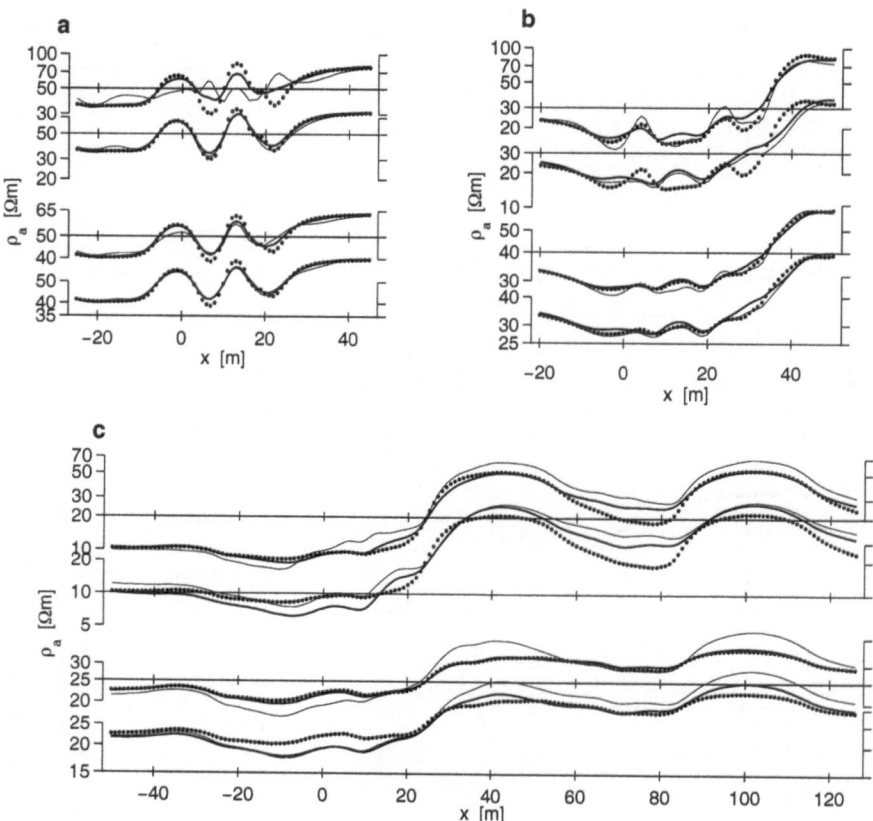

Fig. 7. (a) Zoom on the model responses in Figures 5 and 6 for electrode configuration s1 displayed as the lower and upper curves of the upper pair of curves, respectively. The lower pair of curves display true model response (dotted curve) of a model with lower contrasts (Figure 8a) and responses of model estimates in Figures 8c, d, e, and f. (b) is the same as (a) except that the responses are for electrode configuration s5. (c) is the same as (a) except that the responses are for electrode configuration s8. The lower pair of curves in (a), (b) and (c) are gained up by a factor of two.

3a). The resistive artifacts at the surface of the ground between the resistive blocks in the model estimate in Figure 3f may cause the lack of fit for the model response obtained using a $\nu = 0$.

The difference in data fit between the two cases (Figures 5 and 6) generally increases with increased electrode separation for the rest of the electrode configurations, i.e. the data fit of electrode configuration s8 (Figure 7c) is substantially better for the response of the model estimate obtained using a $\nu = 0$ than for the response of the model estimate obtained using a $\nu = 0.5$.

The responses of the intermediate electrode configurations of the model estimate obtained using $\nu = 0.5$ yields not only a wrong amplitude but even a wrong

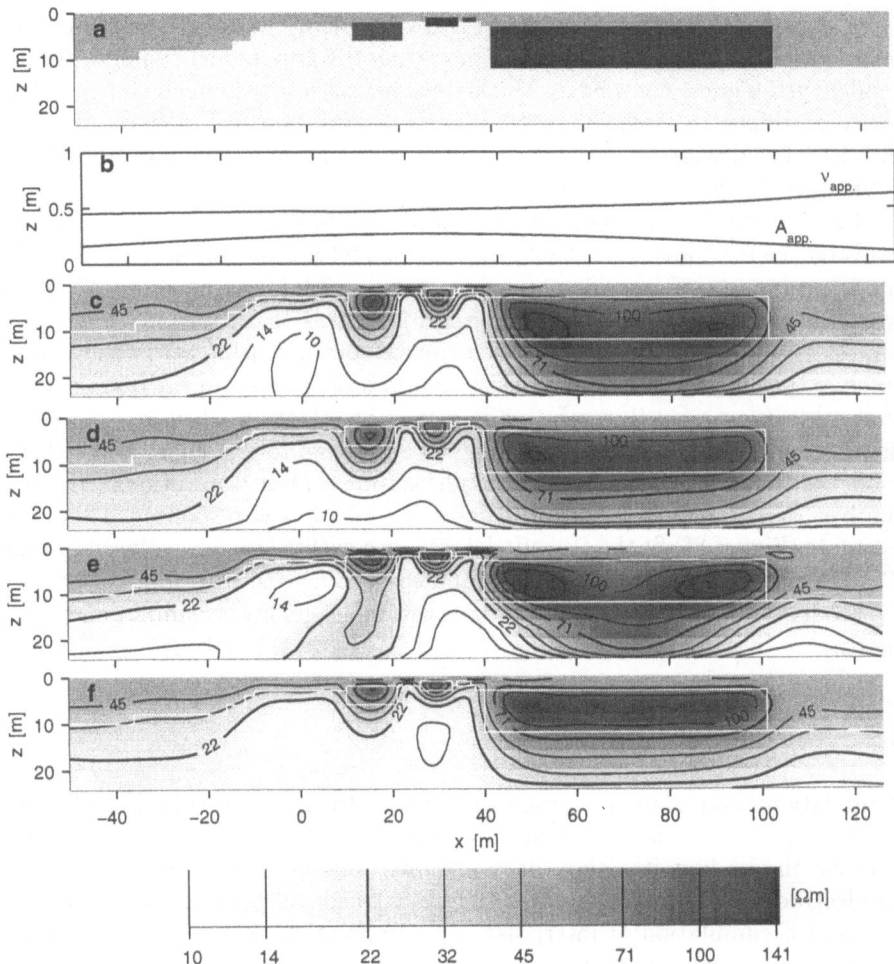

Fig. 8. True model (a) with lower resistivity contrasts. Resistive blocks (111–136 Ωm) are buried in a conductive halfspace of 18 Ωm and overlain by an overburden of 50 Ωm. (b) shows apparent ν and A estimated from data computed for the PACES array. MCD model estimate (c) obtained using a $\nu = 0.5$. Model estimate (d) after 7 iterations with initial $A_{\mathrm{W}} = 0.05$ and using a $\nu = 0.5$. (e) is the same as (c) except that a $\nu = 0$ is used. (f) is the same as (d) except that a $\nu = 0$ is used. The contour interval is marked on the colour bar. Colour bar and contour interval are chosen so that structures are dicertly comparable to the case in Figure 3, i.e. contour lines 22 and 71 Ωm correspond to contour lines 10 and 100 Ωm.

polarity, i.e bumps in the curves are complitely out of phase with bumps in the true model response curve. This is displayed for the interval between -20 m and 50 m in Figure 7b. The model response of the model estimate obtained using $\nu = 0$ suffers a little from the same effect. However, this is not the case for the response (Figure 7b) of a model estimate (Figure 8d) obtained using a $\nu = 0.5$

from data computed for a model (Figure 8a) with the same structures but a lower amplitude of the anomalies. Comparing the true model response from the model with highest contrasts with the true model response from the model with lower contrasts, it is seen that the shape of the responses has changed because of a stronger distortion of the current flow in the ground, which is more difficult to map properly with the approximate Fréchet derivative.

Figure 8 displays MCD model estimates and model estimates obtained after convergence for a lower resistivity contrast model similar to Figure 3. The MCD model estimate in Figure 8c obtained using a $\nu = 0.5$ and an $A = 0.25$ (chosen from the estimated ν and A in Figure 8b) resolves the low contrast anomalies almost as well as the iterated model estimate obtained using a $\nu = 0$ (Figure 3f) resolves the high contrast anomalies. It is seen more clearly for the lower contrasts that an improvement in resolution is gained by iterating when $\nu = 0.5$ (Figure 8d). Stepping ν down to 0 and keeping $A = 0.25$ lead to an understabilized MCD model estimate (Figure 8e), but a model estimate after 6 iterations (Figure 8f) resolves the anomalies very well. Almost all over do the responses of the model estimate (Figure 8f) fit the true model response within 5%, as displayed in Figure 7 for electrode configurations s1, s5, and s8.

Model estimates obtained from data with noise look very similar and are not shown.

3.1 Non-Periodic Data

Field data measured along a profile do not fulfil the requirement for the IMCD as it is set up, because field data are not periodic and data gaps occur for various reasons. In the first iteration, data are interpolated to a regular sampling for all electrode configurations and extrapolated using a periodic spline. In order to avoid accumulation of interpolation/extrapolation errors in the subsequent iterations, the data perturbation (6) is modified in the following very simple way

$$\delta d(x, s) = \begin{cases} d_{\mathrm{obs}}(x, s) - \mathrm{g}(\mathbf{m}_n, s) & \text{if } d_{\mathrm{obs}}(x, s) \text{ is measured} \\ 0 & \text{if } d_{\mathrm{obs}}(x, s) \text{ is interpolated/extrapolated.} \end{cases}$$

(30)

The extrapolated part of $\delta\mathbf{d}$ should have a length at least equal to that of the largest electrode configuration used to prevent information from the data to disturb at the other end of the profile.

Figures 9a and 10a display sections of a stochastic resistivity distribution which is a realization of a von Kármán covariance function (15) where $L = 1000$ m, $\nu = 0.5$, and $A = 0.5$. The simulation is carried out in the wavenumber domain resulting in a periodic model with a total profile length of 768 m. Model responses are computed for the PACES array and used as the periodic data. A set of non-periodic data with a profile length of 512 m is made from the periodic data. The same iterative scheme, where the initial $A_{\mathrm{W}} = 0.05$, is used inverting the two data sets, except that (30) substitutes (6) in inversion of the

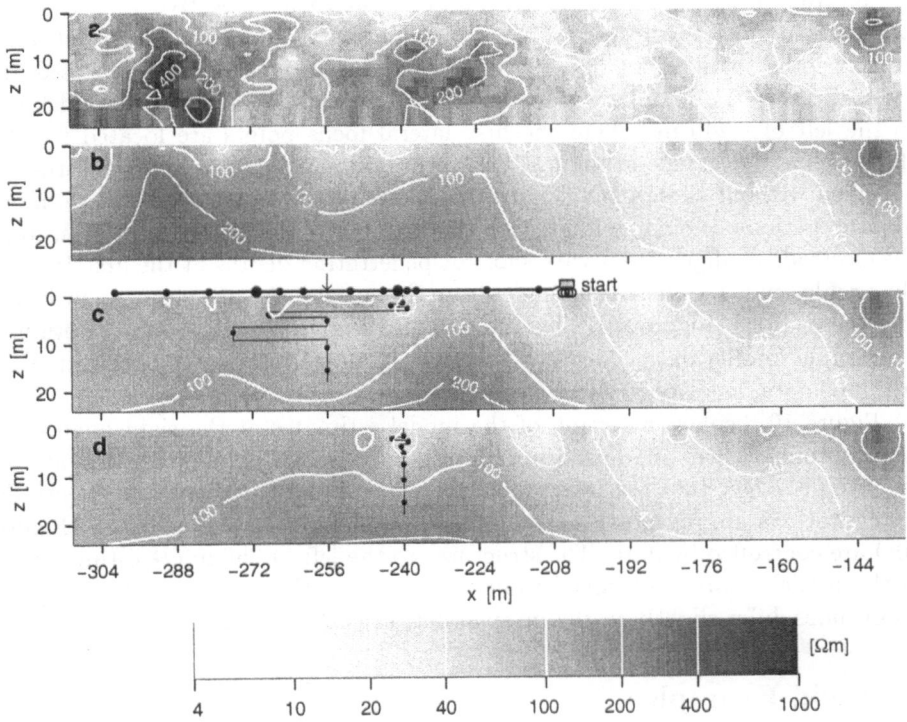

Fig. 9. Non-periodic effects at profile start. (a) is a section of the true model that is a realization of a von Kármán covariance function with $L = 1000$ m, $\nu = 0.5$, and $A = 0.5$. (b) is a section of a model estimate obtained from the periodic data set. (c) is a section of a model estimate obtained from the non-periodic data set with profile start at -256 m. (d) is a section of a model estimate obtained from the non-periodic data set, where the data are trimmed at profile start. The dots connected with lines in (c) and (d) display the first lateral focus point of all electrode configurations used, plotted at their respective focus depths. The periodic data set is the model responses computed for the PACES array and the non-periodic data set is a segment of those. The contour interval is marked on the colour bar.

non-periodic data. The inversions are carried out using covariance matrices with the true $\nu = 0.5$ and true $A = 0.5$.

Figure 9b displays a section of the periodic model estimate obtained after 10 iterations and Figure 9c displays the same section of a model estimate after 10 iterations obtained using the non-periodic data. The left-hand edge of the profile is at -256 m which is the first common reference point for all 8 electrode configurations. The common reference point is the midpoint between the current electrodes (Figure 2). This point is the lateral focus point only for the symmetric electrode configurations (s5, s7 and s8 in Figure 2). The lateral focus points for the pole-pole like electrode configurations (s1, s2, s3 s4, and s6 in Figure 2) are at the midpoints between the two poles (Figure 2). The lateral focus points are plotted on top of the section in Figure 9c at their respective vertical focus depths. The model estimate looks very similar to the periodic model estimate at all depths to the right of −195 m. At that position, all electrodes used are located inside the first lateral focus point. The data points with lateral focus to the left of −240 m, where the first lateral focus points are located for the three shortest electrode configurations, seem to do no good, as a model estimate obtained without these data (Figure 9d) looks rather similar but a little more like the periodic model estimate. Apparently, the absence of information from the electrode configurations with shortest penetration depths at the first 16 m of the profile tends to remove anomalies from the surface and move them downward into the sections (Figure 9c). This artifact, also seen in model estimates obtained from other profile data, does not show up in model estimates obtained from trimmed data, as displayed in Figure 9d.

Figure 10 displays a section of the model estimates at the right-hand edge of the profile, where the last common lateral reference point is located at 256 m. At this edge, the last lateral focus points of the three shortest electrode configurations are outmost, so the upper layers of the model estimate (Figure 10c) are controlled by data. The structures to the left of 256 m are estimated as in the model estimate obtained from periodic data (Figure 10b), although the amplitudes differ slightly in the interval between 224 m and 256 m.

4   Field Examples

4.1   PACES Data from Ølst

In an exploration for smectite-rich clay, more than 30 profile-kilometres of data, covering about 12 km^2, were collected using the PACES method [27, 18]. Upper Palaeocene clay from the Ølst Formation [9] outcrops in hills formed by glaciotectonics at Ølst, Jutland, Denmark. The heavily disturbed clay deposits are surrounded by glaciofluvial sand and gravel deposits and overlain by a thin clayish till [19].

Selected parts of the data set are split up in segments of about 900 m and the segments are padded to obtain a total length of 1280 m. Figure 11a displays a 176 m interval of a data segment as an apparent resistivity pseudosection where data are plotted at the lateral focus point and at the focus depth of the respective electrode configuration. Figure 11c displays the MCD model estimate. The model estimates after 10 iterations, shown in Figures 11d and e, are obtained using an iterative scheme where the initial $A_W = 0.05$. After estimation from the data (equations 20 and 21), ν is chosen to be 0.5 and A to be 0.4, which are the values used obtaining the model estimates displayed in Figures 11c and d, whereas a $\nu = 0$ is used obtaining the model estimate shown in Figure 11e. The three different model estimates agree very well, except for the deeper part of the sections between 440 m and 500 m, where very strong resistivity contrasts occur.

High resistivities above 100 Ωm are interpreted as dry glaciofluvial sand and gravel, low resistivities below 5 Ωm as Palaeocene clay, and the intermediate

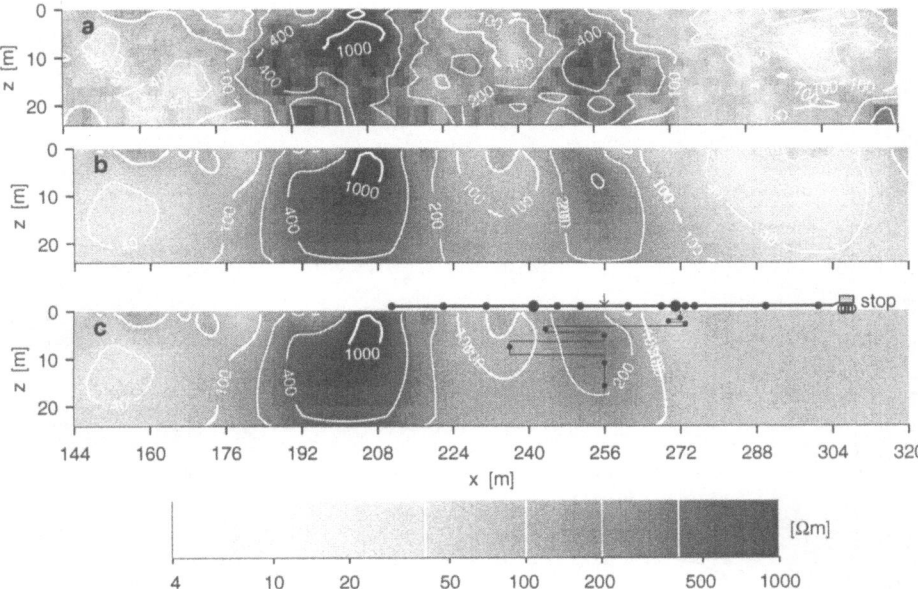

Fig. 10. Non-periodic effects at profile end. (a) is a section of the true model also used in Figure 9. (b) is a section of a model estimate obtained from the periodic data set. (c) is a section of a model estimate obtained from the non-periodic data set with profile end at 256 m. The dots connected with lines in (c) display the last lateral focus point of all electrode configurations used, plotted at their respective focus depths. The contour interval is marked on the colour bar.

resistivities about 30 Ωm a clayish till. The contact between clay and sand can be followed in the adjacent lines supporting the assumption of a 2-D structure with strike direction perpendicular to the profile.

4.2 CVES Data from Grundfør

Profiles of continuous vertical electrical sounding (CVES) data were collected at Grundfør, Jutland, Denmark, as part of a large groundwater project under The Danish Environmental Research Programme. Quaternary aquifers in buried valleys incised into Tertiary clay deposits were mapped and studied in these projects.

Data were collected in Wenner configurations with 10 electrode spacings between 5 m and 120 m using the ABEM Lund Imaging System [1]. Figure 12a displays data from the central profile as an apparent resistivity pseudosection.

After estimation from the data (equations 20 and 21), ν is chosen to be 0.5 and A to be 0.25, which are the values used obtaining the model estimates displayed in Figures 12d and e, whereas a $\nu = 0$ is used obtaining the model estimate shown in Figure 12f. Figure 12d displays the MCD model estimate.

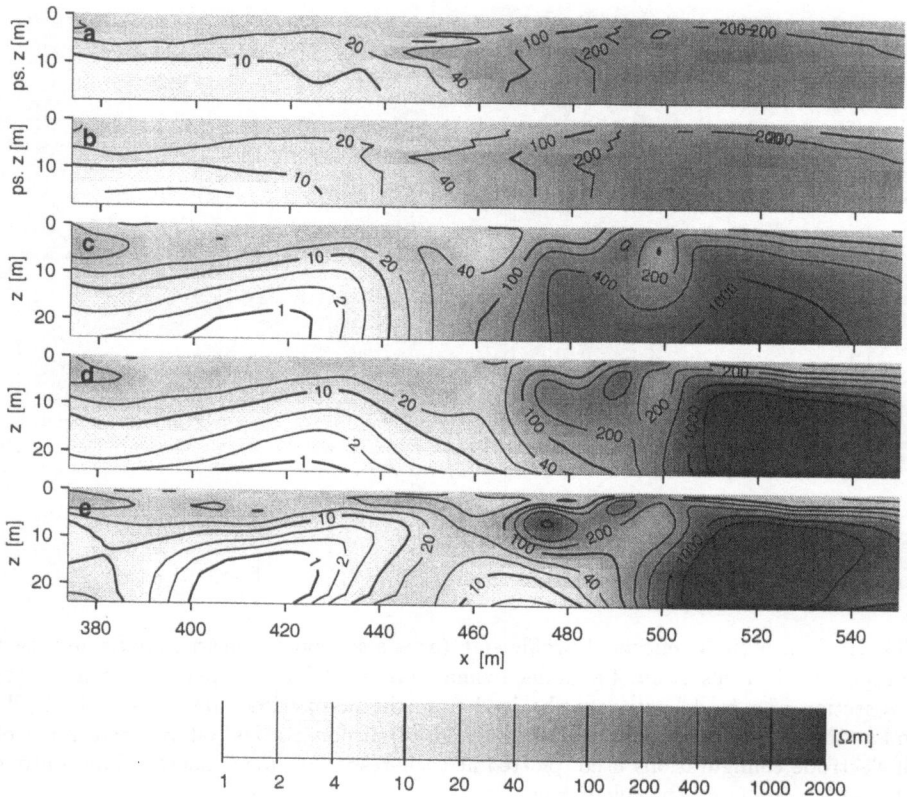

Fig. 11. Data (a) collected with the PACES array are displayed as an apparent resistivity pseudosection with the data located at the lateral focus point and focus depth of the individual electrode configurations. (b) is the model response of the model estimate in (d). MCD model estimate (c) obtained using a $\nu = 0.5$. Model estimate (d) after 10 iterations with the initial $A_W = 0.05$ and using a $\nu = 0.5$. (e) is the same as (d), except that a $\nu = 0$ is used. The contour interval is marked on the colour bar.

Fig. 12. Data (a) collected in 10 Wenner configurations with electrode spacing between 5 m and 120 m are displayed as an apparent resistivity pseudosection. (b) and (c) are responses of the model estimate in (e) and (f), respectively. MCD model estimate (d) obtained using a $\nu = 0.5$. Model estimate (e) after 12 iterations with the initial $A_W = 0.05$ and using a $\nu = 0.5$. (f) is the same as (e) except that a $\nu = 0$ is used. Model estimate obtained by an inversion program using recalculated partial derivatives. Black rectangles in (d), (e), (f), and (g) indicate the location of Ellog drillings, where the in situ resistivity is measured. The resulting resistivity logs are plotted as colour code in the rectangles in (d). The contour interval is marked on the colour bar.

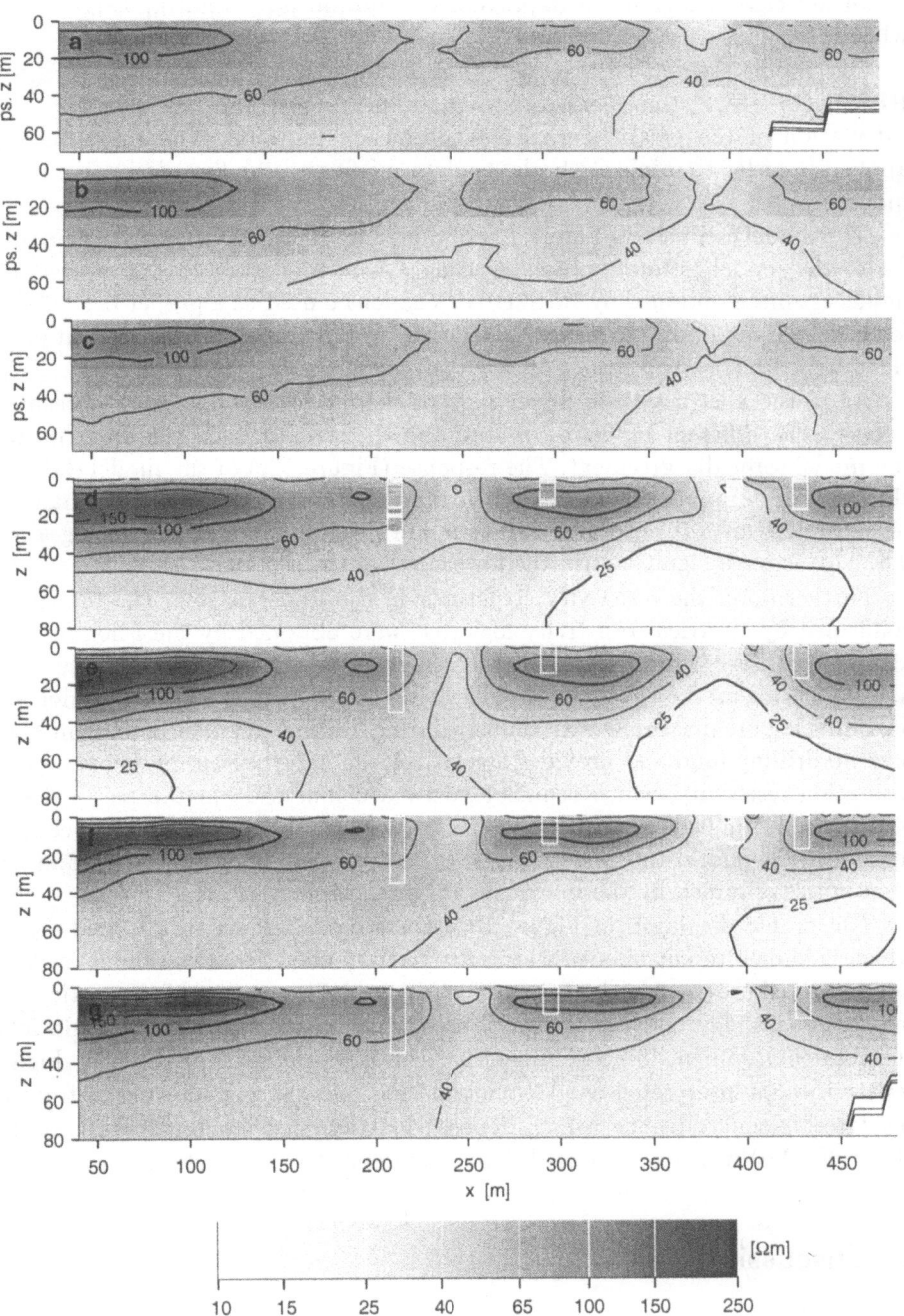

The model estimates after 12 iterations shown in Figures 12e and f are obtained using an iterative scheme where the initial $A_W = 0.05$.

In this case no significant improvement is obtained by iterating (Figure 12e), when a $\nu = 0.5$ is used.

The model estimate in Figure 12g is obtained using the inversion program RES2DINV [13,2] that performs a smoothness constraint 2D inversion with calculation of the partial derivatives after each iteration. The smoothness is applied in order to minimise both the data misfit and the first derivative of the model.

The model estimate in Figure 12g obtained by RES2DINV looks rather similar to the model estimate obtained using a $\nu = 0$ (Figure 12f). However, the model estimate obtained by RES2DINV cannot be used as a proper bench mark for how well approximate partial derivatives work compared to those computed rigorously, because two rather different inversion schemes are used.

As in the Ølst case, the upper part of the model shows a good agreement between the different model estimates, and the structures in the lower part of the model some disagreement. The response (Figure 12c) of the model estimate (Figure 12f) obtained using a $\nu = 0$ fits data (Figure 12a) better than the response (Figure 12b) of the model estimate (Figure 12e) obtained using a $\nu = 0.5$. This is most significant in the lower part of the sections.

Furthermore, the resistivity structures in the upper part of the model are confirmed by electrical resistivity logs that were obtained by the Ellog method at four locations (white rectangles in Figure 12), which were less than 10 m off line. The Ellog method is an auger drilling method, by which an electrical log and a gamma log are measured continuously during drilling [26]. The drilling method uses no drilling mud and provides a relatively undisturbed estimate of the true formation resistivity, which is indicated by a colour code in the rectangles in Figure 12d. Thin beds present in the logs are not resolvable with surface electrical methods. Therefore, the correspondence between the Ellog resistivity and the resistivities obtained by the inversions is quite satisfactory.

The profile displayed in Figure 12 maps a cross section in a buried valley, which is known to cut about 120 m into Tertiary clay. The low (about 25 Ωm) and intermediate resistivities (about 40 Ωm) in the lower part of the section are interpreted as clay and more coarse-grained sediments deposited in a glaciolacustrine environment [8]. The upper part of the section, dominated by higher resistivities, is interpreted as glaciofluvial sand and gravel deposited in front of the latest glacial advance, which deposited a thin incoherent till (resistivities about 40 Ωm) which is seen as the topmost deposit in the section.

5 Discussion

The iterated multichannel deconvolution is a stable inversion procedure that converges to model estimates with a roughness determined by the resistivity amplitude A and shape parameter ν. The differences seen between model estimates with different ν are caused by the difference in ν and not by lack of convergence.

Tests have shown that the inversion procedure seems to be independent of the starting model. Using the true model as well as a stochastic resistivity distribution (as the one in Figure 9a) as starting models, inversions of the data used in Figure 3 resulted in model estimates (not shown) equal to the ones displayed in Figure 3. Only with an inversion scheme that uses the term $\mathbf{C}_m (\mathbf{m}_{\text{prior}} - \mathbf{m}_n)$ (see equation 12 or 13), the result will be independent of the starting model.

Two iterative schemes are tested. The conservative one, which has an extra stabilization term $\mathbf{W}$ at the first iterations, results in model estimates that gain more and more structure and amplitude through the iterations. The other iterative scheme, without $\mathbf{W}$, results after the first iteration, in a model estimate that can be the near optimal model estimate (e.g., Figure 3c) in the linearised approximation or a clearly understabilized model estimate (Figure 3e). Even when the model estimate after the first iteration is clearly understabilized, convergence is obtained towards a model estimate equal to the one obtained by the conservative iteration scheme. Convergence is not necessarily obtained faster by the less stabilized inversion.

Figures 3 and 8 show that $\nu = 0$ leads to a better iteration result than $\nu = 0.5$, which is derived by the estimation of ν from the data. The empirical expressions for estimation of A and ν directly from data were set up on the basis of isotropic covariance functions [17]. The deterministic model (Figures 3 and 8) has anomalies with a horizontal to vertical aspect ratio significantly greater than 1:1, which might explain why $\nu = 0.5$ leads to a poorer resolution. The stochastic resistivity distribution used as true model in Figures 9 and 10 has a horizontal to vertical aspect ratio about 1:1. So, in that case a ν equal to the true ν is the best choice. The ν estimated from data results in reasonable model estimates of the multichannel deconvolution, but it appears that the value of ν should be lower when used in the iterative multichannel deconvolution.

The estimated resistivity amplitude A is rather independent of the shape parameter ν. Tests have shown that A may be varied by at least 30 % without significant changes in the model estimates obtained after convergence. A higher A, the conservative choice, results in model estimates that are smoother, because the nonlinearity error covariance matrix grows as A^4, whereas the model covariance matrix grows as A^2.

The use of A and ν estimated directly from data in the inversion scheme leads to an iterative inversion that converges to reasonable model estimates without the need of line searches. Line searches are often stated as important for the convergence of the iteration scheme [3, 20].

The nonlinearity error examined in [17] is taken into account as a part of the total error covariance matrix, equation (19). The nonlinearity error does not vanish as convergence is reached. Tests (not shown here) indicate that poorer model estimates result from even a moderate reduction in assumed nonlinearity error. Therefore, the full amplitude A is kept at all iterations.

When inverting field data using a 2D inversion procedure it may be wise to keep the error estimates high by choosing the estimated A at its maximum, because the 3-dimensionality of the data acts as errors in a 2D data set.

The inversion procedure, which is periodic in nature, can be modified to work on non-periodic data too, as shown in Figures 9, 10, 11, and 12. Tests have shown that the data set should have a profile length more than 3 times the length of the largest electrode configuration used to obtain reliable model estimates. Padding areas at the ends of the profile, 2–3 times longer than the required length of the largest electrode configuration (for not mixing information from the data at the other end), do also stabilize the inversion.

The nonlinearity of the geoelectrical problem increases with increasing contrast of the resistivity anomalies. The better resolved anomalies in the case with lower resistivity contrast (Figures 8d and f), compared to the anomalies in the high-contrast case (Figures 3d and f), may indicate that the 2D Fréchet derivative of the homogeneous halfspace approximates the partial derivatives the best when the resistivity contrasts are not too high. Sasaki [25] shows, for a theoretical model with resistivity contrast up to a factor of 20, that model estimates obtained by a full 3D inversion are slightly better than model estimates obtained by a 3D inversion using partial derivatives approximated by those of the homogeneous halfspace. Similar results can be drawn from [2] for 2D inversion of theoretical 2D data sets.

The 2D Fréchet derivative of the homogeneous halfspace, which takes relatively long time to compute, is generated once and stored for desired electrode arrays and profile sampling. The covariance matrices are also computed and stored in advance. Because there are no partial derivatives to compute before each iteration and only many small matrices to invert, more than 99% of the total computation time is spent on the forward modelling.

6 Conclusions

A 2D inversion procedure is presented that uses the 2D Fréchet derivative of the homogeneous halfspace as approximate partial derivative. The inversion at each iteration step is solved as a multichannel deconvolution. Thereby, the iterative inversion procedure spends almost all the computation time on the forward modellings.

It appears that the inclusion of nonlinearity errors as a part of the data error covariance matrix is an important reason for the stability of the inversion scheme.

Even for relatively high resistivity contrasts, simple multichannel deconvolution model estimates are fairly good, as shown in [17]. This paper shows that significant improvement results from iterative multichannel deconvolution. In particular, this is the case when the resistivity contrasts are not too high. It is indicated that approximate partial derivatives may have some limitations when resistivity structures become complicated and their amplitudes very high.

Acknowledgements

I wish to thank Doug Oldenburg, The University of British Columbia, for kindly supplying the finite difference code. Kurt Sørensen, University of Aarhus and Aarhus Water Supply have partly financed the development of the iterated multichannel deconvolution. Thanks to Kurt Sørensen and Esben Auken, Aarhus University, for making the field data available. The data from Ølst were collected for Aarhus County. Collection of CVES and Ellog data in Grundfør was carried out under the Groundwater Group of The Danish Environmental Research Programme. I am grateful to Bo Holm Jacobsen for his invaluable ideas for improvements during the development of the inversion procedure and to both Bo Holm Jacobsen and Niels Bøie Christensen for their constructive comments to the manuscript.

References

1. Dahlin, T., 1996, 2D resistivity surveying for environmental and engineering applications, First Break, 14, 275–283.
2. Dahlin, T. and Loke, M. H., 1998, Resolution of 2D Wenner resistivity imaging as assessed by numerical modelling, Journal of Applied Geophysics, 38, 237–249.
3. deGroot Hedlin, C. and Constable, S., 1990, Occam's inversion to generate smooth, two-dimensional models from magnetotelluric data, Geophysics, 55, 1613–1624.
4. Edwards, L. S., 1977, A modified pseudosection for resistivity and IP, Geophysics, 42, 1020–1036.
5. Ellis, R. G. and Oldenburg, D. W., 1994, The pole-pole 3-D DC-resistivity inverse problem: A conjugate gradient approach, Geophysical Journal International, 119, 187–194.
6. Erdélyi, A., ed., 1954, Table of Integral Tranforms, Vol. I, McGraw-Hill Book Company, Inc., New York, Toronto, London.
7. Goff, J. A. and Jordan, T. H., 1988, Stochastic modeling of seafloor morphology: Inversion of sea beam data for second-order statistics, Journal of Geophysical Reseach, 93(B11), 13,589–13,608.
8. Gravesen, P., 1997, Three-dimentional geological model of the complex aquifers in an incised Quartenary valley in Jutland, Denmark, in 'Proceedings', Environmental and Engineering Geophysical Society, European Section, 229–232.
9. Heilmann-Clausen, C., Nielsen, O. B. and Gersner, F., 1985, Lithostatigraphy and depositional environments in the Upper Paleocene and Eocene of Denmark, Bulletin of the Geological Society of Denmark, 33, 287–323.
10. Jackson, D. D., 1979, The use of a priori data to resolve nonuniqueness in linear inversion, Geophysical Journal of the Royal Astronomical Society, 57, 137–157.
11. Li, Y. and Oldenburg, D. W., 1994, Inversion of 3-D DC resistivity data using an approximate inverse mapping, Geophysical Journal International, 116, 527–537.
12. Loke, M. H. and Barker, R. D., 1995, Least-squares deconvolution of apparent resistivity pseudosection, Geophysics, 60, 1682–1690.
13. Loke, M. H. and Barker, R. D., 1996, Rapid least-squares inversion of apparent resistivity pseudosections by a quasi-Newton method, Geophysical Prospecting, 44, 131–152.

14. Maurer, H., Holliger, K. and Boerner, D. E. , 1998, Stochastic regularization: Smoothness or similarity?, Geophysical Reseach Letters, 25, 2889–2892.
15. McGillivray, P. R. , 1992, Forward modeling and inversion of DC resistivity and MMR data, PhD thesis, The University of British Columbia, Vancouver, Canada.
16. Møller, I. and Sørensen, K. I. , 1998, A new approach for fast 2D geoelectrical mapping of near-surface structures, European Journal of Environmental and Engineering Geophysics, 2, 247–261.
17. Møller, I., Jacobsen, B. H. and Christensen, N. B., 1999, Multichannel deconvolution of 2-D geoelectrical data, Geophysics. Submitted.
18. Møller, I., Sørensen, K. I. and Christensen, N. B. , 1998, DC-resistivity multi-electrode profiling in hydrogeological investigations: A comparative study of the pulled array continuous electrical sounding method and a multi-electrode method with fixed electrodes, in 'Proceedings of the Symposium on the Application of Geophysics to Engineering and Environmental Problems, Chicago, Illinois, 1998', 869–874.
19. Nielsen, O. B. , 1973, Eocæne aflejringer i Ølst-området og deres indpasning i områdets Kvartærgeologi (Eocene deposits in the Ølst area and their fit in the quartenary geology of the area), Dansk Geologisk Forening, Årskrift for 1972, 100–110. in danish.
20. Oldenburg, D. W., 1994, Pragtical strategies for the solution of large-scale electromagnetic inverse problems, Radio Science, 29, 1081–1099.
21. Oldenburg, D. W. and Ellis, R. G. , 1991, Inversion of geophysical data using an approximate inverse mapping, Geophysical Journal International, 105, 325–353.
22. Oldenburg, D. W., McGillivray, P. R. and Ellis, R. G. , 1993, Generalized subspace methods for large-scale inverse problems, Geophysical Journal International, 114, 12–20.
23. Panissod, C., Dabas, M., Hesse, A., Jolivet, A., Tabbagh, J. and Tabbagh, A. , 1998, Recent devoloments in shallow-depth electrical and electrostatic prospecting using mobile arrays, Geophysics, 63, 1542–1550.
24. Sasaki, Y. , 1989, Two-dimensional joint inversion of magnetotelluric and dipole-dipole resistivity data, Geophysics, 54, 254–262.
25. Sasaki, Y. , 1994, 3-D resistivity inversion using the finite-element method, Geophysics, 59, 1839–1848.
26. Sørensen, K. I., 1989, A method for measurement of the electrical formation resistivity while auger drilling, First Break, 7, 403–407.
27. Sørensen, K. I., 1996, Pulled Array Continuous Electrical Profiling, First Break, 14, 85–90.
28. Sørensen, K. I. and Sørensen, K., 1995, Pulled Array Continuous Vertical Electrical Sounding, PA-CVES, in 'Proceedings of the Symposium on the Application of Geophysics to Engineering and Environmental Problems, Orlando 1995', 893–898. Environmental and Engineering Geophysical Society, P.O. Box 4475, Englewood, CO 80155, USA.
29. Tarantola, A. and Valette, B. , 1982 a, Generalized nonlinear inverse problems solved using the least squares criterion, Reviews of Geophysics and Space Physics, 20, 219–232.
30. Tarantola, A. and Valette, B. , 1982 b, Inverse Problems: Quest of information, Journal of Geophysics, 50, 159–170.
31. Zhang, J., Mackie, R. L. and Madden, T. R., 1995, 3-D resistivity forward modeling and inversion using conjugate gradients, Geophysics, 60, 1313–1325.

Image Reconstruction as an Inverse Problem

Frank P. Pijpers

Theoretical Astrophysics Center
Institute for Physics and Astronomy, Aarhus University
Ny Munkegade, 8000 Århus C, Denmark
Email: fpp@obs.aau.dk
Home Page: http://www.obs.aau.dk/~fpp

Abstract. A method known as Subtractive Optimally Localized Averages (SOLA) has been adapted to treat the problem of correcting images for known blurring, such as due to imperfect optics. The method constructs a linear transformation between any image detected under the conditions for which this blurring applies and the same image as it would have been for different (improved) conditions. The method has been developed specifically with astronomical applications in mind, and some typical examples are shown. The method is explicitly constrained to preserve flux. Because of its linearity the method will preserve relative orientation of different objects in the field.

1 Concerning Astronomical Inverse Problems

There are very many problems within the field of astronomy which involve solving an inverse problem. In general inversions in astronomy arise when observational (experimental) data are a convolution of some quantity of astrophysical interest and a known or measured effect. The latter can be a known property of the instrument used for the observation, an effect of projection on the sky or, as in helioseismology, a convolution along the ray path of a seismic wave in the Sun. Since the measured data is sampled discretely and suffers from measurement errors of various kinds, it is rare that an exact analytical inversion can be carried out. Furthermore what distinguishes astronomy from most other experimental physical sciences is that both the sampling and the data errors are difficult or impossible to control. A number of numerical inversion techniques are currently in use that try to deal with these difficulties in various ways. A recent review of the subject is by Pijpers (1997b). A useful although somewhat older reference is the book by Craig and Brown (1986) which discusses many techniques and applications of inversions in astronomy. A recent review of image reconstruction techniques, some of which are also in use in astronomy, is by Banham & Katsaggelos (1997), which also contains a number of useful references to other work in the field of image reconstruction.

This paper is concerned with the method of Subtractive Optimally Localized Averages (SOLA) as applied to the problem of image reconstruction. The SOLA method (Pijpers & Thompson, 1992, 1994) was originally developed for the problem of helioseismology. In the mathematical literature a virtually identical method was presented slightly earlier (cf. Louis & Maass, 1990), and there are probably other fields where the same method has been derived independently as well. The method is based on the Backus & Gilbert method (1968, 1969, 1970) but is much faster, which is an essential consideration since the data volume in helioseismology is several orders of magnitude larger than in the typical whole earth geoseismic problem to which the Backus & Gilbert method was originally applied. In so-called structure inversions in helioseismology, quantities such as the behaviour of the sound speed as a function of radius inside the Sun are determined. Such inversions usually involve of the order of 2000 measured frequencies of the resonant oscillations of the Sun. In inversions for the solar rotation as a function of radius and latitude the amount of data involved is usually a factor of 10 larger. The SOLA method is now widely in use within the field of helioseismology and has also found applications in astronomical time series analysis (Pijpers & Wanders, 1994; Pijpers, 1997a).

In the application of SOLA to the problem of image deconvolution there is no difference of principle in the method when compared to its helioseismic application, but there are several practical considerations which need to be taken into account in order to construct an algorithm that is sufficiently fast. The SOLA method is much faster than the Backus & Gilbert method because it only requires one matrix factorization where the Backus & Gilbert method (applied to the same helioseismic problem) would require very many : as many as positions for which a localized estimate of the quantity under the integral sign is desired. For helioseismology this number could easily amount to a few hundred positions. Nevertheless the SOLA method is a slow method in the sense that it requires the factorization of an $N \times N$ matrix which in the most general case will require a number of operations that scales as $\sim N^3$ where N is the number of data available. For most workstations $N \sim 2000$ poses no particular problems and so the structure inversions have been carried out using the SOLA method as originally presented. In the case of the solar rotation problem, it turns out that the integration kernels satisfy certain relations which has the consequence that instead of solving for the full matrix one can reduce the problem to factorizing a small set of matrices that are much smaller than the full original (cf Pijpers & Thompson, 1996). A similar reduction of effort can be achieved using appropriate numerical methods (cf. Larsen & Hansen, 1997).

Most large telescopes in use for astronomy now use CCD detector technology, in which the CCDs can have sizes as large as 1024×1024 pixels, and larger mosaics of CCDs are already being used. It is quite likely that larger monolithic CCDs will become available and so the typical data volume, i.e. recorded fluxes of photons for each pixel, can in principle be in excess of 10^6 items of data. Even one $N \times N$ matrix factorization for $N \sim 10^6$ is a considerable computational effort and thus it is well worthwhile to employ if possible any special properties of the matrix in order to speed up its factorization. As has recently been shown by Pijpers (1999) it is possible under certain conditions to perform the matrix factorization required by the SOLA algorithm in a number of operations that scales as $M^2 \log M$ for $M \times M$ sized CCD images.

In section 2 the SOLA method is briefly described as well as the faster implementation of it that can be used for astronomical image reconstruction. In section 3 the method is applied to a small set of example images to demonstrate the success and limitations of the method. In section 4 follows a discussion of the method, including some additional image processing tools that could be incorporated in the method.

2 The SOLA Method

2.1 General

The SOLA method has been described extensively in the papers by Pijpers and Thompson (1992, 1994) and therefore only a brief description is given here. The strategy of the SOLA method in general is to find a set of linear coefficients c which, when combined with the data, produce a weighted average of the unknown convolved function under the integral sign, where the weighting function is sharply peaked. In the application at hand this means finding the linear transformation between an image recorded at a given resolution and an image appropriate to a different (better) resolution.

The relation between a recorded image D and the actual distribution of flux over the field of view I is:

$$D(x,y) = \int K(x',y'\;;x,y)I(x',y')\,\mathrm{d}x'\mathrm{d}y' \tag{1}$$

where integration kernel K is the point spread function (PSF), which describes how a point source (a Dirac delta function) at position (x',y') is imaged. Of course generally D is not known as a continuous function of (x,y), but instead it is sampled discretely as an image recorded on a CCD is. Thus one has as available data the recorded pixel-by-pixel values of flux $D(x_i,y_j)$. These measured fluxes will usually be corrupted by noise and thus the discretized version of equation (1) is:

$$D_{ij} = \int K_{ij}(x',y')I(x',y')\,\mathrm{d}x'\mathrm{d}y' + n_{ij} \tag{2}$$

where now K_{ij} refers to the PSF appropriate for the pixel at (x_i,y_j) and D_{ij} is the flux value recorded in that pixel, n_{ij} is the contribution from noise.

In the SOLA technique a set of linear coefficients c_l is sought which, when combined with the data, produces a value for the flux R in any given pixel that would correspond to an image recorded with a much narrower PSF. Writing this out explicitly and using (2) yields:

$$R \equiv \sum c_l D_l = \int \left\{ \sum c_l K_l(x',y') \right\} I(x',y')\,\mathrm{d}x'\mathrm{d}y' + \sum c_l n_l \tag{3}$$

in which the double subscript ij has been replaced by a single one l for convenience. Thus one would construct the c_l such that the averaging kernel $\mathcal{K}$ defined by:

$$\mathcal{K}(x', y') \equiv \sum c_l K_l(x', y') \qquad (4)$$

is as sharply peaked as possible. If one does this for all locations on the CCD the collected values R_m are then the fluxes corresponding to the image at this (better) resolution with a (improved) "point spread function" $\mathcal{K}$. The so-called propagated error, the error in the flux R is:

$$\sigma_R^2 \equiv \sum \sum c_l c_m N_{lm} \qquad (5)$$

Here the N_{lm} is the error variance-covariance matrix of the recorded CCD images where both l and m run over all (i, j) combinations of the pixel coordinates. If the errors are uncorrelated between pixels then (5) reduces to:

$$\sigma_R^2 = \sum c_l^2 \sigma_l^2 \qquad (6)$$

which is trivially computed once the coefficients c_l are known.

Ideally one would wish to construct an image corresponding to an infinitely narrow PSF: a Dirac delta function. In practice this cannot be achieved with a finite amount of recorded data. In the deconvolved image the sampling theorem must still be satisfied. A further restriction arises because of the noise term in equation (2). As is well known in helioseismology the linear combination of data corresponding to a very highly resolved measurement usually bears with it a very large propagated error. In order to obtain a flux value for each pixel in the deconvolved image that does not have an excessively large error estimate associated with it, one needs to remain modest in the resolution sought for in the deconvolved image.

Finding the optimal set of coefficients taking these limitations into account can be expressed mathematically in the following minimization problem. One needs to minimize for the coefficients c_l the following:

$$\int [\mathcal{K} - \mathcal{T}]^2 \, dxdy + \mu \sum \sum c_l c_m N_{lm} \qquad (7)$$

Here μ is a free parameter which is used to adjust the relative weight given to minimizing the errors in the deconvolved image and to producing a more sharply peaked kernel $\mathcal{K}$. The higher the value of μ the lower this error but the less successful one will be in producing a narrow PSF. In SOLA one is free to choose the function $\mathcal{T}$. A common choice in SOLA applications is a Gaussian:

$$\mathcal{T} = \frac{1}{f\Delta^2} \exp\left[-\left(\frac{(x - x_0)^2 + (y - y_0)^2}{\Delta^2}\right)\right] \qquad (8)$$

Here (x_0, y_0) is the location for which one wishes to know the flux at the resolution corresponding to the width Δ, and f is a normalization factor chosen such that:

$$\int \mathcal{T} \, dxdy \equiv 1 \qquad (9)$$

although any set of locations (x_0, y_0) can be chosen, a natural choice in the application at hand is to take all original pixel locations (x_i, y_j). If one wishes to deconvolve to sub-pixel scales this can be done by an appropriate choice of the (x_0, y_0) and Δ. Any other target function could be chosen in principle and some other possible choices are discussed in section 4.3.

In terms of an algorithm the problem of minimizing the function (7) can be formulated in various ways. The so-called "normal equation approach" leads to a set of linear equations:

$$A_{lm} c_l = b_m \qquad (10)$$

The elements of the matrix A are given by:

$$A_{lm} \equiv \int K_l(x, y) K_m(x, y) \ dxdy \ + \mu N_{lm} \qquad (11)$$

The elements of the vector b are given by:

$$b_m \equiv \int \mathcal{T}(x, y) K_m(x, y) \ dxdy \qquad (12)$$

Writing out the dependencies on the free parameters explicitly, determining the coefficients c_l results from a straightforward matrix factorization:

$$c_l(x_0, y_0 \ ; \Delta, \mu) = A_{lm}^{-1}(\mu) b_m(x_0, y_0 \ ; \Delta) \qquad (13)$$

It is clear that for each point (x_0, y_0) there is a separate set of coefficients c_l which will depend on the resolution width Δ required and on the error weighting μ. Note that it is not necessary to factorize a matrix for every location (x_0, y_0), which would have been the case with the Backus & Gilbert method, and which would certainly be prohibitive if one wishes to calculate the entire deconvolved image. For a given error weighting μ one needs to factorize A only once. Only the elements of the vector b need be recomputed for different locations or different resolutions.

In order to ensure that at every point in the reconstructed image the summed weight of all measurements is equal and thus a true (weighted) average it is necessary to additionally impose the condition:

$$\sum c_l \equiv 1 \qquad (14)$$

It is this condition that imposes photometric accuracy on the reconstructed image : i.e. the flux is conserved. Using the method of Lagrange multipliers this condition is easily incorporated into the matrix equation (10) by augmenting the matrix A with a row and column of 1's, and a corner element equal to 0. The vector b gains one extra element equal to 1 as well.

2.2 A Faster Algorithm

Although the method described above can work in principle with general PSFs K, the matrix factorization becomes intractable very quickly as the number of pixels increases. For an image of $M \times M$ pixels the number of elements in the matrix A is $M^2 \times M^2$. The matrix A is symmetric but even so a naive matrix factorization routine would require a number of operations scaling as M^6.

As is shown in detail by Pijpers (1999) the matrix A in equation (11) becomes of a special character if one assumes that the PSF is constant over the field of view:

$$K(x', y' ; x, y) \equiv K(x - x', y - y') \tag{15}$$

In this case all elements of A can be determined by doing a 2-dimensional convolution which has an operations count $\mathcal{O}(M^2 \log M)$ if FFT algorithms are employed. Similarly the vectors b in (12) can be evaluated for all locations (x_i, y_j) with a single two-dimensional convolution of K and T, again dominated by the FFT.

If the CCD pixels are assumed to be equally spaced the matrix A for $\mu = 0$ can be constructed in such a way that it becomes of a special type known as symmetric block circulant with circulant blocks (BCCB), for which very fast factorization algorithms exist. Circulant matrices have the property that every row is identical to the previous row, but shifted to the right by one element. The shifting is 'wrapped around' so that the first element on each row is equal to the last element of the previous row. Thus the main diagonal elements are all equal and on every diagonal parallel to the main diagonal of the matrix all elements are equal as well. A BCCB matrix is a matrix that can be partitioned into blocks in such a way that each row of blocks is repeated by shifting (and wrapping around) by one block in the subsequent row of blocks and each individual block is circulant. It can be shown that circulant matrices can be multiplied and factorized using Fourier transforms, and by extension BCCB matrices can be multiplied and factorized using two-dimensional Fourier transforms. The restriction on the matrix N_{lm} is that is must also be a symmetric BCCB matrix for the fast factorization algorithm to work. It is evident that fully optimal results can only be obtained if the full $N^2 \times N^2$ covariance matrix of the errors is used. However, the error correlation function for the pixels is expected to behave similarly to the point spread function in the sense that it is large (in absolute value) for small pixel separations and small for large pixel separations, independently of where on the CCD the pixel is located. It is therefore likely that the error covariance matrix will already be BCCB or be very nearly so. Since its role in the minimization of (7) is to regularize the inversion it is in practice not essential that the exact variance-covariance matrix be used. Experience in using SOLA in other fields has shown that the results of linear inversions are robust to minor inaccuracies in the error matrix. In short the matrix factorization of equation (13) can be carried out in $\mathcal{O}(M^2 \log M)$ operations as well.

Finally, as is shown by Pijpers (1999), the multiplication of the collection of vectors b, obtained through the two-dimensional convolution of K and T, through the inverse of the matrix A can also be carried out with a single two-dimensional convolution using FFTs. The set of linear coefficients obtained can be applied to any (set of) images obtained with the same PSF, using again one two-dimensional convolution per image. Therefore the entire process of deconvolving one $2^L \times 2^L$

CCD image can be done with an operations count of $\mathcal{O}(L4^L)$ where the constant of proportionality is ~ 20. Not all operations need to be repeated for a different choice of target PSF, so that there is relatively little computational effort in varying the target resolution, for instance in order to assess which resolution produces optimal results.

3 Demonstrations of the Method

3.1 Astronomical

As a first example in Figure 1 is shown an artificial field of stars. The angular size on the plane of the sky of nearly every star is so small that for any telescope they will appear unresolved : that is to say that the image of a star is indistinguishable from the image of a delta function. In fact for most astronomical detection systems the way that the PSF is determined is through imaging a star close to the object of interest in order to obtain as closely as possible the same optical path through the atmosphere and the telescope optics. There are many real examples, such as the galactic center or globular clusters, where stars cluster so closely together that their individual images overlap and merge such as has happened in the lower left panel of Figure 1. This image is obtained by convolving the object with a PSF which is the (normalized) sum of two Gaussians : a broader one with a $\Delta = 10$ which has 99.9% of the weight and a narrower one with a $\Delta = 1$ pixels which has 0.1% ot the weight. After this one-stage blurring, a small amount of noise has been added. In astronomical research it is often important to be able to determine the brightness of the individual stars in such a cluster. In order to do this one needs a deconvolution method that is guaranteed to preserve the flux of each star image in the deconvolution. This is referred to as photometric accuracy. Most of the standard methods in use in astronomy, such as maximum likelihood (ML) methods (cf. Richardson, 1972; Lucy, 1974, 1994) or maximum entropy methods (MEM) (cf. Narayan & Nityananda, 1986), do not guarantee photometric accuracy and have been shown in some tests to fail this requirement. Furthermore, in order for measured brightnesses to be useful, good error estimates are needed which usually requires Monte Carlo simulations for the non-linear methods such as ML and MEM. For this reason it is useful to compare the deconvolved image (bottom right in Figure 1) and the original image when convolved with the PSF that was used as target PSF in the deconvolution (top right in Figure 1). To the eye there appears to be no difference. As is shown by Pijpers (1999) when a standard astronomical software package for doing photometry on fields of stars is used, the differences between the two images are consistent with the noise that is present in the deconvolved image through propagation of the (simulated) Gaussian distributed random errors that were added to the panel in the bottom left.

3.2 Non-Astronomical

Although for astronomical purposes the example shown above is of interest, it is a very simple image in the sense that it has only one type of objects in it :

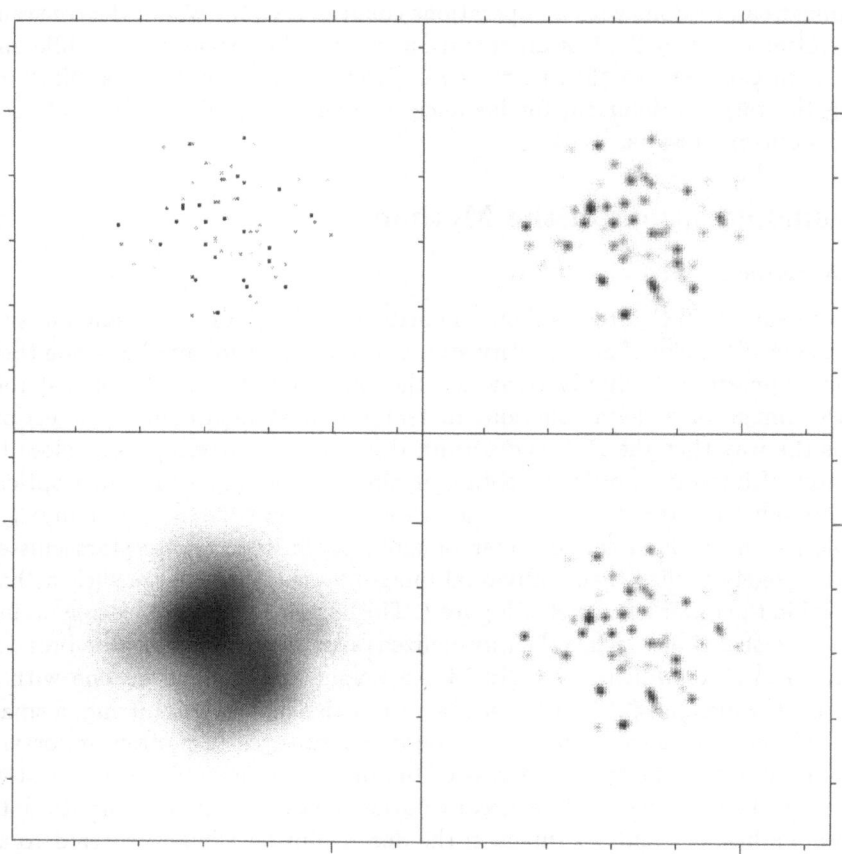

Fig. 1. Top left panel : the original 128 × 128 image. Top right panel : the original convolved with the target PSF : a Gaussian with $\Delta = 1.5$ pixels. Bottom left panel : the original convolved with a PSF which is the sum of a Gaussian with $\Delta = 10$ pixels and an 0.1% contribution from a Gaussian with $\Delta = 1$ pixel. Bottom right panel : the image after SOLA deconvolution of the bottom left image. In all images the grey-scale is linear. In the bottom left image noise is added before deconvolution. In the bottom right image the noise propagated in the deconvolution has an expectation value of ~ 0.5 in arbitrary flux units and the S/N ratio for the brightest pixel is ~ 1000.

point sources. Within the field of radio interferometry in particular a method known as CLEAN (cf. Högbom, 1974; Wakker & Schwarz, 1988) builds up an image by placing point sources in the field and (iteratively) minimizing the difference between the image of this field and the measured radio image. One would therefore not expect any significant advantage of the SOLA method over the CLEAN method for this particular image. However, in real astronomical images it will usually not be the case that there are only point sources present in the field. If the emitting material that is imaged is extended, CLEAN is known not to perform very well without modifications. MEM and ML methods will do better in such cases but may still produce undesirable ringing around strong point sources, effectively transferring flux from the extended object surrounding the point source, into the point source.

Fig. 2. A 512 × 512 image. Otherwise as for Figure 1

In order to demonstrate the method on an image which is evidently not built up out of just point sources or a similarly limited set of building blocks, a photo of a human face was treated in an identical manner as Figure 1. The linear coefficients of the transformation are not identical because this image is much larger than image 1, but the blurring PSF has the same properties as for that image. The results are shown in the same style as in Figure 1. It is evident that the facial features and the fine structure of the hair and the feather in the head band are nearly indistinguishable in the blurred image and recovered very well in the deconvolved image. The most obvious differences are concentrated near the borders of the image. Such edge effects will always occur because in the convolution information has been lost at the edges. Part of the flux that is positioned near the edges of the original image is placed outside the borders of the image after the convolution. Strictly that flux is unknown when doing the deconvolution so that, when padding the image with some value for the flux per pixel, the deconvolution will produce a set of dark and bright stripes parallel to the edges of the image. The strength of these stripes depends strongly on the image under consideration as well as on the ratio of the widths of the PSFs before and after deconvolution.

4 Discussion

4.1 Error Propagation

It is of particular importance for quantitative analysis (photometry) of the brightness of objects within the deconvolved image to have reliable estimates of the error in the flux. It is clear from equations (5) and (6), that it is trivial to compute the full propagated error covariance matrix once the linear transformation has been established. One single number that is illustrative of the behaviour of the errors in the deconvolution is the error magnification factor Λ:

$$\Lambda^2 \equiv \sum c_l^2 \tag{16}$$

If the variance of the errors in the flux in each pixel is identical then the variance of the errors in the flux in the deconvolved image is simply Λ times that variance. Hence Λ is a measure of how much the errors increase in the deconvolution. Of course one must keep in mind that even if the errors are uncorrelated between pixels before the deconvolution, after deconvolution there will be a correlation in the errors among nearby pixels.

In Figure 3 is shown the error magnification as a function of the specified width of the target PSF, for the images of Figure 1. The behaviour for the deconvolution of Figure 2 is very nearly identical. Although the value of Δ of the target is used for the x-axis it is evident that the errors increase only if the resolution of the deconvolved image is better than the resolution of the blurred image. If not then effectively the deconvolution is a smoothing operation and should produce smaller errors. Apart from minor changes of shape this curve should be independent of the blurring PSF, if plotted as a function of the ratio of the second moments (widths) of the PSFs of the deconvolved and blurred images.

4.2 Dealing With Boundary Effects

The issue of boundary effects has been considered by many authors (cf. Banham & Katsaggelos, 1997). Consider for simplicity a 1-dimensional example. Without

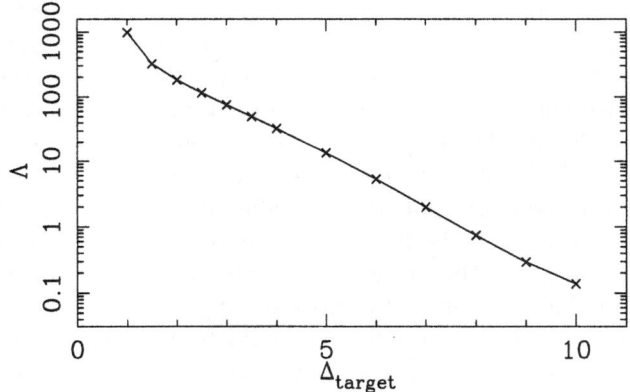

Fig. 3. Error magnification Λ as a function of resolution

any padding the coefficient matrix is Toeplitz (where in the 2-dimensional case it would be block Toeplitz with Toeplitz blocks), so that the corresponding version of equation (11) can be written as

$$Tx = b \tag{17}$$

It is faster to work with circulant matrices instead because then one can use FFTs to implement matrix multiplications. Therefore the matrix T is augmented and so are the vectors x and b:

$$C = \begin{pmatrix} \times & \times & \times \\ \times & T & \times \\ \times & \times & \times \end{pmatrix}, \quad \widehat{x} = \begin{pmatrix} x_1 \\ x_2 \\ x_3 \end{pmatrix}, \quad \widehat{b} = \begin{pmatrix} b_1 \\ b_2 \\ b_3 \end{pmatrix} \tag{18}$$

It is evident that if $x_1 = x_3 = 0$ and $x_2 = x$ then $b_2 = b$. Unfortunately the case here is that x_2 is to be determined and only the b_1 and b_3 can be specified. When computing $\widehat{x} = C^{-1}\widehat{b}$, setting $b_1 = b_3 = 0$ does not guarantee that $x_2 = x$, nor that $x_1 = x_3 = 0$. Therefore, near the edges of the deconvolved image, a significant contribution to the value of the flux comes from fluxes in pixels that lie outside the blurred image. In the implementation presented by Pijpers (1999) the image is embedded in a virtual array that is twice as large and periodic in every dimension in order to obtain b_1 and b_3. The linear coefficients $\widehat{x}$ thus obtained are then combined with the image which is similarly embedded in an array. If the immediate surroundings of the original image were known, there would not be any problem in the reconstruction. Since they are not, it is necessary to use information from the image to simulate the border region outside the image.

Zero padding is the most straightforward way of treating the unknown surroundings of the image. For images that have a similar behaviour towards the edges as image 1 does in that towards the edges there is very little flux, this is probably close to optimal. For images as shown in 2 the edge effects become very pronounced when this is done. Here, padding with the mean value of the flux per pixel in the image works better. For astronomical images mean-value padding is probably not appropriate because the dynamic range is very high. One or a few very bright point sources in the field can substantially increase the mean flux level which is then not at all representative of the typical flux per pixel. Padding with a median value would then be more appropriate but this requires sorting all fluxes which adds computing time to the algorithm.

In any case whatever constant padding value might be chosen, it will always introduce a discontinuity in the flux at the edge which will be propagated into the deconvolved image by the linear transformation of the deconvolution. One can attempt to reduce this effect by extrapolating the behaviour of the image at the edges. A zero order extrapolation would be to keep the flux per pixel constant along rows or columns perpendicular to each edge. As a next step one can then progressively smooth along directions parallel to each edge so that after a certain band of rows or columns the flux has become nearly constant and equal to the mean value on the edges of the image. This is what is used in the image shown in the bottom right panel of Figure 2.

Another way of dealing with this problem is to reduce spatial resolution toward the borders of images, so that proportionally less information from beyond the edges of the image is needed as they are approached. In terms of the algorithm this is more expensive since the calculation of the vector elements b in equation (13) through the inverse of the matrix can then no longer be done using 2D FFTs, nor can the matrix-vector multiplications. A compromise is to decrease the desired resolution in a (small) number of discrete steps for a border-region.

In all cases one should be aware that such solutions are cosmetic in character and may preclude accurate quantitative analysis of structures that are close to the edges of images.

4.3 Incorporating Wavelet-Based Filtering

Although filtering is an issue quite separate from deconvolution in the strictest sense, in many astronomical applications filtering is considered desirable in order to suppress noise. In this way it is hoped to detect even very faint extended structures in the presence of strong sources. Since filtering always implies a smoothing operation it is of course counterproductive to filter uniformly over deconvolved images : one might as well have deconvolved to a less high resolution. Thus what would be required in such cases is to set the resolution of the deconvolved images dynamically : dependent on the structure in the image to be deconvolved. One way in which this problem has been pursued is to use wavelets in order to obtain the multi-resolution support of objects within images (cf. Pantin & Starck, 1994).

In itself adjusting the resolution on the basis of what is in the image to be deconvolved is completely counter to the strategy of SOLA which achieves deconvolution without any bias concerning the character of structures in the images. However, it is quite easy to use SOLA deconvolution in order to obtain the multiresolution support, after which any filtering that is considered desirable can proceed. The multiresolution support of an image is usually obtained by convolving the image with a set of wavelets. The basic wavelet can in principle be any integrable function with zero mean and which has some scaling parameter which sets the resolution. All other wavelets in the same set are derived from this basic wavelet by taking a number of different values for the scaling parameter. In one dimension for instance the second derivative of a Gaussian satisfies these criteria, and the 2-dimensional generalization of this is the Laplacian of a Gaussian. The scaling parameter in this case is evidently the width of the Gaussian. Another option is to take the difference between two normalized Gaussians with different widths. Of course there are very many more choices possible. The convolution of the image with each of these wavelets produces a set of images at different spatial resolutions, which taken together form the multiresolution support. Within the cube of data obtained in this way (stacking up the images in order of increasing spatial resolution) it can in some cases be easier to identify at each resolution level which structures one would wish to filter out. Recombining the datacube into one single image after this filtering process will then produce an image in which only those structures are retained that are of interest for the researcher. One should realize however that in this case the resulting image does **not** have a PSF that is invariant over the image !

In any case, whichever wavelet one would wish to use, it is a quite straightforward generalization of SOLA to use that wavelet as a target function instead of the Gaussian of equation (8). By taking a set of different scaling parameters/resolutions one obtains a set of target functions, and the SOLA algorithm proceeds as before. Of course there is no need to re-factorize a matrix for every individual wavelet, so there is relatively little computational overhead in obtaining a set of images at different resolutions. The appropriate filtering depends very much on the application and so it is not appropriate to address that issue further in this context.

5 Conclusions

It is demonstrated here that the SOLA method is successfully modified to perform fast unbiased deconvolution to correct images for a large class of blurring effects and in the presence of noise. Apart from its speed a clear advantage of this method is that no knowledge is required of the structures in the images to be corrected for blurring. This in combination with the fact that the errors in the deconvolved image are trivial to compute after SOLA deconvolution makes this method particularly appropriate if structures in the deconvolved images are to be subjected to any kind of quantitative analysis.

Acknowledgements

The Theoretical Astrophysics Center is a collaboration between Copenhagen University and Aarhus University and is funded by Danmarks Grundforskningsfond. Per Christian Hansen is thanked for his help in improving the presentation of this paper.

References

1. Backus, G.E., Gilbert, J.F., (1967) : Geophysical Journal of the Royal Astronomical Society **13**, 247
2. Backus, G.E., Gilbert, J.F., (1968) : Geophysical Journal of the Royal Astronomical Society **16**, 169
3. Backus, G.E., Gilbert, J.F., (1970) : Philosophical Transactions of the Royal Society of London A **266**, 123
4. Banham, M.R., Katsaggelos, A.K., (1997) : IEEE Signal Processing Magazine, March, 24
5. Craig, I.J.D., Brown, J.C., (1986) : "Inverse problems in Astronomy : a guide to inversion strategies for remotely sensed data", (Adam Hilger, Bristol & Boston)
6. Högbom, J.A., (1974) : Astronomy & Astrophysics Suppl. Ser.**15**, 417
7. Larsen, R.M., Hansen, P.C., (1997) : Astronomy & Astrophysics Suppl. Ser.**121**, 587
8. Louis, A.K., Maass, P., (1990) : Inverse Problems **6**, 427
9. Lucy, L.B., (1974) : Astronomical Journal **79**, 745
10. Lucy, L.B., (1994) : Astronomy & Astrophysics **289**, 983
11. Narayan, R., Nityananda, R., (1986) : Annual Reviews of Astronomy & Astrophysics **24**, 127
12. Pantin, E., Starck, J.-L., (1994) : Astronomy & Astrophysics Suppl. Ser.**118**, 575
13. Pijpers, F.P., (1997a) : Monthly Notices of the Royal Astronomical Society **289**, 933
14. Pijpers, F.P., (1997b) : "Inversions in Astronomy and the SOLA method", in IMA Volumes in Mathematics and its Applications 90, Inverse problems in wave propagation eds. G. Chavent, G. Papanicolaou, P. Sacks, W. Symes, (Springer, New York), 419
15. Pijpers F.P., (1999) : Monthly Notices of the Royal Astronomical Society **307**, 659
16. Pijpers F.P., Thompson M.J., (1992) : Astronomy & Astrophysics **262**, L33
17. Pijpers F.P., Thompson M.J., (1994) : Astronomy & Astrophysics **281**, 231
18. Pijpers F.P., Thompson M.J., (1996) : Monthly Notices of the Royal Astronomical Society **279**, 498
19. Pijpers F.P., Wanders, I., (1994) : Monthly Notices of the Royal Astronomical Society **271**, 183
20. Richardson, W.H., (1972) : Journal of the Optical Society of America **62**, 55
21. Wakker, B.P., Schwarz, U.J., (1988) : Astronomy & Astrophysics **200**, 312

Resolution Studies of Fluid Flow Models Near the Core-Mantle Boundary Using Bayesian Inversion

Camilla Rygaard-Hjalsted, Klaus Mosegaard, and Niels Olsen

Dept. of Geophysics, Niels Bohr Institute
University of Copenhagen, Juliane Maries Vej 30,
DK-2100 Copenhagen Ø, Denmark

Abstract. We model and analyze the horizontal fluid flow near the core-mantle boundary from geomagnetic observations. The non-linear core motion problem is studied by means of a Bayesian Monte Carlo approach under the frozen flux assumption and subject to a *relaxed* geostrophic bound. In relaxing the geostrophic constraint we allow a non-zero Lorentz force to enter into the force balance of the core. We loosen the geostrophic constraint in steps and present a series of fluid flow models deviating from geostrophy by up to 29%. Models with up to a few percent ageostrophy have structures that are generally well resolved and are clearly under strong influence of the geostrophic constraint. Models of higher ageostrophy have structures that are generally poorly resolved as a result of the relaxation of the geostrophic bound. However, we find a certain well-determined structure of upwelling underneath the east Indian Ocean in our full series of models.

1 Introduction

Fluid flow in Earth's outer core as determined from inversion of geomagnetic data has generally been modeled with the aim of finding an *optimal solution* to the inverse problem (for two extensive review papers, see [7] and [28]). This is commonly done by relying on the frozen flux hypothesis [22], some uniqueness reducing constraint (steadiness of flow [24], [25], steadiness of acceleration [26], toroidal flow [27], [18], or tangentially geostrophic flow [17]), and seeking the solution that conforms with a certain smoothness criterion [7].

In the present study we take a Bayesian view point. We use spherical harmonic representations for the velocity field $\mathbf{u}$, the magnetic field $\mathbf{B}$ and its first time derivative $\partial\mathbf{B}/\partial t$ to obtain the velocity coefficients collected in vector $\mathbf{v}$, the magnetic field coefficients in vector $\mathbf{g}$ and the coefficients of the secular variation in vector $\dot{\mathbf{g}}$. We consider solving for $\mathbf{v}$ near the core-mantle boundary (CMB) from $\mathbf{g}, \dot{\mathbf{g}}$ at epoch 1980, where $\mathbf{g}, \dot{\mathbf{g}}$ will be referred to as *data*. Through spherical harmonic analysis they are downward continuations to the CMB of observations of the magnetic field and its secular variation at or near the Earth's surface. The observations at or near the Earth's surface are necessarily uncertain and there

256

are additional errors involved in the extrapolation to the CMB. We shall refer to the imprecisions in **g**, **ġ** as *uncertainty* in *data.*

In the Bayesian framework we do not look for one optimal solution, but for a whole *family of solutions*, each of which is consistent with data within its uncertainty and certain *a priori* information. Each member of the family is a near-independent sample (see Appendix A) drawn from the *a posteriori* distribution in model space, which is the distribution that combines information held in the data and in the prior.

With this set of solutions we wish to study the non-uniqueness of the problem and the propagation of uncertainties from data to model. With a common term we call this: resolution. Structures in common to many or all members are well resolved, whereas structures that differ in many or all members are poorly resolved [21]. By structures we mean, for example, correlated components of the motion like westward flow or circulations north and south of the geographic equator. In particular we look at the rotation of the fluid flow (vorticity strength) and the horizontal divergence of the horizontal fluid flow (upwelling and downwelling). The existence of many solutions to this inverse problem reflects the fact that it is fundamentally non-unique [1] and that we have non-uniqueness due to a finite amount of inaccurate data. As will become apparent we study both the fundamental non-uniqueness and the non-uniqueness arising from uncertain data.

The plan for the present inversion is that frozen flux is assumed and an approximate tangentially geostrophic constraint is imposed on the governing system of equations. As a new development we loosen the strict geostrophic constraint. In doing so, we indirectly study the forces in the Navier Stokes equation neglected in the geostrophic balance of which the Lorentz force is considered dominant. Due to lack of knowledge of the toroidal part of the magnetic field at the CMB and its radial derivative we are unable to calculate the Lorentz force. In the geostrophic balance the Coriolis force is assumed much larger than the Lorentz force and the latter is set to zero. However, in this paper we allow the Lorentz force to be non-zero. We loosen the geostrophic constraint in steps and produce a series of *ageostrophic* flows which deviate from geostrophy by 1.2% to 29.2%.

The geostrophic solution space is a linear subspace of the whole solution space. As we deviate from the non-uniqueness reducing geostrophic constraint we see a multitude of solutions that are consistent with data, data uncertainty and (new) prior information. The velocity coefficients become increasingly uncertain with the increasing ageostrophy of the flows, but a well-determined structure of upwelling under the east Indian Ocean is present throughout our series of models. This indicates that certain functions of the coefficients do not change significantly under the relaxed geostrophic constraint even though the individual coefficients may vary markedly.

The relationship between the magnetic field **B**, its first time derivative $\partial\mathbf{B}/\partial t$ and the velocity vector field **u** near CMB is expressed in the magnetic induction equation

$$\frac{\partial \mathbf{B}}{\partial t} = \nabla \times (\mathbf{u} \times \mathbf{B}) + \eta \nabla^2 \mathbf{B} \qquad (1)$$

where $\eta = \frac{1}{\mu_0 \sigma_E}$ is the magnetic diffusivity in the fluid outer core near the CMB, with μ_0 being magnetic permeability ($4\pi \times 10^{-7}$ NA^{-2}) and σ_E being electrical conductivity. The two terms on the right hand side of (1) are respectively magnetic advection and diffusion. Throughout this paper we neglect the diffusion term justified by scaling arguments, and thus work under the *frozen flux hypothesis* (see e.g. [7]) which reduces (1) to

$$\frac{\partial \mathbf{B}}{\partial t} = \nabla \times (\mathbf{u} \times \mathbf{B}). \qquad (2)$$

The pair $(\partial \mathbf{B}/\partial t, \mathbf{B})$ depends non-linearly on $\mathbf{u}$. However, with just one exception [13], the problem as stated in (2) has been linearized. In the linearized problem, $\mathbf{B}$ is taken to be known precisely and thereby as a known constant which implies that $\partial \mathbf{B}/\partial t$ becomes linearly dependent on $\mathbf{u}$. Jackson [13] presents a probabilistic approach to incorporating the magnetic field uncertainties into the inversion; he suggests a maximum likelihood estimate with a probability density function (pdf) quite different to and considerably more complicated than the corresponding Gaussian distribution that results from the linearized analysis.

Our procedure solves the full non-linear problem (2) based on an inverse Monte Carlo sampling technique [19]. We perform an extensive sampling of core flow models consistent with data and its uncertainty including uncertainty in the magnetic field at the CMB. Our procedure is one that combines information held in data and in the a priori information probabilistically. As discussed in [21] we can with a modification of the algorithm artificially remove the imprecision in the core magnetic field and thus also perform a sampling of solutions to the linearized inverse problem.

This paper is organized as follows: First we present the theoretical background and the governing equations of the numerical system. In Section 2.1 we look at its non-linearity in more detail. In Section 2.2 we develop the formalism for the relaxed geostrophic constraint and show how to include this constraint as a priori information into the Bayesian inversion. The a posteriori probability density and the inverse Monte Carlo sampling strategy are presented in Section 2.3. Two derivative features of the fluid flow: vorticity strength and up- and downwelling are presented for a selected number of our flow models in Section 3. Finally in Section 4 we draw conclusions from our results from resolution studies and discuss the Monte Carlo inverse sampling strategy. In Appendix A we outline the difficulties in minimizing sampling errors of the Monte Carlo sampling strategy.

2   Background

Assuming a source-free region at the Earth's surface and downwards to the CMB we can represent the magnetic vector field $\mathbf{B}$ at the CMB as the negative gradient of a scalar potential V,

$$\mathbf{B} = -\nabla V. \tag{3}$$

We expand the potential in spherical harmonics for the internal field only in the usual way:

$$V = a \, \Re e \left\{ \sum_{l=1}^{\infty} \sum_{m=0}^{l} \left(\frac{a}{r} \right)^{l+1} g_l^m P_l^m (\cos \vartheta) e^{im\lambda} \right\} \tag{4}$$

where $a = 6371.2$ km is the nominal Earth radius, $\Re e$ denotes the real part, $P_l^m (\cos \vartheta)$ are Schmidt quasi-normalized associated Legendre functions [15], [3], ϑ and λ are geographic colatitude and longitude and the complex Gauss coefficients g_l^m defining the internal field are related to the usual real Gauss coefficients $\hat{g}_l^m$ and $\hat{h}_l^m$ through

$$g_l^m = \hat{g}_l^m - i \, \hat{h}_l^m \, .$$

Likewise the complex coefficients of the secular variation are related to the corresponding real coefficients through

$$\dot{g}_l^m = \dot{\hat{g}}_l^m - i\dot{\hat{h}}_l^m \, .$$

We consider equation (2) immediately beneath the CMB and take the radial part of the flow u_r to vanish there (u_r is identically zero at the CMB) and describe only the horizontal part $\mathbf{u_H}$ of the velocity

$$\frac{\partial B_r}{\partial t} + \nabla_{\mathbf{H}} \cdot (\mathbf{u_H} B_r) = 0, \tag{5}$$

where $\nabla_{\mathbf{H}} = \nabla - \mathbf{r}(\mathbf{r} \cdot \nabla)$ and $\mathbf{H}$ denotes 'horizontal'. $\mathbf{u_H}$ is decomposed for convenience into its toroidal and poloidal parts

$$\mathbf{u_H} = \mathbf{u_H}^{tor} + \mathbf{u_H}^{pol} = \left(\begin{array}{c} \frac{1}{\sin\vartheta} \frac{\partial T}{\partial \lambda} \\ -\frac{\partial T}{\partial \vartheta} \end{array} \right) + \left(\begin{array}{c} -\frac{\partial S}{\partial \vartheta} \\ \frac{1}{\sin\vartheta} \frac{\partial S}{\partial \lambda} \end{array} \right), \tag{6}$$

where T and S are the toroidal and poloidal scalars respectively, expanded in spherical harmonics as

$$T = \Re e \left\{ \sum t_l^m P_l^m (\cos \vartheta) \, e^{im\lambda} \right\}, \quad S = \Re e \left\{ \sum s_l^m P_l^m (\cos \vartheta) \, e^{im\lambda} \right\}. \tag{7}$$

The toroidal and poloidal coefficients t_l^m and s_l^m are complex and have dimension of velocity [e.g. km/yr].

We express (5) in terms of a series of spherical harmonics coefficients, see [29] or [12]

$$\dot{g}_j = \sum j, k \tilde{A}_{i,j,k} \, g_j \, v_k, \tag{8}$$

where following the notation of [13] degree l and order m have been contracted into one index i, j or k and t_l^m and s_l^m are contained in the vector $\mathbf{v}$ with elements v_k. In vector form equation (8) becomes

$$\dot{\mathbf{g}} = \mathbf{A}(\mathbf{g})\mathbf{v}, \tag{9}$$

where $\mathbf{A}(\mathbf{g})$ is the design matrix dependent on $\mathbf{g}$ via the third order matrix $\tilde{\mathbf{A}}$ with elements $\tilde{A}_{i,j,k}$

$$A_{i,k} = \sum_j \tilde{A}_{i,j,k} \, g_k. \tag{10}$$

In the implicit non-linear inverse problem posed in (9) the data are vectors $\mathbf{g}$ and $\dot{\mathbf{g}}$ and the model parameters are vectors $\mathbf{v}$.

Our data are from the *ufm1* model ([8]) at epoch 1980 truncated at degree and order 13, thus we have 195 real coefficients for the magnetic field at the CMB and likewise for the secular variation. The uncertainties in $\dot{\mathbf{g}}$ and $\mathbf{g}$ are parametrized by two Normal probability distributions with variance $(1 \cdot l^{-1}$ $[nT/yr])^2$ for $\dot{\mathbf{g}}$ and $(5 \cdot l^{-1} \, [nT])^2$ for $\mathbf{g}$ using estimates from [15] as a guideline.

2.1 The Non-Linearity

The influence of the non-linearity of this problem was investigated by Jackson [13] analytically. He presents the solution to the problem

$$\dot{\mathbf{g}} = \check{\mathbf{A}}(\mathbf{v}) \, \mathbf{g} + \mathbf{e}, \tag{11}$$

where the vector $\mathbf{e}$ is added to account for the errors in the secular variation $\dot{\mathbf{g}}$ and where the design matrix $\check{\mathbf{A}}(\mathbf{v})$ is now dependent on the velocity coefficients in $\mathbf{v}$. Thus in (11),

$$\check{A}_{i,k} = \sum_j \tilde{A}_{i,j,k} \, v_k. \tag{12}$$

He specifies the form of the probability density $\rho(\mathbf{g})$ of the magnetic field, the prior probability density $\rho(\mathbf{v})$ of the velocity field and the conditional probability $\theta(\mathbf{v}, \mathbf{g})$ of the secular variation given the velocity and the magnetic fields

$$\rho(\mathbf{g}) \propto \exp\left\{-\frac{1}{2}(\mathbf{g} - \mathbf{g}_0)^T \mathbf{C}_\mathbf{g}^{-1} (\mathbf{g} - \mathbf{g}_0)\right\} \tag{13}$$

$$\rho(\mathbf{v}) \propto \exp\left\{-\frac{1}{2}(\mathbf{v} - \mathbf{v}_0)^T \mathbf{C}_\mathbf{v}^{-1} (\mathbf{v} - \mathbf{v}_0)\right\} \tag{14}$$

$$\theta(\dot{\mathbf{g}} \mid \mathbf{v}, \mathbf{g}) \propto \exp\left(-\frac{1}{2}\mathbf{e}^T \mathbf{C}_\mathbf{e}^{-1} \mathbf{e}\right), \tag{15}$$

where the magnetic field and the velocity field have Gaussian distributions around $\mathbf{g}_0$ and $\mathbf{v}_0$ and the uncertainties on the secular variation are also taken to be Gaussian. $\mathbf{C}_\mathbf{g}, \mathbf{C}_\mathbf{v}$ and $\mathbf{C}_\mathbf{e}$ are the appropriate covariance matrices. The a posteriori pdf is found to be

$$\sigma(\mathbf{v}) \propto \exp\left\{-\frac{1}{2}\left(\dot{\mathbf{g}}\mathbf{C}_\mathbf{e}^{-1}\dot{\mathbf{g}} + \mathbf{g}_0^T \mathbf{C}_\mathbf{g}^{-1}\mathbf{g}_0 + \right.\right.$$
$$\left.\left. (\mathbf{v} - \mathbf{v}_0)^T \mathbf{C}_\mathbf{v}^{-1} (\mathbf{v} - \mathbf{v}_0) - \mathbf{r}^T \mathbf{N}^{-1}\mathbf{r} + \log[\det \mathbf{N}]\right)\right\} \tag{16}$$

where

$$\mathbf{N} = \left(\check{\mathbf{A}}(\mathbf{v})\right)^T \mathbf{C}_\mathbf{e}^{-1} \check{\mathbf{A}}(\mathbf{v}) + \mathbf{C}_\mathbf{g}^{-1} \tag{17}$$

and

$$\mathbf{r} = \mathbf{C_g^{-1} g_0} + \left(\check{\mathbf{A}} \left(\mathbf{v} \right) \right)^T \mathbf{C_e^{-1} \dot{g}}. \tag{18}$$

In this study we *sample* the probability distribution (16), which is of rather complicated form and not susceptible to analytical treatment. Our a priori pdf on $\mathbf{v}$ is a Gaussian distribution centered in the geostrophic subspace with a spread described by $\mathbf{C_v}$.

2.2 The Geostrophic Constraint and its Relaxation

The starting point for describing geostrophic motion is the set of equations governing the dynamics of the fluid flow in the core; the Navier-Stokes equation in the Boussinesq approximation; see, e.g., [7]

$$\rho_0 \left(\frac{\partial \mathbf{u}}{\partial t} + \mathbf{u} \cdot \nabla \mathbf{u} + 2\mathbf{\Omega} \times \mathbf{u} \right) = -\nabla p + \rho' \mathbf{f_g} + \mathbf{J} \times \mathbf{B} - \rho_0 \nu \nabla^2 \mathbf{u} \tag{19}$$

where ρ_0 and ρ' are the hydrostatic density and departure from hydrostatic density, respectively, $\mathbf{u}$ is the velocity vector, $\mathbf{\Omega}$ is the Earth's rotation vector, p is the non-hydrostatic part of the pressure, $\mathbf{f_g}$ is the gravitational acceleration vector, ν is kinematic viscosity and $\mathbf{J}$ and $\mathbf{B}$ are the current density and magnetic field vectors, respectively.

In the geostrophic momentum equation the Coriolis force on the left hand side of (19) balances the buoyancy forces on its right hand side (all other terms in (19) are considered negligibly small), see e.g. [16]:

$$2\rho_0 \left(\mathbf{\Omega} \times \mathbf{u} \right) = -\nabla p + \rho' \mathbf{f_g}. \tag{20}$$

By curling (20) we get the thermal wind equation

$$2\rho_0 \left(\mathbf{\Omega} \cdot \nabla \right) \mathbf{u} = \mathbf{f_g} \times \nabla \rho'. \tag{21}$$

Its radial component gives us the tangentially geostrophic motion or geostrophic constraint

$$\nabla_{\mathbf{H}} \cdot \left(\mathbf{u_H} \cos \vartheta \right) = 0. \tag{22}$$

Relaxing Geostrophy. A simple way of relaxing the geostrophy bound would be to replace the equality (20) by an inequality

$$0 < \| \nabla_{\mathbf{H}} \cdot \left(\mathbf{u_H} \cos \vartheta \right) \| \leq \varepsilon, \tag{23}$$

where the quantity $\varepsilon > 0$ describes the deviation from strict geostrophy. However, for algorithmic reasons we have followed a slightly different route. We write $\nabla_{\mathbf{H}} \cdot \left(\mathbf{u_H} \cos \vartheta \right)$ as a function of colatitude ϑ and longitude λ, which we in turn separate into a matrix $\mathbf{F}$ times the solution vector $\mathbf{v} = (\mathbf{t}, \mathbf{s})$, thus

$$\nabla_{\mathbf{H}} \cdot \left(\mathbf{u_H} \cos \vartheta \right) = F \left(\vartheta, \lambda \right) = \mathbf{F} \cdot \mathbf{v}, \tag{24}$$

where $\nabla_{\mathbf{H}} \cdot (\mathbf{u_H} \cos \vartheta)$ is evaluated on a grid of 5 by 5 degrees over the globe. Substituting (24) into (23) would give

$$0 < \mathbf{v}^T \cdot \mathbf{F}^T \mathbf{F} \cdot \mathbf{v} \le \varepsilon^2, \tag{25}$$

but we choose to replace (25) with a similar soft bound [2] expressed by the prior

$$\rho(\mathbf{v}) = k_1 \cdot \exp\left(-\frac{1}{2} \mathbf{v}^T \cdot \frac{\mathbf{F}^T \mathbf{F}}{\varepsilon^2} \cdot \mathbf{v}\right), \tag{26}$$

where k_1 is a normalization constant, and $\rho(\mathbf{v})$ is our a priori information on $\mathbf{v}$. As can be seen in (26), ε^2 can be regarded as a "variance". Since the uncertainty distribution on the magnetic field coefficients and the noise distribution on the velocity field coefficients are independent, we can multiply their probability distributions and their joint priori probability distribution $\rho(\mathbf{g}, \mathbf{v})$ becomes

$$\rho(\mathbf{g}, \mathbf{v}) = k_2 \left[\exp\left(-\frac{1}{2} \mathbf{g}^T \mathbf{C_g^{-1}} \mathbf{g}\right) \exp\left(-\frac{1}{2} \mathbf{v}^T \mathbf{C_v^{-1}} \mathbf{v}\right) \right], \tag{27}$$

where k_2 is a normalization constant and $\mathbf{C_g}$ is the (diagonal) magnetic field covariance matrix and the (full) covariance $\mathbf{C_v}$ is the inverse of $\frac{\mathbf{F}^T \mathbf{F}}{\varepsilon^2}$ (see equation 26). In our study we have looked at a range of flows with three values of ε, $\varepsilon = 10^{-2}$, $\varepsilon = 10^{-1}$ and $\varepsilon = 1$.

2.3 Bayesian Inversion

Equation (9) enables us to predict the secular variation $\dot{\mathbf{g}}$ from $\mathbf{g}$ and $\mathbf{v}$. Since the uncertainty distribution on the magnetic field coefficients and the prior distribution on the velocity field coefficients are Gaussian we can define a likelihood function for $(\mathbf{g}, \mathbf{v})$ treating $\mathbf{g}$ as model parameters in the calculations (see [21] for further details)

$$L(\mathbf{g}, \mathbf{v}) = k_3 \exp\left\{-\frac{1}{2} [\dot{\mathbf{g}} - \mathbf{A}(\mathbf{g}) \mathbf{v}]^T \mathbf{C_{\dot{g}}^{-1}} [\dot{\mathbf{g}} - \mathbf{A}(\mathbf{g}) \mathbf{v}]\right\}, \tag{28}$$

where k_3 is a normalization constant and $\mathbf{C_{\dot{g}}}$ is the secular variation covariance matrix. We follow [23] in which the joint a posteriori probability density $\sigma(\mathbf{g}, \mathbf{v})$ combines the information held in $L(\mathbf{g}, \mathbf{v})$ and $\rho(\mathbf{g}, \mathbf{v})$

$$\sigma(\mathbf{g}, \mathbf{v}) = k_4 \rho(\mathbf{g}, \mathbf{v}) L(\mathbf{g}, \mathbf{v}), \tag{29}$$

where k_4 is a normalization constant. The solution to the inverse problem is the a posteriori probability density of the velocities alone and so we integrate (29) over the data space $\mathcal{D}_1$ of the magnetic field coefficients

$$\sigma(\mathbf{v}) = \int_{\mathcal{D}_1} \sigma(\mathbf{g}, \mathbf{v}) \, d\mathbf{g}. \tag{30}$$

The complexity of Jackson's posterior (16) prevents us from calculating $\sigma(\mathbf{v})$ directly. The aim of the inverse sampling technique is to *sample* the posterior (30), the idea being that a large number of near-independent samples drawn from the posterior distribution can be used as a representation of the distribution. To complete the task of sampling $\sigma(\mathbf{v})$ we need an algorithm that samples according to $\rho(\mathbf{g}, \mathbf{v})$, and a forward algorithm that calculates $L(\mathbf{g}, \mathbf{v})$ [21]. Given a random function V, designed such that a random walk performed by iterative application of V

$$(\mathbf{g}, \mathbf{v})^{(n+1)} = V(\mathbf{g}, \mathbf{v})^{(n)} \tag{31}$$

would sample the prior $\rho(\mathbf{g}, \mathbf{v})$, our inverse Monte Carlo sampling algorithm proceeds by iterative application of the random function W

$$W[(\mathbf{g}, \mathbf{v})] = \begin{cases} V(\mathbf{g}, \mathbf{v}) & \text{if} \quad U(0,1) \le \min\left[1, \frac{L(V(\mathbf{g}, \mathbf{v}))}{L(\mathbf{g}, \mathbf{v})}\right], \\ (\mathbf{g}, \mathbf{v}) & \text{else} \end{cases} \tag{32}$$

where $U(0,1)$ is a random number uniformly distributed on the interval $[0,1]$. This algorithm can be described as one that in each iteration selects a trial model according to the prior, and accepts the move if it leads to a higher likelihood, whereas a move leading to a lower likelihood is accepted with a probability proportional to a ratio of the likelihoods.

3   Results and Discussion

Regularization versus a posteriori filtering The usual way to avoid problems with non-uniqueness and instability of solutions to inverse problems is through regularization. Regularization (a priori) reduces the subset of permitted solutions to the inverse problem through minimization of a given penalty function or norm. In core flow studies the vorticity norm is an important example of this.

In our study we avoid regularization, because it introduces a non-physical bias to the solution. We do, however, present smooth solutions, but for a different reason, and produced in a different way than described above. The advantage of inverse Monte Carlo sampling is that, in contrast to techniques using regularization, we need not put nonuniqueness-reducing constraints on our solutions a priori since we are not searching for only *one* solution. In fact our prior, a relaxed (soft) geostrophic bound, is incapable of producing a unique solution to the inverse problem. Instead, the Monte Carlo sampler generates a multitude of solutions, all compatible with prior information and data, and all containing structure that is both well resolved and poorly resolved by this information. The point is now that poorly resolved structure fluctuates strongly from one posterior model to the other, in contrast to well resolved structure that is preserved in most of these models. In our analysis of the posterior models it is our task to find well resolved structure, and this may not be an easy task. It turns out that one of the best ways to discover such structure is to replay a large number of a posteriori models as a movie (Mosegaard and Tarantola, 1995) and in this way spot with the eye the stable, and hence well resolved, features in the models.

Figs. 1-4 are snapshots from such a movie. Practically however, even inspection of the a posteriori movie may not allow discovery of well resolved structure, simply because fluctuations of poorly resolved features will dominate the picture. We therefore filter away most of these strongly fluctuating features before investigating the movie. In our problem, strong fluctuations occur in small scale structures of the flow, so a high-cut filter is useful. The resulting models are rather smooth, but they should not be confused with regularized models. One important difference is that our smoothing is applied a posteriori, in contrast to regularization which is an a priori operation. The consequence of posterior filtering is that whereas all unfiltered models fit data, the filtered models generally do not. Another difference is that our smoothing procedure may not necessarily remove all structure that is considered "unnecessarily busy" in a regularization context.

Vorticity strength. Being only interested in large-scale structures of the flow the following models are filtered a posteriori from degree 4 to degree 13 with an almost linearly decreasing function such that half the signal is retained for the degree 8 harmonics. We present a series of twelve randomly selected samples from the posterior pdf for $\varepsilon = 10^{-2}$ and hence under strong influence of geostrophy (Fig. 1). We first study the quantity $-\hat{\mathbf{r}}(\nabla \times \mathbf{u_H}) = r\nabla^2 T$ which we call the *vorticity strength*, where $\hat{\mathbf{r}}$ is the unit vector in the radial direction. The vorticity strength is the toroidal counterpart to the poloidal upwelling and downwelling $r\nabla^2 S$ (se below). These models have well resolved vortices. A well resolved structure is one that exists in many (or all) samples and is similar in location and intensity. For example, all models have a high intensity clockwise (red) vortex under South America, one clockwise vortex of somewhat lower intensity under South Africa and a large anticlockwise (blue) 'bone' structure in the southern hemisphere extending from the south east Pacific to the southern Indian Ocean. In most models this structure encapsulates the South American patch. Its stable eastern extent might be a manifest of the south Indian Ocean gyre. We can identify westward flow in the western hemisphere along the equator by the clockwise patch north of the equator and anticlockwise patch south of the equator in the east Pacific. When comparing with flows from previous published inversions we find that our models are missing the 'jets' from the poles at 90 degrees east and that we resolve a clockwise vortex under South America which is a feature that, to our knowledge, has not been reported in the past.

In contrast, we see considerable dissimilarity in the twelve models for $\varepsilon = 10^{-1}$ (Fig. 2). The 'bone' has dissolved and structures differ from one model to the other indicating poor resolution. We observe a *transition* from generally good to generally poor resolvability in our models at $\varepsilon \approx 2 \cdot 10^{-2}$.

Upwelling/downwelling. Figs. 3 and 4 show $r\nabla^2 S$ of the same twelve models. Likewise, these models exhibit well resolved structures of horizontal divergence $\nabla_{\mathbf{H}} \cdot \mathbf{u_H} = r\nabla^2 S$ for $\varepsilon = 10^{-2}$. All models have upwelling under South Africa, which previous studies have shown to be an upwelling area e.g. [5], [6] and under the East Indian Ocean. Since both the vorticity strength and the horizontal

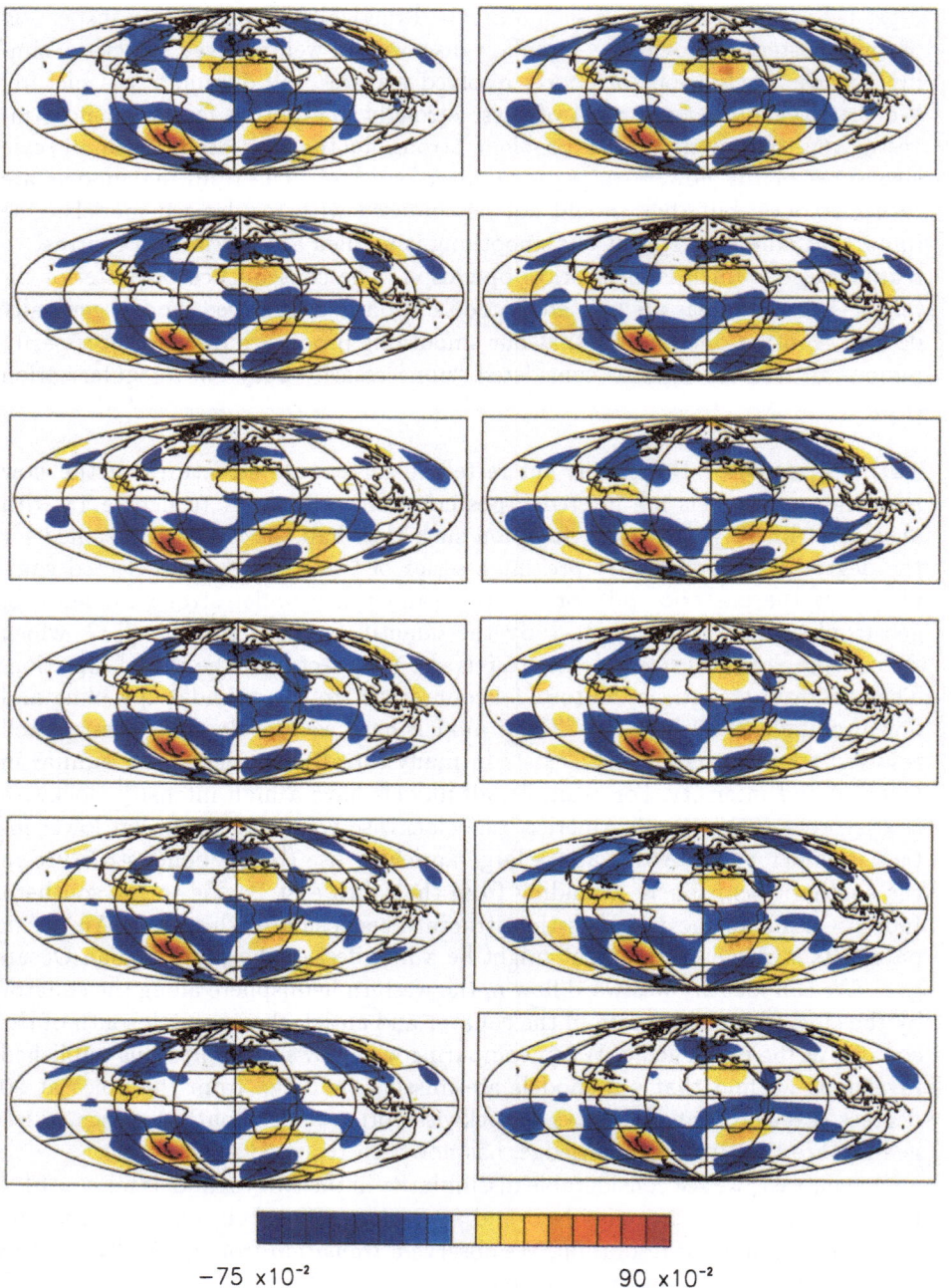

Fig. 1. Vorticity strength in $[yr^{-1}]$ from non-linear Monte Carlo inversion for $\varepsilon = 10^{-2}$. Positive values (yellow/red colors) represent clockwise vortices and negative values (blue colors) represent anticlockwise vortices. These flows are quite similar; they have many well resolved structures in common and are clearly under strong influence of the geostrophic constraint.

265

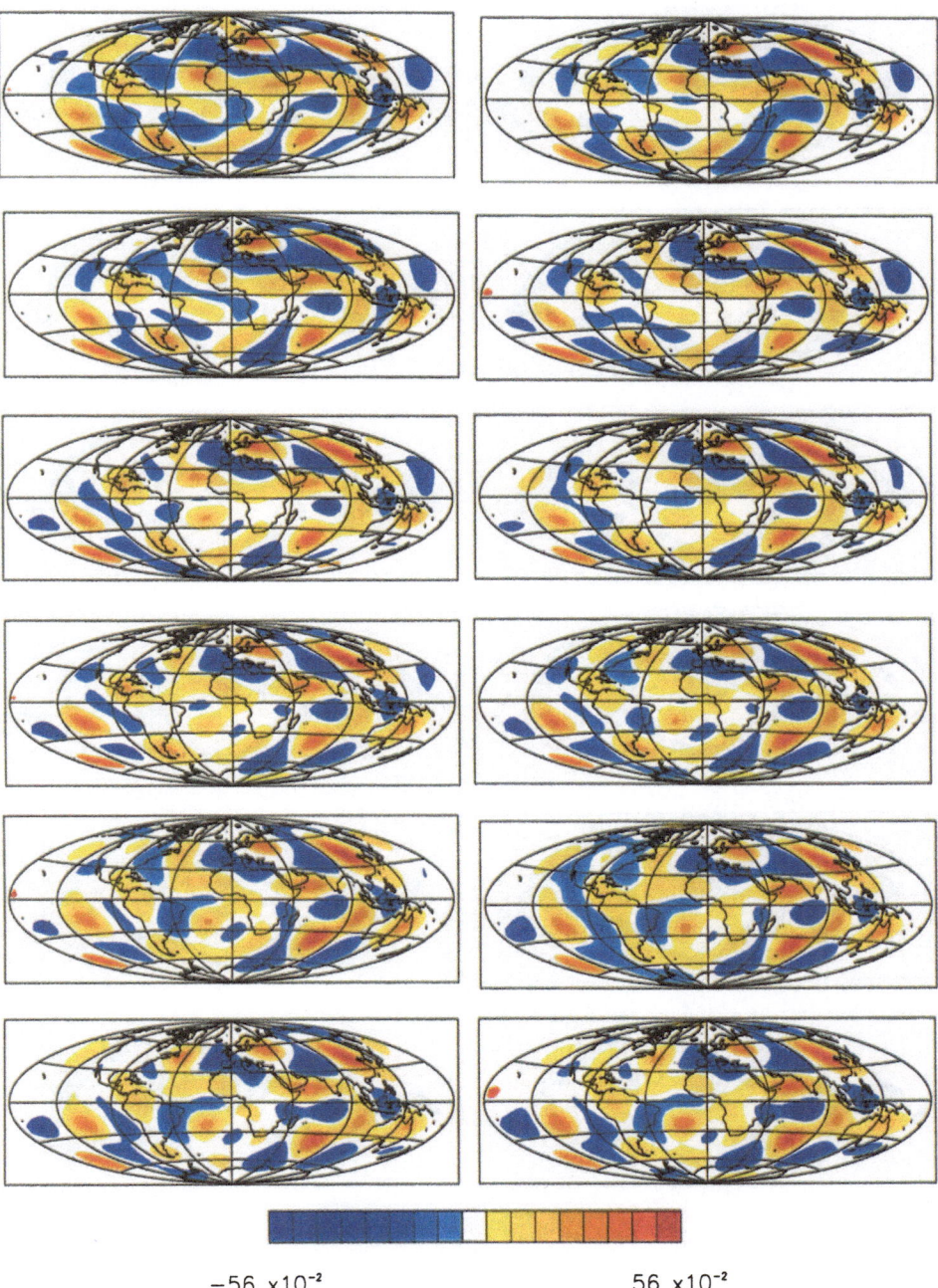

Fig. 2. Vorticity strength in $[yr^{-1}]$ from non-linear Monte Carlo inversion for $\varepsilon = 10^{-1}$. Positive values (yellow/red colors) represent clockwise vorticity strengths and negative values (blue colors) represent anticlockwise vorticity strengths. Notice that the color scale is different from Fig. 1. These flows are markedly different from those in Fig. 1 and they vary individually which indicates generally poor resolution of the structures.

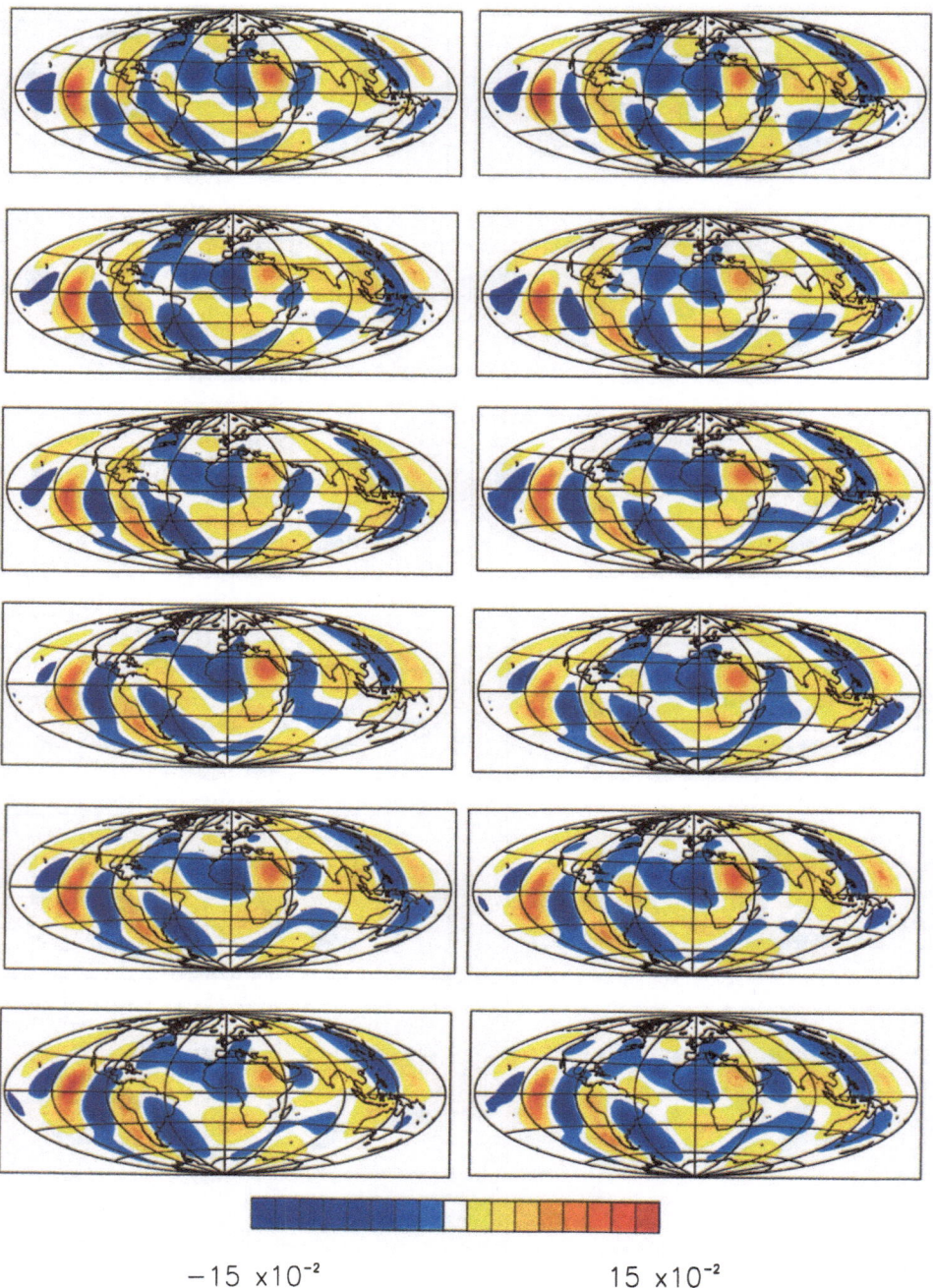

Fig. 3. Divergence of horizontal fluid flow in $[yr^{-1}]$ from non-linear Monte Carlo inversion for $\varepsilon = 10^{-2}$. Positive values (yellow/red colors) represent upwelling and negative values (blue colors) represent downwelling. These flows are very similar; they have many well resolved structures in common and are clearly under strong influence of the geostrophic constraint.

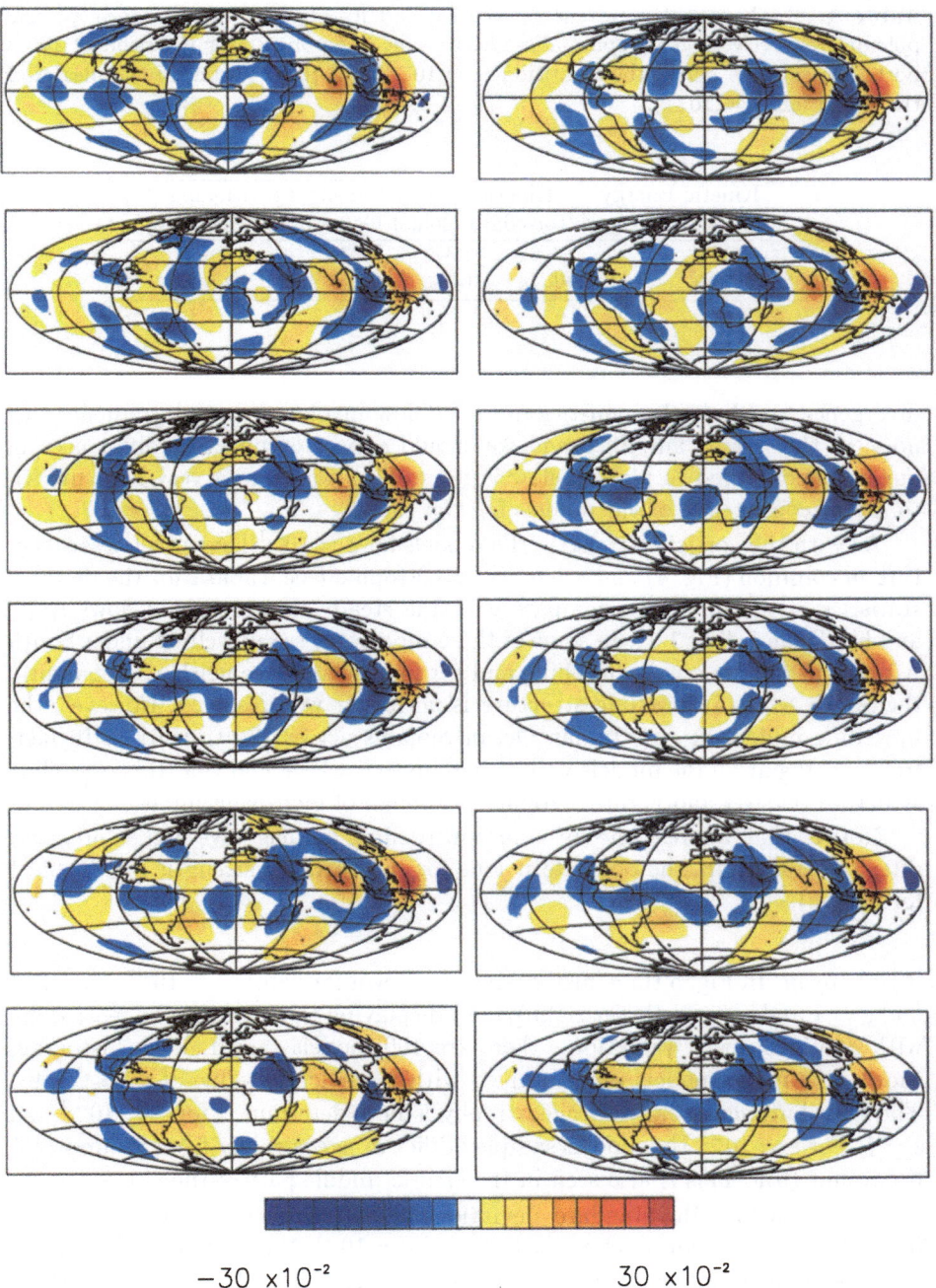

Fig. 4. Divergence of horizontal fluid flow in $[yr^{-1}]$ from non-linear Monte Carlo inversion for $\varepsilon = 10^{-1}$. positive values (yellow/red colors) represent upwelling and negative values (blue colors) represent downwelling. Notice that the color scale is different from Fig. 3. These flows are markedly different from those in Fig. 3 and they vary individually which indicates generally poor resolution of the structures.

Table 1. Total mean kinetic energy density and its partitioning into toroidal and poloidal respectively geostrophic and ageostrophic components for inversion of 100 a posteriori samples before the transition ($\varepsilon = 10^{-2}$) and 100 a posteriori samples after the transition ($\varepsilon = 10^{-1}$).

ε	Kinetic Energy Density [km/yr]^2	Kinetic Energy Density Partitioning [%]			
		Toroidal	Poloidal	Geostrophic	Ageostrophic
10^{-2}	6624.6	95.66	4.34	98.33	1.66
10^{-1}	7043.3	85.44	14.56	85.18	14.82

divergence are derived quantities from the flow involving spatial gradients, the noise in the flow is amplified and we should expect emphazised differences in our plots compared to plots of flow vectors. This points to a very high model-to-model consistency of our models for $\varepsilon = 10^{-2}$.

As with the vorticity strength, the models for $\varepsilon = 10^{-1}$ have much less structure in common (Fig. 4) and are more ageostrophic (see Table 1 for the detailed statistics of the flows and compare with the steady unconstrained flows in [9] and his Table 1, flow 1-1). In some of these models we see upwelling under South Africa, in some we see downwelling and yet in others we see neither. Apparently the South African upwelling structure is no longer well resolved. However, the upwelling under the east Indian Ocean remains across the transition. In fact, from looking at all the models with values up to $\varepsilon = 1$ we find that this upwelling structure persists and is rather robust (although of increasing intensity).

Notice, that whereas the up- and downwelling models have higher intensity after the transition, the opposite it true for the vorticity strength (see Table 1 for details).

Ageostrophy. In Fig. 5 three histograms are shown for values $\varepsilon = 10^{-2}$, $\varepsilon = 10^{-1}$ and $\varepsilon = 1$ and a curve (lower right panel) displaying the increase of ageostrophy with ε. Within each value for ε there are 100 models shown. Each histogram shows how many models were sampled with a particular value of ageostrophy. The distributions reveal that ageostrophy is best determined for $\varepsilon = 10^{-2}$ and $\varepsilon = 1$. For $\varepsilon = 10^{-1}$ the models sampled differ in ageostrophy from around 11% to around 19%. This is also seen in the erratic middle part of the curve.

Fig. 6 displays the standard deviations of horizontal divergence estimated from 554 posterior samples for $\varepsilon = 10^{-2}$ and $\varepsilon = 10^{-1}$. We see both greater areal extent and higher intensity of the standard deviations after the transition. In Fig. 7 we see four examples of this underneath Cape Town, Athens, Sandwich Islands and Sri Lanka. Table 2 summarizes the statistics of the distributions shown in fig. 7. Under Cape Town, Sandwich Islands and Sri Lanka we see that while the distributions are narrow and tall before the transition ($\varepsilon = 10^{-2}$, gray bars) they are wide and short after the transition ($\varepsilon = 10^{-1}$, white bars). Under Athens the resolvability is good both before and after the transition.

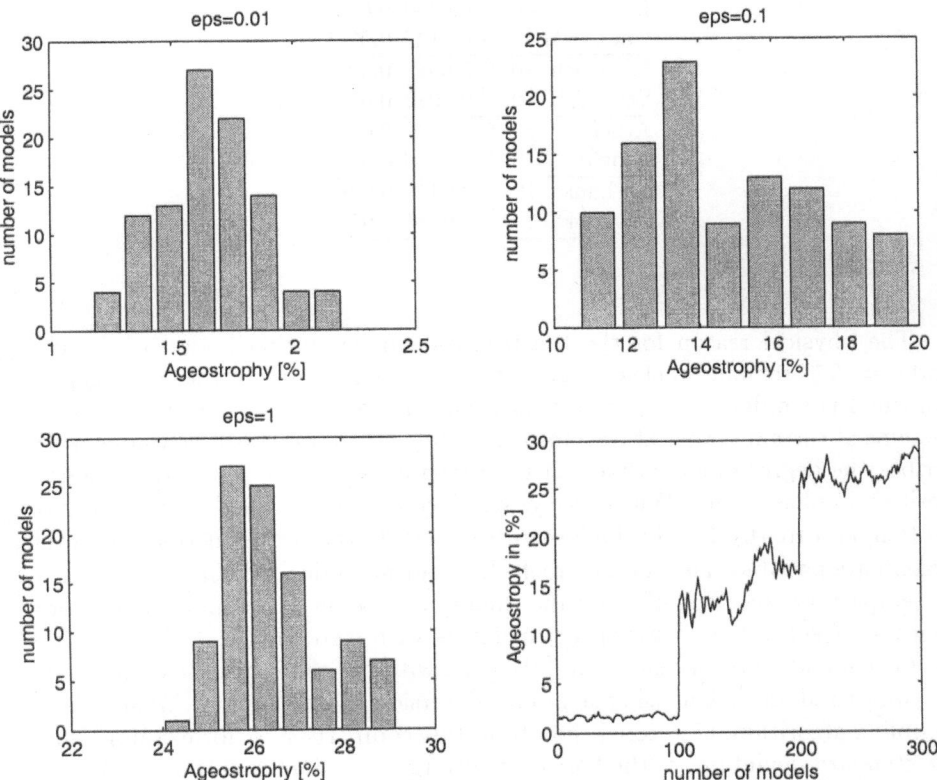

Fig. 5. Ageostrophic part in [%] of the kinetic energy from non-linear Monte Carlo inversion for $\varepsilon = 10^{-2}; \varepsilon = 10^{-1}; \varepsilon = 1$. For each value of ε there are 100 a posteriori samples. The curve in the lower right panel shows the three levels for ε in one plot.

270

Table 2. Means and standard deviations for divergence of horizontal flow in $[yr^{-1}]$ (positive values represent upwelling and negative values represent downwelling) for 554 models at four locations before ($\varepsilon = 10^{-2}$) and after ($\varepsilon = 10^{-1}$) the transition displayed in Figure 7

Location	ϵ	mean	st.dev.
Capetown	10^{-2}	0.0368	0.0055
Capetown	10^{-1}	0.0127	0.0275
Sandwich	10^{-2}	0.0042	0.0038
Sandwich	10^{-1}	0.0038	0.0208
Athens	10^{-2}	-0.0058	0.0085
Athens	10^{-1}	-0.0054	0.0071
Sri Lanka	10^{-2}	0.0383	0.0082
Sri Lanka	10^{-1}	0.0964	0.0307

The physical reason for the the transition in resolution between $\varepsilon = 10^{-2}$ and $\varepsilon = 10^{-1}$ remains unclear. However, it may be speculated whether this phenomenon is similar to the phase transitions observed in statistical mechanical systems at certain temperatures. In such systems the temperature parameter describes the degree of smoothness of the probability distribution of the system's position in phase space. When the system during heating exceeds a certain critical temperature the distribution becomes so smooth that large fluctuations of the system are possible. The system "melts". In our numerical system ε is similar to a "temperature parameter" since the smoothness of the a posteriori distribution over model space (our phase space) increases with increasing ε.

Figs. 6 and 7 are also shown in [21] along with the values for the corresponding linearized inversion using a modified version of the inverse Monte Carlo sampling algorithm. One conclusion from this comparison is, for example, that the standard deviations of the horizontal divergence are up to 6 times higher for linear inversion.

4 Conclusion

Based on an analysis of the horizontal fluid flow beneath the CMB we study resolution properties of the flow under a relaxed geostrophic bound described by the parameter ε, which measures deviation from strict geostrophy. In particular, we examine the varying model resolution depending on the degree of deviation from a strict geostrophy bound.

We increase ε in steps and display the flow structures for values $\varepsilon = 10^{-2}$, $\varepsilon = 10^{-1}$ and $\varepsilon = 1$. For flows with $\varepsilon = 10^{-2}$ we observe well resolved structures, clearly under strong influence of the geostrophic constraint. For higher values of ε structures are generally poorly resolved as a result of the relaxed constraint. We find a *transition* from generally good resolution to generally poor resolution in the models for $\varepsilon \approx 2 \times 10^{-2}$. However, a well-determined structure of upwelling

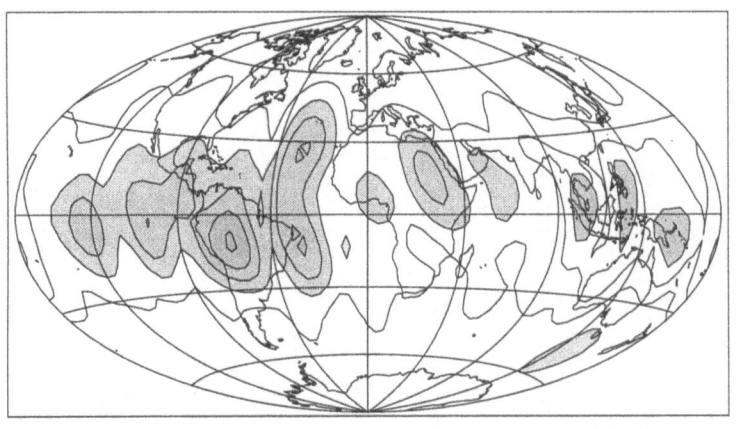

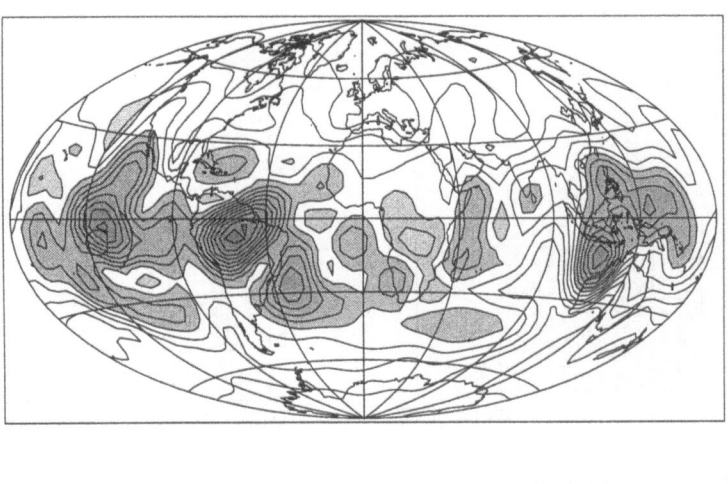

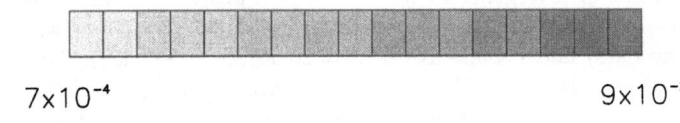

7×10^{-4} 9×10^{-2}

Fig. 6. Standard deviation of divergence of horizontal fluid flow in $[yr^{-1}]$ for (top) $\varepsilon = 10^{-2}$ and (bottom) $\varepsilon = 10^{-1}$.

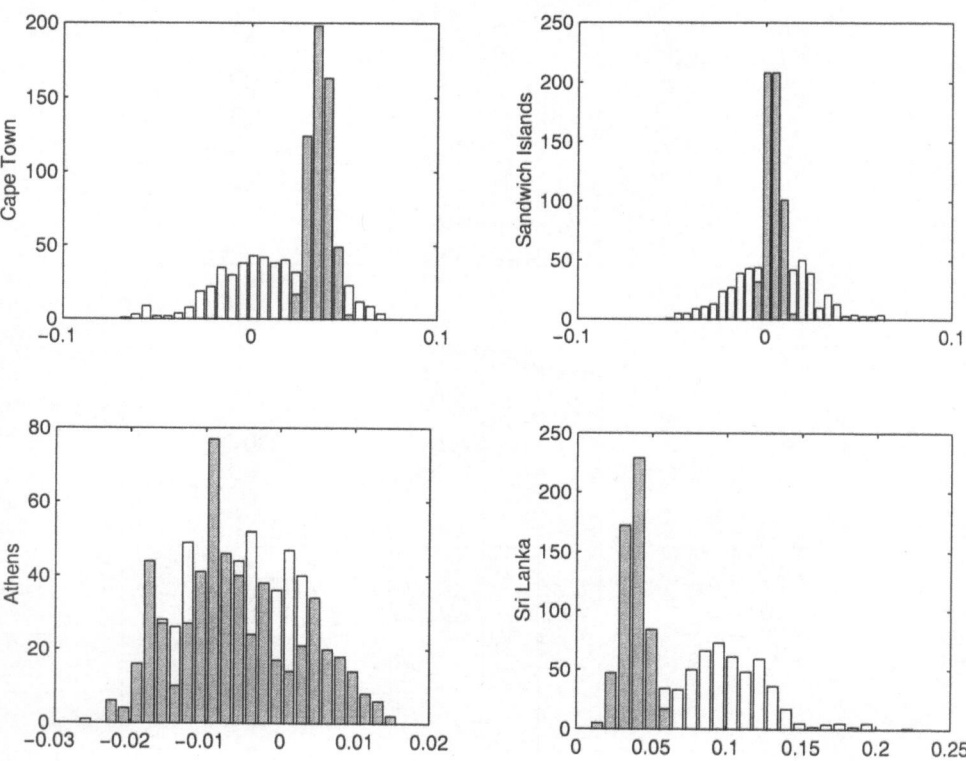

Fig. 7. Histograms of divergence in $[yr^{-1}]$ of 554 a posteriori samples (upwelling are positive values and downwelling are negative values) for $\varepsilon = 10^{-1}$ (white bars) and $\varepsilon = 10^{-2}$ (gray bars) under Cape Town ($\lambda = 20\,\mathrm{deg}, \phi = -33\,\mathrm{deg}$), Sandwich Islands ($\lambda = 355\,\mathrm{deg}, \phi = -54\,\mathrm{deg}$), Athens ($\lambda = 25\,\mathrm{deg}, \phi = 38\,\mathrm{deg}$) and Sri Lanka ($\lambda = 85\,\mathrm{deg}, \phi = 7\,\mathrm{deg}$, where λ, ϕ are longitude and latitude). Notice the different horizontal and vertical scales.

underneath the east Indian Ocean is conserved throughout our series of models indicating that certain functions of the flow are almost unchanged even though the resolving power of the individual coefficients decreases. The non-uniqueness studied for non-zero ε is the fundamental non-uniqueness of the inverse problem, addressed by [1].

Acknowledgments We thank Kathy Whaler, Richard Holme and Jeremy Bloxham for their encouragement and useful suggestions. A program written by Jeremy Bloxham to calculate the geostrophic partitioning of the flow and the statistics displayed in Table 1 was kindly made available to us.

Appendix A

Minimizing Sampling Errors

The Monte Carlo method used in this study belongs to the class of so-called Markov-Chain Monte Carlo (MCMC) method. These types of methods are well-established in Bayesian computation see e.g. [4], but were adapted to analysis of inverse problems by [19]. The method can be demonstrated to sample the posterior probability density asymptotically, that is, the correlation between samples taken with a separation of n iterations converges to zero as n goes to infinity [10]. In practice (as for instance in this study) the method is used to produce approximately independent samples from the posterior by using only a subset of all the models generated by the algorithm, namely models that are separated by more than N iterations, where N is a large number.

The question is, however, how large N should be. Unfortunately, no theory for MCMC methods exists that allows estimation of N given an allowed maximum correlation between samples. We are therefore reduced to using a number of test procedures to ensure approximate sample independence, see [11] or [20]. The procedures (all of which have been used in this study) are:

- Choose step lengths as large as possible, but still short enough to keep the acceptance probability high (30-50%).
- Start sampling in a region of high posterior probability (in a model that is reasonably consistent with data and prior information).
- Observe selected parameters (model parameters, predicted data, data misfit, etc.) as the sampling proceeds. These (finite) time series of parameters must attain stationarity before the output samples from the algorithms are used.

None of the above procedures will mathematically guarantee independence of the samples obtained by the algorithm. However, for many problems they will provide convincing arguments for the approximate correctness of the analysis.

For comparison it should be noted that iterative algorithms used for estimating solutions to weakly non-linear problems suffer from convergence problems that are quite similar to those of MCMC methods. Perfect convergence to the optimal solution will not be obtained in a finite number of iterations, and the only way to ensure approximate convergence is to observe the behavior of, e.g., the data misfit as the iterations proceed.

References

1. Backus, G. E., Kinematics of geomagnetic secular variation in a perfectly conducting core, Phil. Trans. Roy. Soc. Lond. A., 1968, 263, 239-266
2. Backus, G. E., Comparing hard and soft bounds in geophysical inverse problems, Geophys. J., 1988, 94, 249-261
3. Backus, G. E. and Parker, R. L. and Constable, C. G., Foundations of geomagnetism, Cambridge University Press, 1996, New York, N.Y.
4. Bernardo, J. M. and Smith, A. F.., Bayesian Theory, Wiley Series in Probability and Mathematical Statistics, 1997, New York, NY.
5. Bloxham, J., The dynamical regime of fluid flow at the core surface, Geophys. Res. Lett., 1988, 15, 6, 585-588.
6. Bloxham, J., The determination of fluid flow at the core surface from geomagnetic observations, Mathematical Geophysics, D. Reidel Publishing Company, 1988, Vlaar, N. J. and Nolet, G. and Wortel, M. J. R. and Cloetingh, S. A. P. L., 1, University of Utrecht, The Netherlands
7. Bloxham, J. and Jackson, A., Fluid flow near the surface of Earth's outer core, Rev. Geophys. Space Phys., 1991, 29, 97-120
8. Bloxham, J. and Jackson, A., Time-dependent Mapping of the Magnetic Field at the Core-Mantle Boundary, J. Geophys. Res., 1992, 97, 19537-19563
9. Bloxham, J., The Steady Part of the Secular Variation of the Earth's Magnetic Field, J. Geophys. Res., 1992, 97, B13, 19.565-19.579
10. Feller, W., An Introduction to Probability Theory and its Applications, John Wiley, 1970, New York
11. Hastings, W. K., Monte Carlo sampling methods using Markov chains and their applications, Biometrica, 1970, 57, 1, 97-109
12. Jackson, A. and Bloxham, J., Mapping the fluid flow and shear near the core surface using the radial and horizontal components of the magnetic field, Geophys. J. Int., 1991, 105, 199-212
13. Jackson, A., An approach to estimation problems containing uncertain parameters, Phys. Earth Planet. Inter., 1995, 90, 145-156
14. Jackson, A., Kelvin's theorem applied to Earth's core, 1996, The Royal Society
15. Langel, R. A., The main field, Geomagnetism, Academic Press, 1987, Jacobs, J. A., 4, Orlando, Fl.
16. Le Mouël, J. -L., Outer-core geostrophic flow and secular variation of the Earth's magnetic field, Nature, 1984, 311, 734-735
17. Le Mouël, J. -L. and Gire, C. and Madden, T., Motions at the core surface in the geostrophic approximation, Phys. Earth Planet. Inter., 1985, 39, 270-287
18. Lloyd, D. and Gubbins, D., Toroidal fluid motion at the top of the Earth's core, Geophys. J. Int., 1990, 100, 455-467
19. Mosegaard, K. and Tarantola, A., Monte Carlo sampling of solutions to Inverse problems, J. Geophys. Res., 1995, 100, 12431-12447
20. Mosegaard, K., Resolution Analysis of General Inverse Problems through Inverse Monte Carlo Sampling, Inverse Problems, 1998, In press
21. Mosegaard, K. and Rygaard-Hjalsted, C., Bayesian analysis of implicit inverse problem, Inverse Problems , 1998, Submitted
22. Roberts, P. H. and Scott, S., On Analysis of the Secular Variation, 1. A Hydrodynamic Constraint: Theory, J. Geomagn. Geoelectr., 1965, 17, 137-151
23. Tarantola, A. and Valette, B., Inverse Problems = Quest for Information, J. Geophys., 1982, 50, 159-170

24. Voorhies, C.V., Steady flows at the top of Earth's core derived from geomagnetic field models, J. Geophys. Res., 1986, 91, 12444-12466
25. Voorhies, C.V., Steady surficial core motions: An alternate method, Geophys. Res. Lett., 1986, 13, 1537-1540
26. Voorhies, C. V., Time-varying fluid flow at the top of the earth's core derived from definitive reference field models, J. Geophys. Res., 1995, 100, B7, June, 10029-10039
27. Whaler, K.A., Does the whole of the Earth's Core convect?, Nature, 1980, 287, 528-530
28. Whaler, K.A. and Davis, R. G., Probing the Earth's Core with Geomagnetism, Earth's Deep Interiour, Gordon and Breach Science Publishers, 1997, Crossley, D. J., 7, 5, 115-166
29. Whaler, K.A., Geomagnetic evidence for fluid upwelling at the core-mantle boundary, Geophys. J. R. Astr. Soc., 1986, 86, 563-588

Inversion of GPS Occultation Data for Atmospheric Profiling

Stig Syndergaard

Atmosphere Ionosphere Remote Sensing Division,
Danish Meteorological Institute,
Copenhagen, Denmark
Email: ssy@dmi.dk

1   Introduction

The GPS (Global Positioning System) radio occultation technique can provide valuable information about the atmospheric refractive index related to the temperature, pressure, and moisture, as well as the ionospheric electron density. Generally, the radio occultation method has been widely used in various planetary missions the last 30–35 years, and has given unique information about the atmospheres of the other planets in our solar system (e.g., Kliore et al. [1965]; Fjeldbo et al. [1971]; Tyler [1987]). Just recently, with the launch of the American GPS/MET (GPS/Meteorology) experiment in April 1995, the technique has been successfully applied to the Earth's atmosphere as well (e.g., Ware et al. [1996]; Kursinski et al. [1996]; Rocken et al. [1997]).

The basis of the GPS/MET occultation experiment is the GPS constellation, originally designed for precise positioning. The GPS constellation consists of 24 satellites in 6 orbital planes at a distance of about 20,200 km from the Earth's surface. These satellites continuously transmit electro-magnetic waves at two frequencies: $f_1 = 1.57542\,\text{GHz}$ and $f_2 = 1.22760\,\text{GHz}$. In the GPS/MET experiment the signals are received at a Low Earth Orbit (LEO) satellite about 750 km above the surface. On their way the signals pass through the Earth's ionosphere and neutral atmosphere (Fig. 1).

In the ionosphere and neutral atmosphere the signals are distorted because of the refractive index of the medium. In the geometrical optics approximation, the result is that a signal received at the LEO has been subject to a small bending. If we know the precise positions and velocities of the satellites (in principle obtained by simultaneous observations of different GPS satellites), we are able to measure the bending angle, α. Due to the satellite motions the whole atmosphere from top to surface is scanned, obtaining a set of bending angles related to different heights in the atmosphere. Using the Abel transform (e.g., Fjeldbo et al. [1971]), the measured bending angles can be inverted to an atmospheric refractive index profile. In the case that one is only interested in the neutral atmosphere, the scanned region is restricted to the lowest ~ 100 km of the atmosphere. In this paper we shall concentrate on the neutral atmosphere occultation experiment.

The refractive index in the neutral atmosphere is related to the temperature, T [K], the total pressure, p [hPa], and the partial pressure of water vapor,

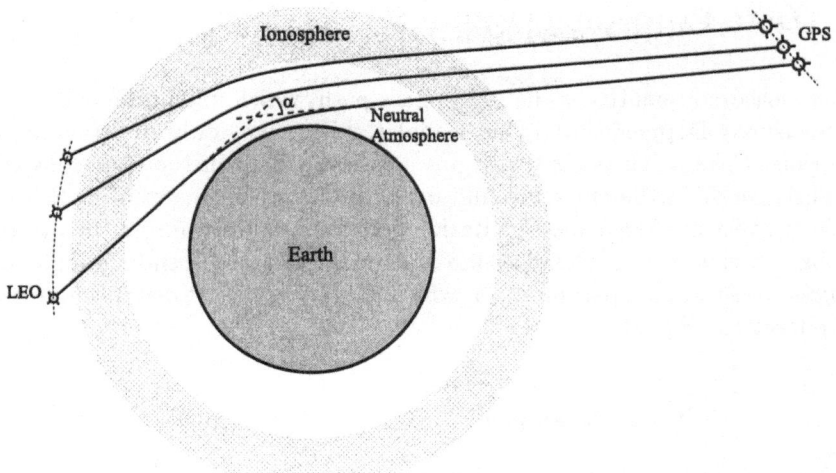

Fig. 1. Illustration of the occultation geometry. Signals transmitted by a GPS satellite are refracted by the ionosphere and the neutral atmosphere. When received at a LEO satellite outside the main part of the ionosphere the signals have been subject to bending.

p_w [hPa], as (e.g., Smith and Weintraub [1953])

$$\mu = 1 + 77.60 \cdot 10^{-6} \frac{p}{T} + 0.373 \frac{p_w}{T^2} \ . \tag{1}$$

The term related to the partial pressure of water vapor can sometimes be neglected, being a fair assumption at high latitudes. Generally, the water vapor in the lower troposphere can be neglected in regions colder than 250 K (Kursinski et al. [1996]). In the tropical troposphere the water vapor term can usually not be neglected and the derived temperature ignoring this term in (1) is therefore often denoted as the *dry temperature*.

In Sect. 2 we outline the inversion procedure, going from bending angles to the dry temperature profiles. This outline is only meant as an overview for the reader not familiar with the method, and does not take into account problems involving differentiation of noisy phase data to obtain the bending angles, ionospheric residual effects, diffraction effects, and model extrapolation dealing with infinite integration limits. For a closer inspection of these difficulties the reader is referred to (Syndergaard [1999]) and references therein. Section 3 is dedicated to a random error propagation analysis related to the Abel transform. In Sect. 4 some results of inverted data from the GPS/MET experiment are shown and compared to other data sources like radiosondes and numerical weather analyses. A case study, testing the spherical symmetry assumption is carried out in Sect. 5, and finally, conclusions and remarks are given in Sect. 6.

278

2 Data Processing Overview

The measured quantity is the GPS phase path, which upon differentiation gives a measured Doppler shift. This Doppler shift is not only due to the relative motion of the satellites but also due to the bending in the atmosphere which slightly modifies the emergent and incident angles of the signal to the satellite velocity vectors. From the occultation geometry, and assuming the atmosphere to be spherically symmetric, we are able to calculate the bending angle, α, then related to an impact parameter, a, which is the asymptotic distance to the center of refraction (Fig. 2).

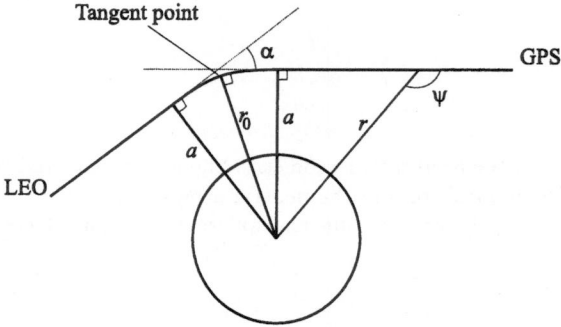

Fig. 2. The geometry of the occultation is assumed to be spherically symmetric. The impact parameter, a, is defined as the perpendicular distance between either of the ray asymptotes and the center of refraction. At any point along the ray, ψ is the angle between the wave normal and the radius vector. The tangent point is the closest approach of a ray to the Earth's surface.

When only interested in the contribution from the neutral atmosphere to the bending of the signal, the ionosphere contribution (overshadowing the contribution from the neutral atmosphere for rays with tangent points above some 45 km) has to be eliminated. This is possible, at least to a large extent, using a dual frequency combination. To eliminate most of the ionospheric effects a linear combination of the bending angles, α_1 and α_2, for each of the two GPS signals can be applied as

$$\alpha(a) = \frac{f_1^2 \alpha_1(a) - f_2^2 \alpha_2(a)}{f_1^2 - f_2^2} . \tag{2}$$

This combination was first proposed by Vorob'ev and Krasil'nikova ([1994]) and has shown to give better results than the traditional combination of phases usually applied in ground based GPS measurements. A detailed theoretical analysis of the ionosphere calibration in GPS occultation measurements can be found in (Syndergaard [1999]).

In a spherically layered medium, Snell's law of refraction results in Bouger's law given by

$$\mu(r)r\sin\psi = a \ , \tag{3}$$

where r is the radial distance and ψ is the angle between the wave normal and the radius vector at points along the ray (Fig. 2). The quantity $\mu(r)r\sin\psi$ is an invariant along the ray. Based on Bouger's law, Fjeldbo et al. ([1971]) derived an integral transform relating $\alpha(a)$ and $\mu(r_0)$:

$$\alpha(a) = -2a \int_{r_0}^{\infty} \frac{d\ln\mu/dr}{\sqrt{\mu^2 r^2 - a^2}} dr \ , \tag{4}$$

and the inverse relation:

$$\mu(r_0) = \exp\left(\frac{1}{\pi} \int_{a}^{\infty} \frac{\alpha(a')}{\sqrt{a'^2 - a^2}} da'\right) \ . \tag{5}$$

In (5), r_0 is the radial distance from the center of refraction to the tangent point (Fig. 2). Equations (4) and (5) can be regarded as a special case of the Abel transform, first derived by the nineteenth century Norwegian mathematician, Niels Henrik Abel. These equations have been derived under the assumption of local spherical symmetry, and as we shall see later, this is a very good assumption for the neutral part of the atmosphere in most cases. Having calculated a set of bending angles as a function of impact parameters, (5) directly gives us the refractive index at the tangent point altitudes. The tangent point altitude for a given impact parameter can then be derived from Bouger's law for $\psi = 90°$, giving

$$r_0 = \frac{a}{\mu(r_0)} \ . \tag{6}$$

For convenience the refractivity, N, is here defined as

$$N = \mu - 1 = 77.60 \cdot 10^{-6} \frac{p}{T} \ , \tag{7}$$

ignoring the water vapor term. Close to the surface the refractivity reach a value of about $3 \cdot 10^{-4}$ under dry conditions. Under wet conditions the refractivity (including the water vapor term) may reach values of about $4.5 \cdot 10^{-4}$.

If we assume ideal gas behavior, it follows that the dry air density, ρ, is proportional to the dry air refractivity given by (7). Therefore, applying (5) for a range of impact parameter values—and knowing the radius of the Earth—we obtain a dry air density profile. From the dry air density profile, pressure, p, as a function of altitude, h, is generated by applying the hydrostatic equilibrium assumption:

$$p(h) = \int_{h}^{\infty} g(h')\rho(h')dh' \ , \tag{8}$$

where $g(h)$ is the gravitational acceleration profile at the tangent point location. Finally the equation of state is applied once again to obtain the temperature:

$$T(h) = \frac{p(h)}{R_d \rho(h)} \ , \tag{9}$$

R_d being the gas constant for dry air.

The temperature obtained in this way is a very good estimate of the real temperature in regions where the water vapor term has only little influence on the total refractivity. This is not generally the case in tropical regions at tropospheric heights. Then the dry temperature will underestimate the real temperature. As a consequence, in moist regions the water vapor pressure may be found if auxiliary information of the real temperature is available. From (1) it follows that

$$p_w = T^2 \frac{N - 77.60 \cdot 10^{-6} p/T}{0.373} \ , \tag{10}$$

where T denotes the auxiliary temperature while N is the refractivity measured from the occultation data. The total pressure, p, may be obtained either from auxiliary sources also, or the hydrostatic equation may be invoked to get the relation between temperature, pressure and moisture. This latter approach becomes an iterative procedure, and a method for water vapor retrieval using such a scheme has been elaborated by Gorbunov et al. ([1996]). In this paper we shall consider dry air retrievals only.

3 Random Error Propagation Analysis

As seen in Sect. 2 we have an analytic formulation of the inverse problem. Nevertheless it is illuminating to take (4) and discretize it in order to do a random error propagation analysis using singular value decomposition. For this we choose to discretize the atmosphere into 100 layers, each of thickness 1 km. This is a representative value of the inherent vertical resolution of the method, which is limited by the first Fresnel diameter (cf Sect. 5).

We will first put (4) into a form ready for discretization. This can be done by a change of variable, $x = \mu r$, while at the same time using the approximation $\ln(\mu) \approx N$, to obtain

$$\alpha(a) = \int_{a_1}^{b} K(x, a) m(x) dx \ , \tag{11}$$

with

$$K(x, a) = \begin{cases} 0 & \text{for } x < a \\ \frac{2a}{\sqrt{x^2 - a^2}} & \text{for } x \geq a \end{cases} , \qquad m(x) = -\frac{dN}{dx} \ . \tag{12}$$

At ~ 100 km altitude and above we will assume the refractivity gradients to be negligible, so that the upper integration limit can be approximated by $b = a_1 + 100$ km, where $a_1 = 6371$ km is a representative value of the lowest impact parameter close to the Earth's surface. Splitting up the integration into $n = 100$ sub-integrals in which $m(x) = m_j$ ($j = 1, \ldots, n$) are assumed to be constants, we get

$$\alpha(a) = \sum_{j=1}^{n} \int_{a_j}^{a_{j+1}} K(x, a) m_j dx \ , \tag{13}$$

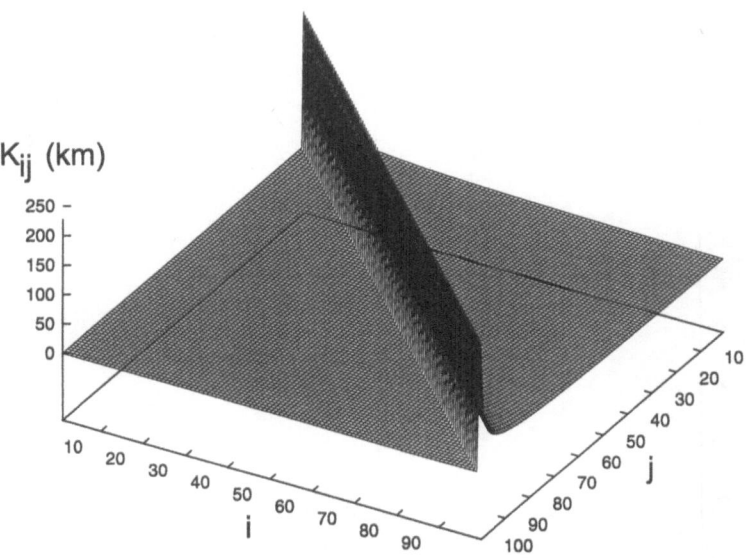

Fig. 3. The 100×100 matrix related to the kernel of the inverse problem.

where $a_{n+1} = b$ and $a_{j+1} - a_j = \Delta a = 1\,\text{km}$. The integration of the kernel, $K(x, a)$, can be evaluated analytically in each subinterval to give

$$
\int_{a_j}^{a_{j+1}} K(x, a) dx = 2a \ln \left(\frac{a_{j+1} + \sqrt{a_{j+1}^2 - a^2}}{a_j + \sqrt{a_j^2 - a^2}} \right) \quad , \qquad a_j \geq a \ . \tag{14}
$$

Choosing $\alpha_i = \alpha(a_i)$ $(i = 1, \ldots, n)$ to be n data points of bending angles with impact parameters $a_i = a_1 + (i-1)\Delta a$, we arrive at a discretized formulation of (11):

$$
\boldsymbol{\alpha} = \boldsymbol{K}\boldsymbol{m} \tag{15}
$$

with

$$
K_{ij} = \begin{cases} 0 & \text{for} \quad j < i \\ 2a_i \ln \left(\frac{a_{j+1} + \sqrt{a_{j+1}^2 - a_i^2}}{a_j + \sqrt{a_j^2 - a_i^2}} \right) & \text{for} \quad j \geq i \end{cases} . \tag{16}
$$

In (15), $\boldsymbol{\alpha}$ is our data vector of bending angles and $\boldsymbol{m}$ our model vector of the refractivity gradients. Figure 3 is a plot of the resulting 100×100 dimensional $\boldsymbol{K}$ matrix. It is an upper triangular matrix, and due to the shape of the kernel the diagonal elements are far the largest ones.

Doing the singular value decomposition (e.g., Scales and Smith [1996]),

$$
\boldsymbol{K} = \boldsymbol{U}\boldsymbol{\Sigma}\boldsymbol{V}^{\mathrm{T}} \ , \tag{17}
$$

we find, not surprisingly, that the condition number is low ($= 9.67$), meaning that this is a very well-posed problem. Assuming uncorrelated errors on the

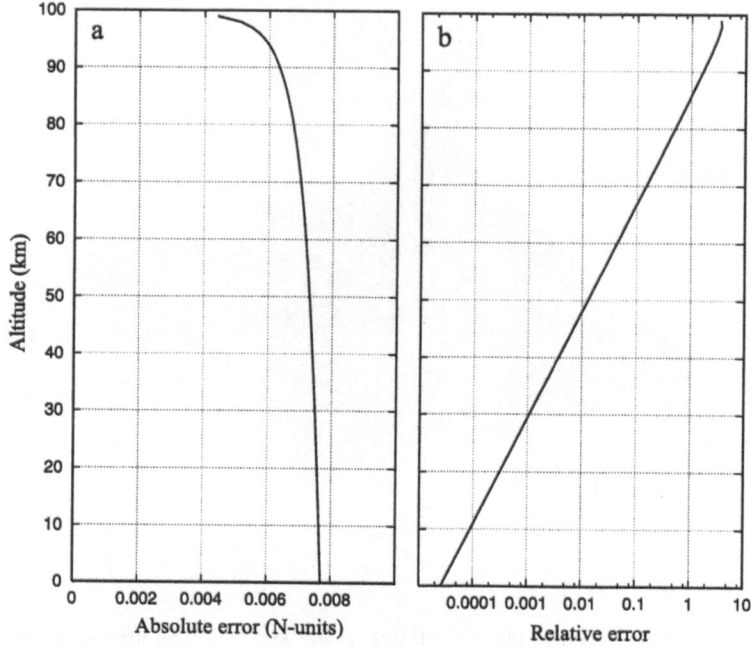

Fig. 4. Standard deviation on retrieved refractivity. a) In N-units ($1\,\text{N-unit} = 10^{-6}$). b) Relative to a background refractivity taken as $N = 3 \cdot 10^{-4} \exp(-(a_k - a_1)/8\,\text{km})$.

bending angles[1], with a constant standard deviation σ_α, the error covariance matrix for the refractivity gradients becomes

$$C_m = \sigma_\alpha^2 V \Sigma^{-2} V^{\mathrm{T}} \ . \tag{18}$$

From the error covariance matrix of the refractivity gradients we obtain the resulting standard deviation on the refractivity, $N_k = N(a_k)$ $(k = 1, \ldots, n)$, at different heights:

$$\sigma_{N_k}^2 = (\Delta a)^2 \sum_{i,j \geq k} C_{ij} \ . \tag{19}$$

For $\sigma_\alpha = 10^{-6}\,\text{rad}$ (which is not so far from what can be obtained from the GPS/MET data with this technique), Fig. 4a shows how the refractivity standard deviation decreases slightly with altitude in the atmosphere. However, since the atmospheric refractivity is decreasing almost exponentially with altitude, the relative errors in refractivity increases with altitude (Fig. 4b). At altitudes above $\sim 70\,\text{km}$ the air is so thin that the bending angle signal becomes less than the

---

[1] Since the bending angles are derived from the phase measurements, involving numerical differentiation, the bending angle errors will not be perfectly uncorrelated. A more elaborate random error propagation analysis for the whole problem going from phase to temperature can be found in (Syndergaard [1999]).

bending angle noise. For the retrieved refractivity the relative error exceeds 20 %
above 70 km. The error behavior seen in Fig. 4b maps into a similar behavior on
the temperature standard deviation, which is the main reason that this method
is only capable of providing good temperature estimates (better than 1–2 °C)
below some 30–40 km altitude. In the estimations above we have made a small
approximation, equalizing the impact height with the altitude. In fact the impact
height and the actual tangent height differs by a few kilometers at the lowest
altitudes.

The result in Fig. 4b is similar to earlier results given by Kursinski et al.
([1997]). However, it should be emphasized that the resulting standard deviation of the refractivity—besides the standard deviation of the bending angle—
depends on the number of discretization levels and the spacing between them as
well as the vertical correlation between the bending angle errors (Syndergaard
[1999]).

4 Results

The method outlined in Sect. 2 has been used to obtain the dry temperature
from the GPS/MET phase data in a few cases. Figure 5 and 6 compares inversion
results with nearby radiosonde data and numerical analyses from the European
Centre for Medium-range Weather Forecasts (ECMWF) and the U. S. National
Centers for Environmental Prediction (NCEP).

In Fig. 5 we see a good agreement between the GPS/MET and the radiosonde
temperature. Both measurement methods catch some of the same temperature
variations in the tropopause and stratosphere up to the altitude where the radiosonde stops giving data. The differences are only a few °C. The inversion
result also agrees quite well with the numerical analyses up to an altitude where
we expect the inversion accuracy to become poor (above 35 km). Below some
4 km the results are not good because of increasing moisture in the lower troposphere. Generally, it should be noted that the correlative data (radiosonde data
and the analyses data) are separated in time and space from the occultation
data, since the correlative data are only available at fixed times and locations
while the temporal and spatial distribution of the occultation events is almost
random. This fact may also account for some of the discrepancies between the
occultation measurements and the correlative data.

In the lower troposphere moisture may cause sharp vertical gradients in the
refractivity which influences the signal tracking performance of the receiver. If
the defocusing of the signal becomes to large, the signal becomes so weak that
the receiver looses the signal. In extreme cases super-refraction may occur, which
is the situation when the bending at the tangent point becomes larger than the
Earth's curvature (Kursinski et al. [1997]). In such cases the signal will never
reach the receiver. Even if the signal is tracked correctly, atmospheric multipath propagation, which will occur in regions of sharp vertical gradients, makes
it more difficult to retrieve the refractivity correctly. In such situations there is
no unique value of the bending angle connected to the Doppler shift at the LEO

284

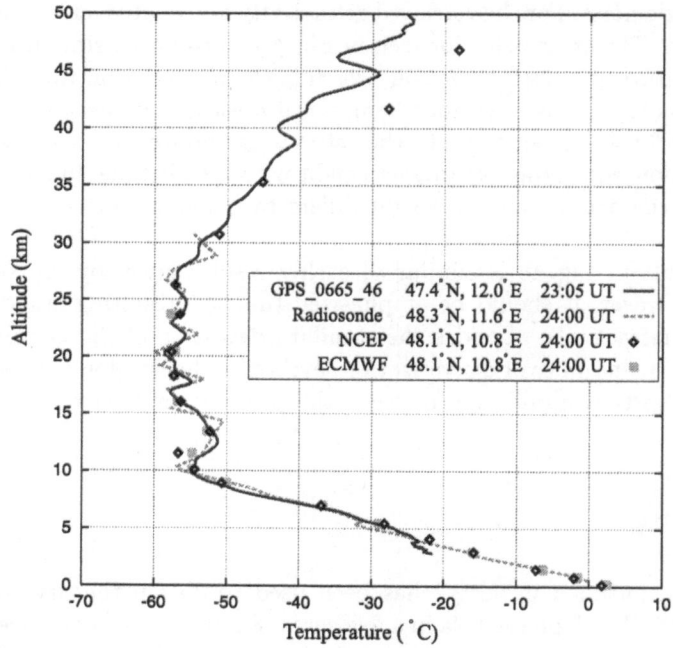

Fig. 5. GPS/MET sounding (occ. no. 665, day 46, 1997) compared with a nearby radiosonde measurement and numerical analyses from ECMWF and NCEP.

satellite. Instead, more advanced methods like back-propagating the electromagnetic field to a region nearer the tangent point, by means of wave theory, should be applied (Karayel and Hinson [1997]; Gorbunov and Gurvich [1998]). The back-propagation method also corrects for diffraction effects, theoretically improving the vertical resolution in the troposphere by a factor of about 2–5, depending on the altitude (Mortensen et al. [1999]). In any case, the moist and the dry contribution to the refractivity in the lower troposphere cannot be distinguished without auxiliary information of either temperature or water vapor (Kursinski et al. [1995]; Ware et al. [1996]).

In Fig. 6 the radiosonde and the GPS/MET sounding match extremely well. This is a high latitude occultation where the air is mostly dry all the way down to the altitude where the receiver looses track of the signal. Also the numerical analyses data agree well with the GPS/MET sounding up to about 20 km where they suggest a warmer stratosphere than the inversion result. However, it should be noticed that the ECMWF and NCEP analyses are more or less driven by radiosonde data, and in this case the nearest radiosonde stops giving data at 17–18 km altitude. Therefore it is most likely that the GPS/MET sounding in this case is closer to the truth than the NCEP results in the range 20–35 km.

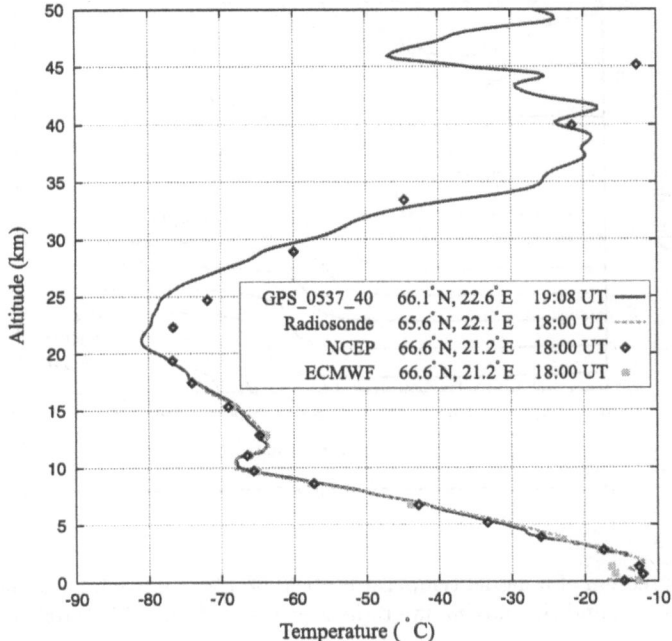

Fig. 6. GPS/MET sounding (occ. no. 537, day 40, 1997) compared with a nearby radiosonde measurement and numerical analyses from ECMWF and NCEP.

At present, serious efforts are made to develop techniques for assimilation of the occultation data, either bending angle or refractivity, into numerical weather prediction models (e.g., Zou et al. [1999]).

5 Testing the Spherical Symmetry Assumption

It should be remembered that the measurement is an integrated effect over the entire path from the GPS satellite to the LEO satellite. However, most of the bending occurs over approximately 700 km of the path, centered at the tangent point (Høeg et al. [1996]). Taking into account the sphericity of the atmosphere, the first Fresnel diameter, z_F, can be used to define a horizontal resolution at the tangent point as

$$D_F = 2\sqrt{2r_0 z_F} \ . \tag{20}$$

If $z_F \ll r_0$, D_F is approximately the horizontal width of a spherical shell having the thickness equal to the first Fresnel diameter. The first Fresnel diameter is the inherent vertical resolution limit when using the geometrical optics approximation, and for a GPS receiver in LEO it varies from about 1.5 km in the stratosphere to about 0.5 km in the lower troposphere (Melbourne et al. [1994]). Roughly speaking, components of all rays that pass through the first Fresnel

286

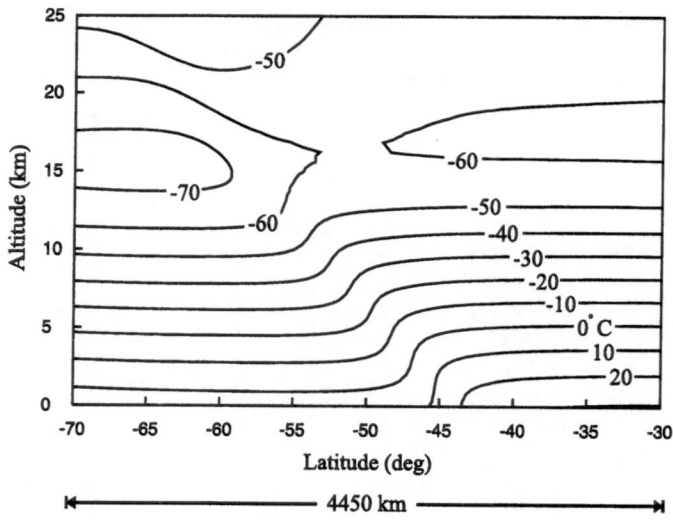

Fig. 7. Cross section of a model temperature field of a severe frontal system. The slope of the front is 1 %, and the maximum gradient at the surface is 5 °C/100 km. Isotherms are in °C.

diameter zone will add constructively to some degree, while the contributions of those outside this zone will cancel (Kursinski [1994]). For $z_F = 1$ km we get $D_F \approx 220$ km. During the occultation the tangent points are drifting horizontally due to the satellite motion, giving different tangent point locations at different heights. This horizontal drift may become as large as 300 km during the lowest 60 km of ray path descend (Kursinski et al. [1997]).

The inversion described in Sect. 2 is based on the assumption of local spherical symmetry, which, if being true, would make our concerns about the horizontal resolution superfluous. However, as we know, the atmosphere is not spherically symmetric, and larger errors than the ones in Fig. 4 are expected below some 40 km, and especially in regions of severe horizontal gradients. To assess this, a small case study has been carried out, including 3D ray tracing through a frontal system to obtain synthetic phase data.

Figure 7 is the cross section of the model temperature field of a severe frontal system. For simplicity, in the simulations the front is directed so that the temperature gradient lies in a meridian plane. To simulate a worst case scenario, the difference between the warm and the cold sides of the front at the surface has been set to about 25 °C. Such differences might be found at mid latitudes over continents during the winter (Hardy et al. [1994]). The model has been constructed from an analytical expression of the refractivity given by Syndergaard ([1999]). The model is based on calculations excluding the water vapor contribution. Hardy et al. ([1994]) performed a similar case study using a more realistic model including the water vapor, but only presenting a single result.

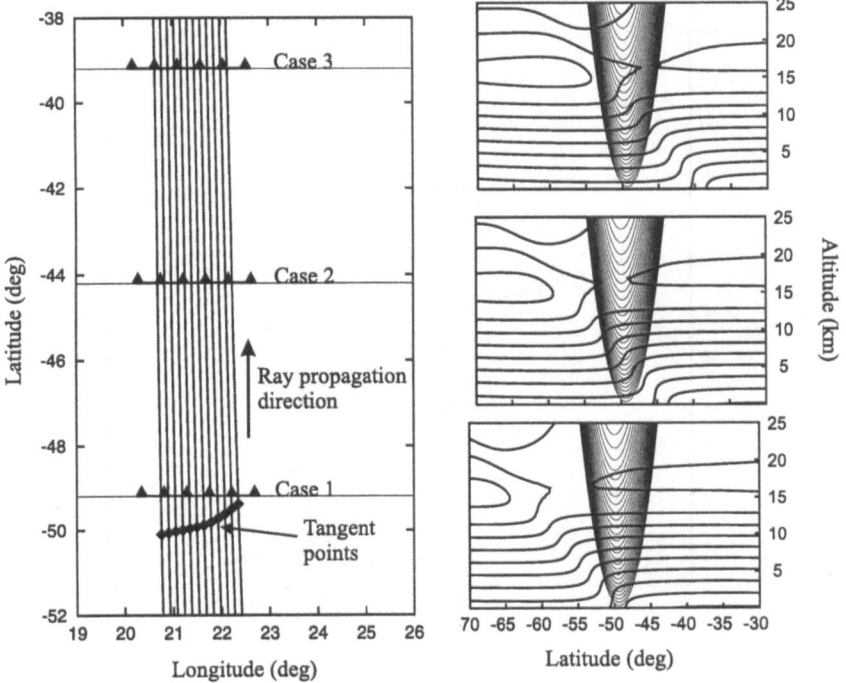

Fig. 8. Left) Ray paths in a longitude-latitude plot. The simulation of rays are based on a real set of GPS/MET orbit ephemerides. Only a few rays out of about 3000 are shown. The left most ray corresponds to a ray tangent altitude of about 100 km, and the right most ray has its tangent point close to the surface. The drift of the tangent points are shown as diamonds. The three cases of placing the base of the front are indicated as lines with triangles. Right) Each of the three cases in a latitude-height plot, with rays (every 50th ray) superimposed on the temperature fields.

Here we shall go a bit further and look at different cases. Excluding the water vapor is no serious violation in favor of the method. On the contrary, in a real frontal system the refractivity may not vary much across the front because the dry and the moist refractivity terms tend to cancel, understood in the way that the cold side is denser but contains less water vapor than the warm side (Hardy et al. [1994]). In Fig. 7 the temperature gradients are produced solely by the dry refractivity term (7).

In the forward ray-tracing modeling, three different cases of occultations going through the frontal system have been simulated, aligning the rays approximately along the meridian: Case 1, where the base of the front has been placed right in front of the tangent points. Case 2, where the base of the front has been placed 5 degrees north of the tangent points. Case 3, where the base of the front has been placed 10 degrees north of the tangent points. In Fig. 8 the placements of the front in each of these three cases are illustrated. The left figure shows the

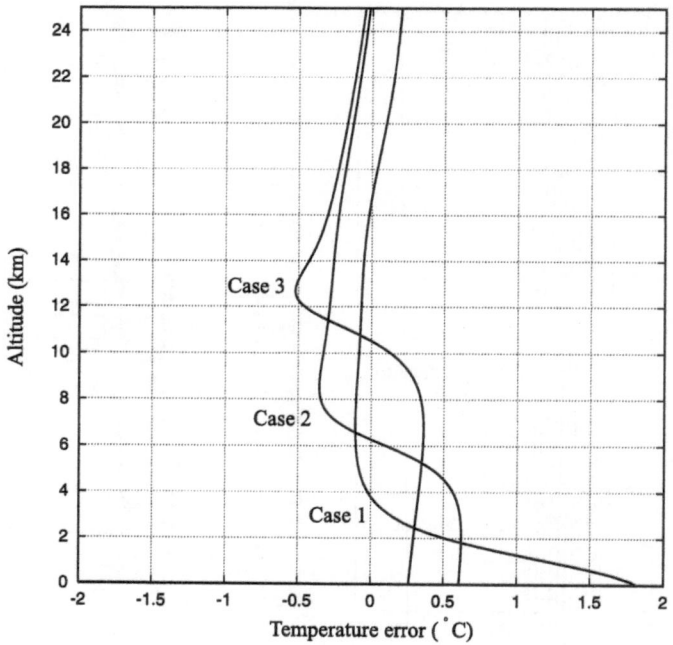

Fig. 9. Retrieval errors for each of the three cases as described in the text.

ray paths and the frontal bases in a longitude-latitude plot. The three figures at the right show the situations in latitude-height plots. Because of the Earth's curvature, the almost straight rays (the bending angle close to the surface is about 1 degree and bending towards the Earth) become convex curves in the three small panels to the right in Fig. 8. Such figures are also an excellent illustration of the spatial resolution of the method. In principle, the horizontal resolution defined by (20) corresponds to twice the horizontal distance between the lowest point on a ray and the neighboring ray, except that the vertical distance between rays in Fig. 8 is not exactly the same as the first Fresnel diameter (see also (Kursinski et al. [1997])).

For each of the three cases, about 3000 samples of synthetic phase data (50 Hz sampling rate) where calculated, covering the altitude range 0–100 km. Bending angles and the corresponding impact parameters were obtained, and the procedure outlined in Sect. 2 was then used to calculate a retrieved temperature profile. A model temperature profile was extracted from the corresponding model field of Fig. 8. The location of the model profile was taken as the latitude and longitude of the point closest to the Earth's center on that line connecting the GPS- and the LEO satellite which was tangent to the Earth's surface. This gives a profile close to the set of tangent points shown in the left part of Fig. 8. The retrieved temperature profile and the model temperature profile was compared and the differences in each of the three cases are plotted in Fig. 9. It is seen that the errors are of the order of 0.5 °C, and in the worst case, which turned out to be case 1, the error is still less than 2 °C.

The retrieved temperature profiles were also compared with profiles extracted from the model field at the exact positions of the tangent points. The observed differences were of the same order as those shown in Fig. 9, although case 2 showed an error as large as case 1, being about 1.5 °C. In the above numerical experiments the frontal slope was 1 % which is a typical value in the real atmosphere. However, close to the surface the slope may become larger (Gurvich and Sokolovskiy [1985]). Increasing the frontal slope to 5 % in case 1 resulted in somewhat larger errors (mostly above 4 km, though still less than 2 °C), confirming similar results by Gurvich and Sokolovskiy ([1985]).

Since the temperature profiles are derived via the hydrostatic integration of the refractivity, errors in refractivity, pressure, and temperature do not map as a one to one relationship. There will be differences in the error behavior as a function of height. Also, in a weather front the strict assumption of hydrostatic equilibrium may be questionable. Here we have only concentrated on the temperature errors, though pressure and refractivity errors may be of equal relevance. The order of magnitude of the relative errors, however, is generally the same.

6   Conclusive Remarks

In this paper the inverse problem for the radio occultation method has been presented to which there exist a unique solution connecting measured bending angles to the atmospheric refractive index through an Abel integral transform. Nevertheless, the problem was discretized to investigate the random error propagation using the singular value decomposition. For a constant value of the bending angle error of 1 μrad, relative errors in refractivity increase nearly exponentially with height, exceeding 1 % above 50 km altitude.

The inversion method was applied to real data to obtain temperature profiles, and the results were validated against radiosonde measurements and numerical analyses results. The agreements were within a few °C.

A case study was carried out testing the method in situations of horizontal inhomogeneities such as a severe frontal system. Errors obtained were less than 2 °C.

Other important retrieval errors are due to residual ionospheric effects and uncertainties in upper boundary conditions in the Abel and hydrostatic integrations.

In the future it is the plan that the occultation method shall provide data to be assimilated into numerical weather prediction models, hopefully improving the initial conditions, and thereby the predictions.

The occultation technique can also be applied to the ionosphere, then obtaining electron density profiles (Hajj and Romans [1998]).

Acknowledgments

I would like to thank the GPS/MET team at UCAR for providing the occultation data as well as the correlative radiosonde- and numerical weather analyses data. Part of the research was carried out at the University of Arizona, Tucson, Arizona, and I would like to thank Benjamin Herman and David Flittner at the University of Arizona for useful discussions regarding the frontal model. The research was funded by a grant from the Danish Space Board.

References

[1971] Fjeldbo, G., Kliore, A. J., and Eshleman, V. R. (1971). The neutral atmosphere of Venus as studied with the Mariner V radio occultation experiments. *The Astronomical Journal*, 76(2), 123–140.

[1998] Gorbunov, M. E. and Gurvich, A. S. (1998). Algorithms of inversion of Microlab-1 satellite data including effects of multipath propagation. *International Journal of Remote Sensing*, 19(12), 2283–2300.

[1996] Gorbunov, M. E., Sokolovsky, S. V., and Bengtsson, L. (1996). Space refractive tomography of the atmosphere: Modeling of direct and inverse problems. Report No. 210, Max-Planck-Institute for Meteorology, Hamburg, Germany.

[1985] Gurvich, A. S. and Sokolovskiy, S. V. (1985). Reconstruction of a pressure field by remote refractometry from space. *Izvestiya, Atmospheric and Oceanic Physics*, 21(1), 7–13.

[1998] Hajj, G. A. and Romans, L. J. (1998). Ionospheric electron density profiles obtained with the Global Positioning System: Results from the GPS/MET experiment. *Radio Science*, 33(1), 175–190.

[1994] Hardy, K. R., Hajj, G. A., and Kursinski, E. R. (1994). Accuracies of atmospheric profiles obtained from GPS occultations. *International Journal of Satellite Communications*, 12, 463–473.

[1996] Høeg, P., Hauchcorne, A., Kirchengast, G., Syndergaard, S., Belloul, B., Leitinger, R., and Rothleitner, W. (1996). Derivation of atmospheric properties using a radio occultation technique. Scientific Report 95-4, Danish Meteorological Institute, Copenhagen, Denmark.

[1997] Karayel, E. T. and Hinson, D. P. (1997). Sub-Fresnel-scale vertical resolution in atmospheric profiles from radio occultation. *Radio Science*, 32(2), 411–423.

[1965] Kliore, A., Cain, D. L., Levy, G. S., Eshleman, V. R., Fjeldbo, G., and Drake, F. D. (1965). Occultation experiment: Results of the first direct measurement of Mars's atmosphere and ionosphere. *Science*, 149, 1243–1248.

[1996] Kursinski, E. R., Hajj, G. A., Bertiger, W. I., Leroy, S. S., Meehan, T. K., Romans, L. J., Schofield, J. T., McCleese, D. J., Melbourne, W. G., Thornton, C. L., Yunck, T. P., Eyre, J. R., and Nagatani, R. N. (1996). Initial results of radio occultation observations of Earth's atmosphere using the global positioning system. *Science*, 271, 1107–1110.

[1995] Kursinski, E. R., Hajj, G. A., Hardy, K. R., Romans, L. J., and Schofield, J. T. (1995). Observing tropospheric water vapor by radio occultation using the global positioning system. *Geophysical Research Letters*, 22(17), 2365–2368.

[1997] Kursinski, E. R., Hajj, G. A., Schofield, J. T., Linfield, R. P., and Hardy, K. R. (1997). Observing Earth's atmosphere with radio occultation measurements using the Global Positioning System. *Journal of Geophysical Research*, 102(D19), 23,429–23,465.

[1994] Kursinski, R. (1994). Monitoring the Earth's atmosphere with GPS. *GPS World*, 5(3), 50–54.

[1994] Melbourne, W. G., Davis, E. S., Duncan, C. B., Hajj, G. A., Hardy, K. R., Kursinski, E. R., Meehan, T. K., Young, L. E., and Yunck, T. P. (1994). The application of spaceborne GPS to atmospheric limb sounding and global change monitoring. JPL-Publication 94-18, Jet Propulsion Laboratory, California Institute of Technology, Pasadena, California.

[1999] Mortensen, M. D., Linfield, R. P., and Kursinski, E. R. (1999). Vertical resolution approaching 100 m for GPS occultations of the Earth's atmosphere. *Radio Science*. Accepted for publication.

[1997] Rocken, C., Anthes, R., Exner, M., Hunt, D., Sokolovskiy, S., Ware, R., Gorbunov, M., Schreiner, W., Feng, D., Herman, B., Kuo, Y.-H., and Zou, X. (1997). Analysis and validation of GPS/MET data in the neutral atmosphere. *Journal of Geophysical Research*, 102(D25), 29,849–29,860.

[1996] Scales, J. A. and Smith, M. L. (1996). *Introductory Geophysical Inverse Theory: Part I*. Samizdat Press, Center for Wave Phenomena, Department of Geophysics, Colorado School of Mines, Golden, Colorado. Available via FTP from landau.mines.edu or 138.67.12.78 or via the WWW from http://landau.mines.edu/~samizdat.

[1953] Smith, Jr., E. K. and Weintraub, S. (1953). The constants in the equation for atmospheric refractive index at radio frequencies. *Journal of Research of the National Bureau of Standards*, 50(1), 39–41.

[1999a] Syndergaard, S. (1999a). On the ionosphere calibration in GPS radio occultation measurements. *Radio Science*. Submitted.

[1999b] Syndergaard, S. (1999b). *Retrieval Analysis and Methodologies in Atmospheric Limb Sounding Using the GNSS Radio Occultation Technique*. PhD thesis, Niels Bohr Institute for Astronomy, Physics and Geophysics, University of Copenhagen, Copenhagen, Denmark.

[1987] Tyler, G. L. (1987). Radio propagation experiments in the outer solar system with Voyager. *Proceedings of the IEEE*, 75(10), 1404–1431.

[1994] Vorob'ev, V. V. and Krasil'nikova, T. G. (1994). Estimation of the accuracy of the atmospheric refractive index recovery from Doppler shift measurements at frequencies used in the NAVSTAR system. *Physics of the Atmosphere and Ocean*, 29(5), 602–609.

[1996] Ware, R., Exner, M., Feng, D., Gorbunov, M., Hardy, K., Herman, B., Kuo, Y., Meehan, T., Melbourne, W., Rocken, C., Schreiner, W., Sokolovskiy, S., Solheim, F., Zou, X., Anthes, R., Businger, S., and Trenberth, K. (1996). GPS sounding of the atmosphere from low Earth orbit: Preliminary results. *Bulletin of the American Meteorological Society*, 77(1), 19–40.

[1999] Zou, X., Vandenberghe, F., Wang, B., Gorbunov, M. E., Kuo, Y.-H., Sokolovskiy, S., Chang, J. C., Sela, J. G., and Anthes, R. (1999). A raytracing operator and its adjoint for the use of GPS/MET refraction angle measurements. *Journal of Geophysical Research - Atmosphere*. Accepted for publication.

A Limited Memory BFGS Method for an Inverse Problem in Atmospheric Imaging

Curtis R. Vogel

Department of Mathematical Sciences
Montana State University
Bozeman, MT 59717-0240 USA.
e-mail: vogel@math.montana.edu
http://www.math.montana.edu/~vogel

Abstract. Multiple time frame phase diversity image data is used to simultaneously estimate the object (i.e., the true image) and the phases, or wavefront profiles, of light that has propagated through the atmosphere. Tikhonov regularization is applied to deal with the instability of this estimation problem. The resulting large scale unconstrained minimization problem is solved numerically using a limited memory BFGS method. Some preliminary computational results obtained with simulated atmospheric image data are presented.

1 Introduction

As light rays propagate through the earth's atmosphere, they are bent because of variations in the index of refraction due to differences in air temperature. This causes image distortion, or blurring. An undesirable consequence of this blurring is the limited spatial resolution of space objects viewed through ground-based telescopes [6]. The Hubble Space Telescope successfully avoids this difficulty because it is above the atmosphere. With ground-based telescopes two approaches can be taken. With adaptive optics [9], a deformable mirror is used to "straighten", or phase conjugate, the bent light rays. A second approach is to formulate a mathematical model (called the *forward model*) relating the image data to the object and the atmospheric distortion and then to solve the *inverse problem*, i.e., to estimate the object and certain features of the atmosphere given observed image data and the forward model.

A simple forward model (see [6]) is the convolution integral equation

$$d(x, y) = \int \int s(x - x', y - y') f(x', y') \, dx'dy' + \eta(x, y) \tag{1}$$
$$= (s \star f)(x, y) + \eta(x, y).$$

Here d represents image data, f represents the *object*, or true image, η represents noise in the data, and s is known as the *point spread function*, or PSF. The PSF is the image that would result from an idealized point object, and it characterizes atmospheric blurring effects. In this context, the inverse problem is to determine both the PSF s and the object f given the image data d. This inverse problem, which is referred to as *blind deconvolution*, is ill-posed in several respects. First, since convolution is symmetric (i.e., $s \star f = f \star s$), one cannot uniquely determine both s and f from a single observation of d. Additional information is needed.

Even if the nonuniqueness difficulty in the model (1) could be overcome, an additional difficulty called *instability* may arise. To illustrate, suppose the PSF s were known. The process of estimating f from d is then called *deconvolution*. If no error is present in the data, one can apply the convolution theorem to obtain the Fourier transform of the object at a given spatial frequency vector $\omega = (\omega_1, \omega_2)$,

$$\mathcal{F}(f)(\omega) = \frac{\mathcal{F}(d)(\omega)}{\mathcal{F}(s)(\omega)}. \tag{2}$$

The object can then be reconstructed by applying the inverse Fourier transform to the right hand side of (2). However, if the PSF s is smooth, then its Fourier transform tends to zero at high spatial frequencies (large ω). If the data d contains error whose Fourier transform does not decay as rapidly, then the right hand side becomes large. This is *error amplification*. Stability can be restored by filtering out components of the estimated solution for which $\mathcal{F}(s)$ is "small", e.g., by replacing the right hand side of (2) by zero whenever the magnitude of the denominator falls below a certain threshold. Note that certain spatial frequency information will be lost when this is done. This phenomena is referred to as *truncation*.

The term *regularization* is used for schemes which restore stability in a manner which provides a good approximation to the true solution. Regularization schemes yield solutions which depend on quantities known as *regularization parameters*. An example is the threshold level in the filtering scheme discussed above. Regularization parameters quantify the tradeoff between error amplification due to instability and truncation due to regularization.

The atmospheric imaging problem to be considered in this paper has the form (1) with some additional special structure. The PSF can be described in terms of a function known as the *phase*, or wavefront profile. A physical process known as *phase diversity* [3, 5, 10] can be applied to generate additional data to partially overcome the nonuniqueness difficulties in blind deconvolution. From several phase diversity images, one then seeks to estimate both the object f and the phase. Additional information can be obtained from multiple time frames, i.e., by capturing a sequence of images over a period of time. Object estimation based on multiple time frame phase diversity data is called *phase-diverse speckle imaging* [7]. A forward model for multiple time frame phase diversity data is presented in the next section. It is assumed in this model that the object is fixed, but the phases vary with time.

The inverse problem of determining the phases and the object, given the forward model and multiple time frame phase diversity data, is unstable with respect to perturbations in the data. To restore stability, a technique known as Tikhonov regularization (see [2]) is applied. To implement this technique, a pair of regularization, or stabilization, functionals must be selected. In addition to restoring stability, these functionals allow the incorporation of a priori information about the phases and the object. Details appear in Section 3.

From Tikhonov regularization, one obtains a cost functional which must be minimized to estimate the phases and the object. The unknown phases and object must first be discretized. Using a pixel, or nodal, representation, we obtain an unconstrained minimization problem with a very large number of unknowns (with $n_x \times n_y$ pixel image arrays and T time frames, we have $n_x n_y (T + 1)$ unknowns. In our numerical simulations, $n_x = n_y = 128$ and $T = 8$, yielding more than 10^5 unknowns). As in [3, 5, 10], we eliminate the object to reduce the number of unknowns. A derivation of the resulting reduced cost functional is sketched in Section 4. A formula for the gradient of this reduced cost functional is also given in this section.

A limited memory BFGS method for the minimization of the reduced cost functional is the topic of Section 5. The BFGS method is a minimization technique which combines low computational expense (only the cost functional and its gradient are computed at each BFGS iteration) with rapid convergence. The limited memory variant has low storage requirements. Finally in Section 6, we present computational results obtained with simulated atmospheric image data.

2 The Forward Model

We will assume that recorded image data can be accurately represented as an $n_x \times n_y$ array of pixel intensities with components

$$[\mathbf{d}]_{ij} = (s \star f)(x_i, y_j) + [\eta]_{ij}, \quad 1 \le i \le n_x, \ 1 \le j \le n_y, \tag{3}$$

where f denotes the true image, or object, s denotes the point spread function, or PSF, the $[\eta]_{ij}$ represent noise in the data, and $\star$ denotes 2-D convolution product. The PSF s quantifies the blurring effects of the atmosphere. We will assume dependence on the phase, or wavefront profile, $\phi(x, y)$ via

$$s[\phi] = |\mathcal{F}^{-1}\{pe^{\imath\phi}\}|^2, \tag{4}$$

where $\mathcal{F}$ denotes the 2-D Fourier transform, $\imath = \sqrt{-1}$, and $p = p(x, y)$ denotes the *pupil*, or aperture, function. For ground-based telescopy, p is an indicator function whose support is determined by the extent of the telescope mirror. See [6] and the references therein for a detailed discussion of atmospheric imaging models.

The nonuniqueness difficulties for the model (3)-(4) can be at least partially resolved with phase diversity data. In its simplest form, one collects a second image,

$$\mathbf{d}' = s[\phi + \theta] \star f + \eta', \tag{5}$$

where θ represents a known phase perturbation. In practice, this data is generated by splitting the beam of light collected from the telescope's primary mirror. See Fig. 1 for an illustration. From one beam, the conventional image $\mathbf{d}$ is formed.

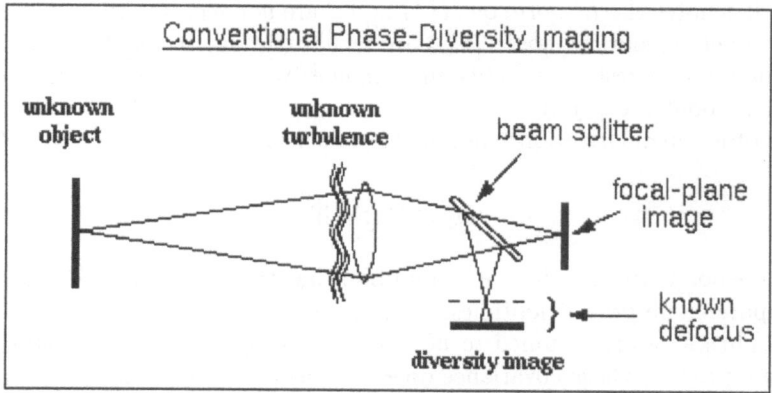

Fig. 1. Phase diversity image formation.

An out-of-focus image $\mathbf{d}'$ is formed from the second beam. This corresponds to a quadratic phase perturbation,

$$\theta(x, y) = c(x^2 + y^2),$$

where the constant c depends on the defocus length.

More phase diversity channels can be added, e.g., using additional beam splitters with different amounts of defocus. Furthermore, images can be captured at different instants in time. Assuming the object f is fixed, but the phase varies with time, we obtain data

$$d_{t,k} = s[\phi_t + \theta_k] \star f + \eta_{t,k}, \tag{6}$$

where $k = 1, \ldots, K$, with K denoting the number of phase diversity channels, and $t = 1, \ldots, T$, with T denoting the number of time frames.

3 Regularization

Phase diversity can greatly reduce the nonuniqueness difficulties. (It should be noted that constant offsets in ϕ still cannot be resolved, since $s[\phi + c] = s[\phi]$. Whether other ambiguities exist is an open question.) One must also deal with instability with respect to perturbations in the data. This can be overcome by Tikhonov regularization [2], or penalized least squares. Given data (6), we minimize the joint cost functional

$$T_{\gamma, \alpha}[f, \phi_1, \ldots, \phi_T] = \sum_{t=1}^{T} \sum_{k=1}^{K} \|s[\phi_t + \theta_k] \star f - d_{t,k}\|^2$$
$$+ \ \gamma J_{obj}[f] + \alpha J_{phase}[\phi_1, \ldots, \phi_T]. \tag{7}$$

Here $||\cdot||$ denotes the L^2 norm on R^2, γ and α are nonnegative scalar regularization parameters, and J_{obj} and J_{phase} are regularization, or penalty, functionals. These functionals restore stability in a manner which incorporates a priori information about the object and phase.

The object regularization functional is taken to be

$$J_{obj}[f] = \frac{1}{2}||f||^2. \tag{8}$$

This corresponds to the minimal assumption that the object has finite intensity. It incorporates no prior smoothness assumptions.

The phases ϕ_t are assumed to be realizations from a wide-sense stationary stochastic process whose covariance operator has a Von Karman spectrum [6],

$$\Lambda(\omega) = C\,(\omega_1^2 + \omega_2^2 + L_0^{-2})^{-11/12}. \tag{9}$$

In numerical simulations, we compute phase realizations

$$\phi = \mathcal{F}^{-1}(\Lambda\,\mathcal{F}(W)), \tag{10}$$

where W is a realization of white noise. The phase regularization operator is picked to have a Fourier representation

$$J_{phase}[\phi_1, \ldots, \phi_T] = \frac{1}{2}\sum_{t=1}^{T}\mathcal{F}^{-1}\left(\frac{|\mathcal{F}(\phi_t)|^2}{\Lambda}\right). \tag{11}$$

4 The Reduced Cost Functional and Its Gradient

The notation and many of the technical details in this section are taken from [10]. The only significant modification is the additional phases. Let upper case letters denote Fourier transformed variables. The Tikhonov cost functional (7) has the representation

$$T_{\gamma,\alpha}[F, \phi_1, \ldots, \phi_T] = \frac{1}{2}\sum_{t=1}^{T}\sum_{k=1}^{K}||S[\phi_t + \theta_k]F - D_{t,k}||^2$$
$$+ \frac{\gamma}{2}||F||^2 + \alpha J_{phase}[\phi_1, \ldots, \phi_T]. \tag{12}$$

Setting $\frac{\partial T_{\gamma,\alpha}}{\partial F} = 0$ yields

$$F[\phi_1, \ldots, \phi_T] = \frac{\sum_{t=1}^{T}\sum_{k=1}^{K}S[\phi_t + \theta_k]^* D_{t,k}}{\gamma + \sum_{t=1}^{T}\sum_{k=1}^{K}|S[\phi_t + \theta_k]|^2}. \tag{13}$$

Here the superscript $*$ denotes complex conjugate, and $|\cdot|$ denotes magnitude of a complex quantity. Given estimates for $\phi_1, \ldots, \phi_T$, one can compute the inverse Fourier transform in (13) to obtain an estimate for the object f. Note that the positive object regularization parameter γ in the denominator of (13) induces stability by preventing division by very small quantities or by zero.

By substituting $F = F[\phi_1, \ldots, \phi_T]$ from (13) back into (12), one obtains the reduced cost functional, which we define by

$$J[\phi_1, \ldots, \phi_T] = \frac{1}{2} \sum_{t=1}^{T} \sum_{k=1}^{K} \|D_{t,k}\|^2 - \frac{1}{2} \int \int_{R^2} \frac{|\sum_{t=1}^{T} \sum_{k=1}^{K} S[\phi_t + \theta_k]^* D_{t,k}|^2}{\gamma + \sum_{t=1}^{T} \sum_{k=1}^{K} |S[\phi_t + \theta_k]|^2}$$
$$+ \alpha J_{phase}[\phi_1, \ldots, \phi_T]. \tag{14}$$

Note that by minimizing (14) to obtain $\phi_1, \ldots, \phi_T$ and then substituting into (13) and inverse Fourier transforming to obtain f, one need not get the same estimates for phases and object as would be obtained by minimizing the joint Tikhonov cost functional (7).

The gradient of the reduced cost functional has a block representation

$$\boldsymbol{g} = \begin{bmatrix} \frac{\partial J}{\partial \phi_1} \\ \vdots \\ \frac{\partial J}{\partial \phi_T} \end{bmatrix} = \begin{bmatrix} g[\phi_1] \\ \vdots \\ g[\phi_T] \end{bmatrix}, \tag{15}$$

where for $t = 1, \ldots, T$,

$$g[\phi_t] = -2 \sum_{k=1}^{K} \text{Imag}(H_k^*[\phi_t] \mathcal{F}(\text{Real}(h_k[\phi_t] \mathcal{F}^{-1} V_k[\phi_t]))) + \alpha \, g_{phase}[\phi_t], \tag{16}$$

with

$$H_k[\phi] = p e^{i(\phi + \theta_k)}, \qquad h_k[\phi] = \mathcal{F}^{-1}(H_k[\phi]), \tag{17}$$
$$s_k[\phi] = |h_k[\phi]|^2, \qquad S_k[\phi] = \mathcal{F}(s_k[\phi]), \tag{18}$$
$$V_k[\phi] = F^*[\phi] D_k - |F[\phi]|^2 S_k[\phi], \tag{19}$$

and from (11),

$$g_{phase}[\phi] = \text{Real}\left(\mathcal{F}^{-1}\left(\frac{\mathcal{F}(\phi)}{\Lambda}\right)\right). \tag{20}$$

The cost of evaluating $J[\phi_1, \ldots, \phi_T]$ is dominated by the computation of the $S[\phi_t + \theta_k]$'s. These require 2 Fourier transform/inverse transform computations per (k, t) pair, cf., (17) and (18). From (11), an transform/inverse transform pair for each t is required to evaluate $J_{phase}[\phi_1, \ldots, \phi_T]$. Thus the total number of transforms needed to evaluate the reduced cost functional is $T(2K + 2)$. If intermediate results are saved, then evaluation of the gradient $\boldsymbol{g}$ requires $T(2K + 1)$ additional Fourier transform/inverse transform computations, cf., (16) and (20). Hence each gradient evaluation is slightly less expensive than each reduced cost functional evaluation.

5   The Limited Memory BFGS Method

Much of the technical background required to understand the material in this section can be found in [1] or [4]. To simplify notation, let $\boldsymbol{\phi} = (\phi_1, \ldots, \phi_T)$. To obtain a (local) minimizer $\boldsymbol{\phi}_*$ of the reduced cost functional $J[\boldsymbol{\phi}]$, we apply quasi-Newton methods, combined with line search step size control. These yield approximations of the form

$$\boldsymbol{\phi}_{\nu+1} = \boldsymbol{\phi}_\nu + \boldsymbol{s}_\nu, \quad \nu = 0, 1, \ldots,$$

where $\boldsymbol{s}_\nu = \overline{\mu} \boldsymbol{d}_\nu$, $\overline{\mu}$ is a positive step length parameter, and the quasi-Newton step $\boldsymbol{d}_\nu$ solves

$$H_\nu \boldsymbol{d} = -\boldsymbol{g}[\boldsymbol{\phi}_\nu], \tag{21}$$

with H_ν a symmetric positive definite approximation to the true Hessian $H[\boldsymbol{\phi}_\nu]$. Positive definiteness guarantees that $\boldsymbol{d}_\nu$ is a descent direction for J at $\boldsymbol{\phi}_\nu$, i.e., $J[\boldsymbol{\phi}_\nu + \mu \boldsymbol{d}_\nu] < J[\boldsymbol{\phi}_\nu]$ for some $\mu > 0$, provided the gradient $\boldsymbol{g}[\boldsymbol{\phi}_\nu]$ is nonzero. However, simply decreasing the cost functional at each iteration is not enough to guarantee that the iterates $\boldsymbol{\phi}_\nu$ converge to a local minimizer. An additional criterion known as the Armijo rule is needed. We employ a backtracking cubic line search algorithm [4, p. 43] to obtain step lengths for which the Armijo rule is satisfied.

If H_ν in (21) is taken to be the true Hessian $H[\boldsymbol{\phi}_\nu]$, then one obtains Newton's method. This method has the advantage of quadratic convergence near a local minimizer. Unfortunately, the computation, storage, and inversion of the Hessian may be prohibitively expensive. For instance, in the test problem presented in the next section, the Hessian has more than 1.7×10^{10} entries. Moreover, far from a local minimizer the Hessian need not be positive definite, and hence the Newton step need not be a descent direction.

One alternative to Newton's method is the BFGS method. Given an initial Hessian approximation H_0, it generates a sequence of Hessian approximates via the rank-two update

$$H_{\nu+1} = H_\nu + \frac{1}{\boldsymbol{y}_\nu^T \boldsymbol{s}_\nu} \boldsymbol{y}_\nu \boldsymbol{y}_\nu^T - \frac{1}{\boldsymbol{s}_\nu^T H_\nu \boldsymbol{s}_\nu} (H_\nu \boldsymbol{s}_\nu)(H_\nu \boldsymbol{s}_\nu)^T, \tag{22}$$

where $\boldsymbol{s}_\nu = \boldsymbol{\phi}_{\nu+1} - \boldsymbol{\phi}_\nu$ is the current step and $\boldsymbol{y}_\nu = \boldsymbol{g}[\boldsymbol{\phi}_{\nu+1}] - \boldsymbol{g}[\boldsymbol{\phi}_\nu]$ is the difference between current and previous gradients. If H_ν is positive definite and $\boldsymbol{y}_\nu^T \boldsymbol{s}_\nu$ is positive, then $H_{\nu+1}$ is guaranteed to be positive definite. In the finite dimensional case, under standard assumptions (i.e., J is smooth, $\boldsymbol{\phi}_0$ is sufficiently close to a local minimizer $\boldsymbol{\phi}_*$, H_0 is sufficiently close to $H[\boldsymbol{\phi}_*]$, and $H[\boldsymbol{\phi}_*]$ is strictly positive; see [4] for details), the BFGS method is guaranteed to be superlinearly convergent. In the infinite dimensional case one needs the additional assumption that $H_0 - H[\boldsymbol{\phi}_*]$ is a compact operator to prove a superlinear convergence rate.

A recursive formula for the inverses of the matrices in (22) is

$$H_{\nu+1}^{-1} = \left(I - \frac{1}{\boldsymbol{y}_\nu^T \boldsymbol{s}_\nu} \boldsymbol{s} \boldsymbol{y}_\nu^T \right) H_\nu^{-1} \left(I - \frac{1}{\boldsymbol{y}_\nu^T \boldsymbol{s}_\nu} \boldsymbol{y}_\nu \boldsymbol{s}_\nu^T \right) + \frac{1}{\boldsymbol{y}_\nu^T \boldsymbol{s}_\nu} \boldsymbol{s}_\nu \boldsymbol{s}_\nu^T. \tag{23}$$

This recursion can be used to solve equation (21) with a BFGS Hessian approximation H_ν. To do so requires storage of ν vectors $\boldsymbol{s}$ and $\boldsymbol{y}$ (i.e., all the previous steps and gradient differences), inner product computations involving the gradient $\boldsymbol{g}[\boldsymbol{\phi}_\nu]$ and the $\boldsymbol{s}$ and $\boldsymbol{y}$ vectors, and the computation of $H_0^{-1}\boldsymbol{v}$ for some vector $\boldsymbol{v}$. With the limited memory BFGS method, only a fixed number of the $\boldsymbol{s}$ and $\boldsymbol{y}$ vectors are retained. As new vectors are added to storage, the oldest vectors are discarded. This can substantially reduce the storage requirements of the method and the cost of computing the quasi-Newton steps. Unfortunately, this may also slow convergence to a linear rate.

From (16) and (20) the true Hessian has the form

$$H[\phi] = K[\phi] + \alpha H_{phase}. \tag{24}$$

Here $\alpha > 0$ is the phase regularization parameter, and H_{phase} is the Hessian of the phase regularization functional J_{phase}. From (15), (16), and (20), H_{phase} is block diagonal with T constant diagonal block operators, which we denote by $\mathcal{R}$, i.e., one can write

$$H_{reg} = \text{diag}(\mathcal{R}, \ldots, \mathcal{R}). \tag{25}$$

From (9), $\mathcal{R}$ is a symmetric positive definite linear operator whose spectrum increases at a rate asymptotically proportional to $|\omega|^{11/6}$ as $|\omega| \to \infty$. The linear operator $K[\phi]$ in (24) is derived from the second term on the right hand side of (14). We suspect that it is a compact operator (whose spectrum tends to 0 for large ω), although this property has yet to be verified. If this is indeed the case, then $H[\phi]$ should be well approximated by αH_{reg}. This motivates our choice of the initial Hessian

$$H_0 = \alpha H_{reg}. \tag{26}$$

From (25), the action of the inverse of the initial on a vector $\boldsymbol{v} = (v_1, \ldots, v_T)$ is given by

$$H_0^{-1}v = \left(\frac{1}{\alpha}\mathcal{R}v_1, \ldots, \frac{1}{\alpha}\mathcal{R}v_T\right), \tag{27}$$

where for each $i = 1, \ldots, T$,

$$\mathcal{R}v_i = \text{Real}(\mathcal{F}^{-1}(\Lambda\mathcal{F}(v_i))). \tag{28}$$

6 Computational Results

Eight simulated atmospheric phases, $\phi_1, \ldots, \phi_8$, were generated on a 128×128 grid according to the model (9)-(10). Note from equation (4) that only the portion of the phases within the support of the pupil affects the data. Hence, only this portion of the phases is plotted. The first three phases are shown in the top row of Fig. 2. The annular region comprising the support of the pupil can clearly be seen in each of the subplots in the top row. The upper left subplot in Fig. 4 also shows the first phase, ϕ_1.

Simulated phase diversity image data was then generated according to the model equation (6). The true object $f = f_{true}(x, y)$ was taken to be a computer

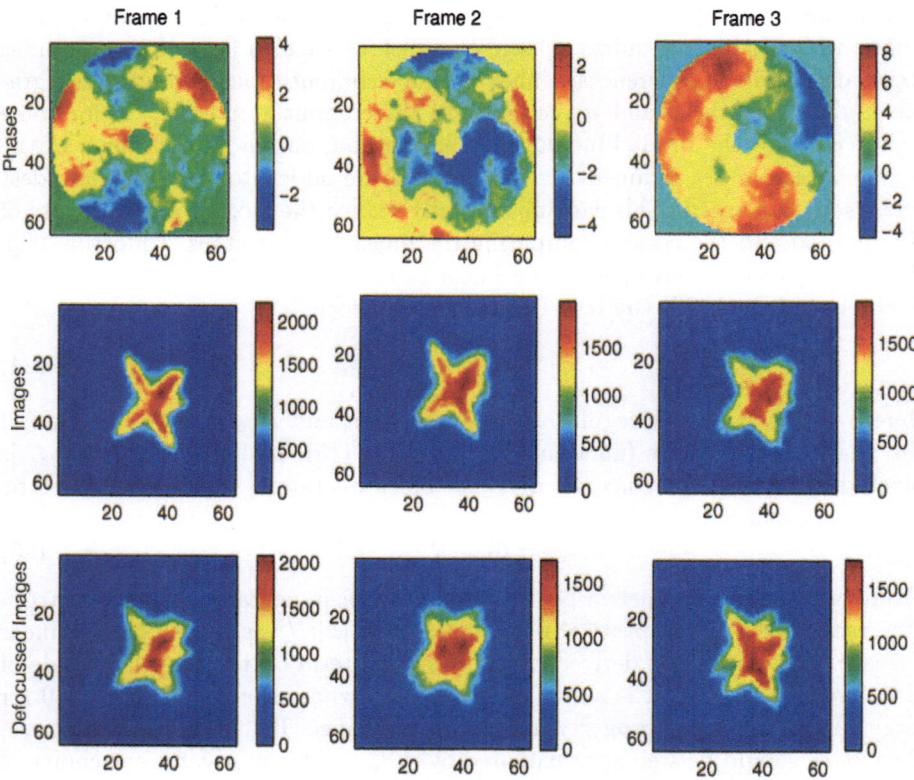

Fig. 2. Phases and image data. The top row shows the first three phases $\phi_t(x, y)$, $t = 1, 2, 3$. The subscript t corresponds the to the time frame number. The second row shows the corresponding conventional images $d_{1,t}$, $t = 1, 2, 3$. The third row shows the corresponding defocussed images $d_{2,t}$, $t = 1, 2, 3$.

rendering of a satellite, and is shown at the upper left of Fig. 3. This same object was used in the quantitative study by Tyler et al [8]. The PSF's $s = s[\phi_t + \theta_k]$, $t = 1, \ldots, 8$, $k = 1, 2$, were generated from the phases via equations (17)-(18). Blurred but noise-free images were generated by computing convolution products $s \star f_{true}$. To simulate measurement noise, Poisson and Gaussian errors were generated. To each pixel (i, j), $1 \leq i, j \leq 128$, a realization of a Poisson pseudo-random variable with mean $(s \star f_{true})(x_i, y_j)$ was computed. To this was added a realization of a Gaussian random variable with mean zero and standard deviation equal to 5. The conventional images (without defocus, i.e., $\theta_1 = 0$ in (6)) are shown in the second row of Fig. 2, while the third row shows the images formed with quadratic phase perturbation θ_2, corresponding to defocus blur. The defocus amount, characterized by the difference in θ_2 between the edge of the aperture and the center, is $0.7 \times 2\pi$.

To solve the inverse problem in a stable manner, we applied Tikhonov regularization with the regularization functionals presented in Section 3. The reduc-

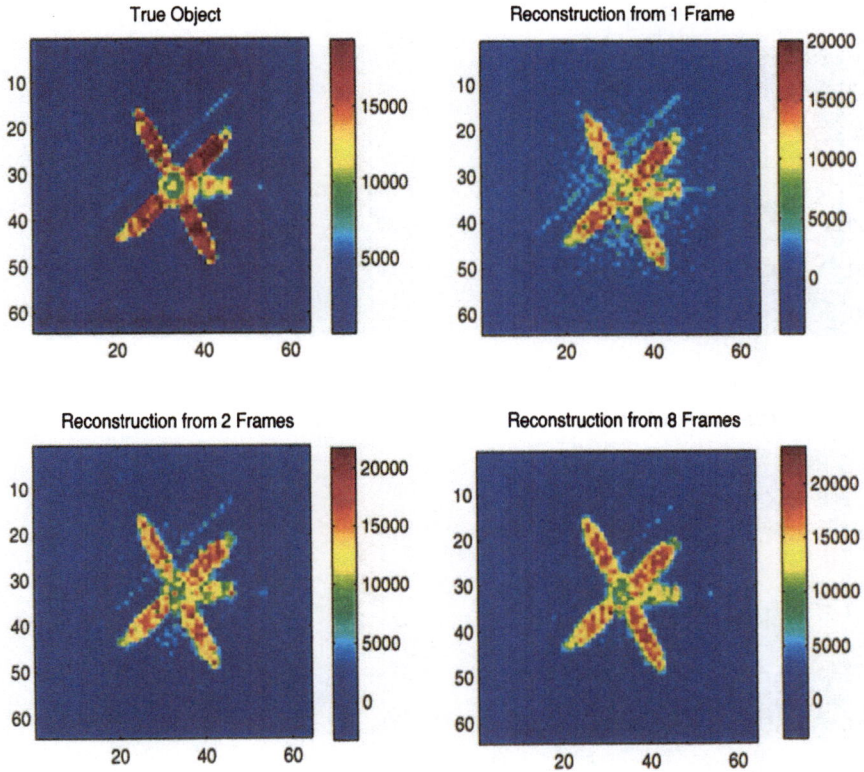

Fig. 3. True and reconstructed objects. Upper left subplot shows true object. Upper right subplot shows object reconstruction obtained with a single time frame. Lower left subplot shows object reconstruction obtained with two time frames. Lower right subplot shows object reconstruction obtained with eight time frames.

tion scheme of Section 4 was employed to eliminate the unknown object. The reduced cost functional was minimized using the implementation of the limited memory BFGS method described in Section 5. Regularization parameter values ($\gamma = 10^{-5}$ and $\alpha = 10^7$) were selected to provide good phase reconstructions. The true and reconstructed objects are shown in Fig. 3, while Fig. 4 shows the true phase ϕ_1 associated with the first time frame, and the corresponding reconstructions of ϕ_1. When we say that a reconstruction was obtained from T time frames, we mean that the data used was $d_{k,t}$, $k = 1, 2$, $t = 1, \ldots, T$, cf. (6).

Fig. 3 clearly shows that the object reconstructions improve as one increases the number of time frames. The situation with the phase reconstructions is somewhat more complicated. A comparison of the upper right and lower left subplots in Fig. 4 shows some improvement in the reconstruction of ϕ_1 when one increases the amount of data from one time frame to two frames. However, an examination of the lower right subplot, which was obtained with eight time frames, shows a slight decrease in the quality of the reconstruction of ϕ_1. It

302

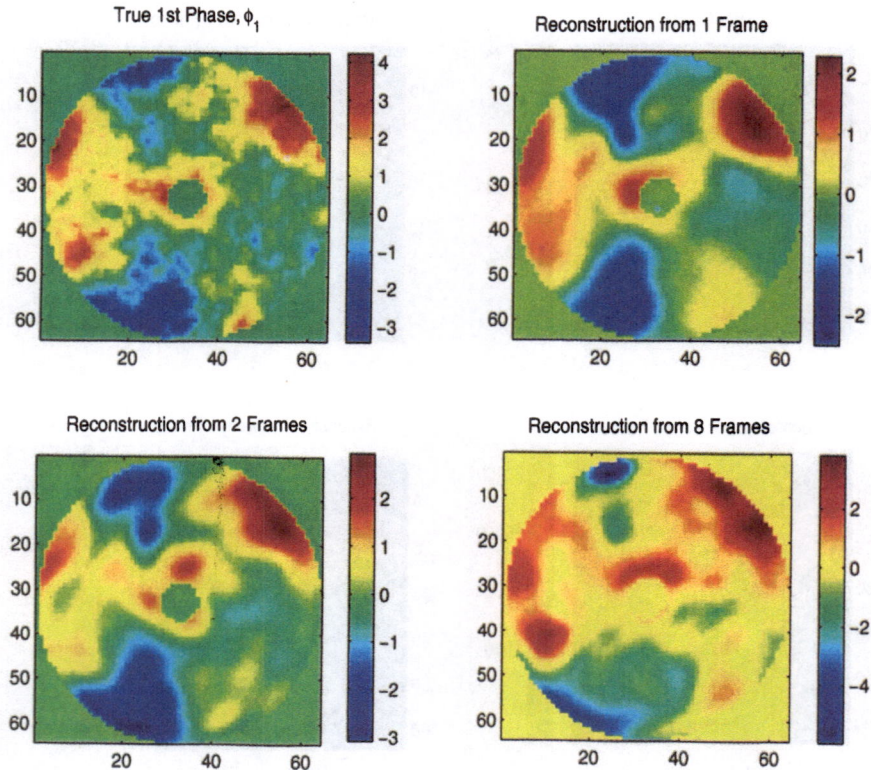

Fig. 4. ϕ_1 and its reconstructions. The upper left subplot shows the true phase ϕ_1. The upper right subplot shows a reconstruction of ϕ_1 obtained with single time frame data. The lower left subplot shows a reconstruction of ϕ_1 obtained with two time frames. The lower right subplot shows a reconstruction of ϕ_1 obtained with eight time frames.

should be noted that when eight frames were used, eight separate phases were simultaneously reconstructed. Although ϕ_1 was estimated less accurately, enough additional information was obtained from the other seven phases to improve the corresponding object reconstruction, cf., equation (13).

Reconstruction quality will also vary with other changes in the data. For instance, increasing the parameter C in equation (9) increases the variability of the phases. This in turn increases the amount of blur, which decreases the quality of the reconstructions. Similar degradations in reconstruction quality occur when the signal to noise ratio is decreased. For a quantitative study of these effects, see [8].

The performance of the limited memory BFGS algorithm depends on a number of factors, e.g., the number of vectors s and y saved, and the values of the regularization parameters γ and α. The data itself may also profoundly influence the numerical performance. Our primary interest here is in understanding the relationship between numerical performance and the number of saved vectors.

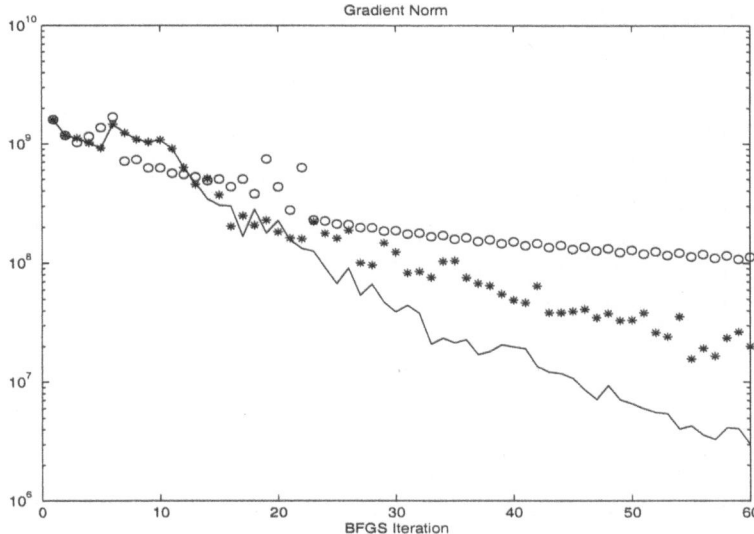

Fig. 5. Gradient norm vs. BFGS iteration count when only one time frame is used in the reconstructions. The circles (o) denote scaled steepest descent (no vectors saved). The stars (*) denote limited memory BFGS with 8 vectors saved, while the solid line denotes that 32 vectors were saved.

If no vectors are saved, the algorithm reduces to a scaled version of the steepest descent method. On the other hand, if all the vectors are saved, the usual BFGS scheme results. Figs. 5 and 6 summarize convergence results as the number of saved vectors changes.

In Fig. 5, one time frame was used to reconstruct the first phase, ϕ_1. Numerical performance, as measured by the norm of the gradient of the reduced cost functional, improves as more vectors are saved. The convergence rates appear to be linear for each of the three cases (no saved vectors, 8 saved vectors, and 32 saved vectors), since the three separate line types in Fig. 5 display a linear trend. This is consistent with the convergence theory for limited memory BFGS.

In Fig. 6, four time frames are used to reconstruct the first four phases, $\phi_1, \ldots, \phi_4$. The results in this case are somewhat puzzling. There is no clear computational advantage to be gained by saving more vectors. Moreover, no matter how many vectors are saved, the numerical performance is quite poor compared to the performance in the single frame case. Why this happens is an open question. The data corresponding to Frame 3 in Fig. 2 suggest that the 4-frame reconstruction problem may be much more difficult than the corresponding single frame problem. Note the phase variation for Frame 3 is much larger than that for Frame 1. In addition, Frame 3 images are more blurred than are Frame 1 images.

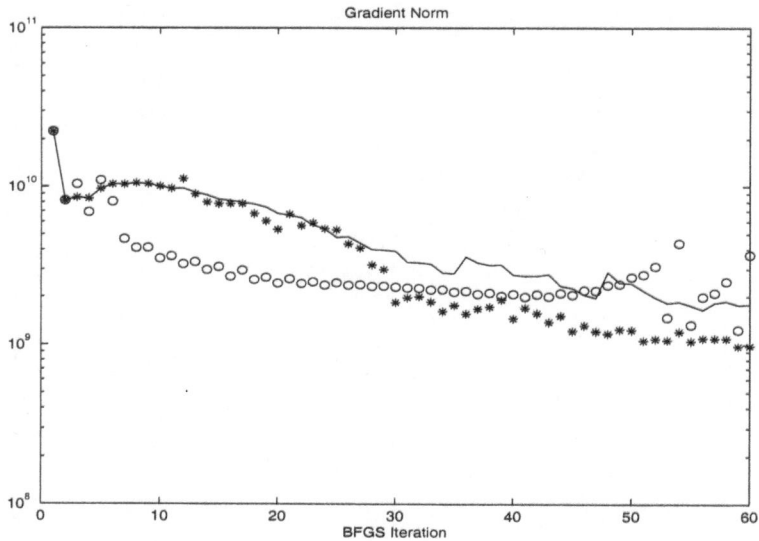

Fig. 6. Gradient norm vs. BFGS iteration count when four time frames are used in the reconstructions. The circles (o) denote scaled steepest descent (no vectors saved). The stars (∗) denote limited memory BFGS with 8 vectors saved, while the solid line denotes that 32 vectors were saved.

References

1. J. E. Dennis, Jr., and R. B. Schnabel, *Numerical Methods for Unconstrained Optimization and Nonlinear Equations*, SIAM Classics in Applied Mathematics Series, **16**, 1996.
2. H. Engl, M. Hanke, and A. Neubauer, *Regularization of Inverse Problems*, Kluwer Academic Publishers, Dordrecht, 1996.
3. R. A. Gonsalves, "Phase retrieval and diversity in adaptive optics," *Optical Engineering*, **21**, pp. 829-832, 1982.
4. C. T. Kelley, *Iterative Methods for Optimization*, SIAM, 1999.
5. R. G. Paxman, T. J. Schulz, and J. R. Fineup, "Joint estimation of object and aberrations by using phase diversity," *J. Optical Soc. Am.*, **9**, pp. 1072-1085, 1992.
6. M. C. Roggeman and B. Welsh, *Imaging Through Turbulence*, CRC Press, 1996.
7. J. H. Seldin and R. G. Paxman, "Phase-diverse speckle reconstruction of solar data", in *Image Reconstruction and Restoration*, T. J. Schulz and D. L. Snyder, eds., Proc. SPIE **2302**, pp. 268-280 (1994).
8. D. W. Tyler, S. D. Ford, B. R. Hunt, R. G. Paxman, M. C. Roggemann, J. C. Rountree, T. J. Schulz, K. J. Schulze, J. H. Seldin, D. G. Sheppard, B. E. Stribling, W. C. Van Kampen, and B. M. Welsh, "Comparison of image reconstruction algorithms using adaptive optics instrumentation", in *Adaptive Optical System Technologies*, Proc. SPIE **3353** (1998), pp. 160-171.
9. R. K. Tyson, *Principles of Adaptive Optics, 2nd Edition*, Academic Press, 1998.
10. C. R. Vogel, T. Chan, and R. Plemmons, "Fast algorithms for phase diversity-based blind deconvolution", in *Adaptive Optical System Technologies*, Proc. SPIE **3353** (1998), pp. 994-1005.

Lecture Notes in Earth Sciences

For information about Vols. 1–19
please contact your bookseller or Springer-Verlag